Handbook of Experimental Pharmacology

Volume 292

The *Handbook of Experimental Pharmacology* is one of the most authoritative and influential book series in pharmacology. It provides critical and comprehensive discussions of the most significant areas of pharmacological research, written by leading international authorities. Each volume in the series represents the most informative and contemporary account of its subject available, making it an unrivalled reference source.

HEP is indexed in Medline/PubMed and Scopus.

Andreas Papapetropoulos • Csaba Szabo
Editors

Hydrogen Sulfide: A Toxicant with Healing Potential

Springer

Editors
Andreas Papapetropoulos
Department of Pharmacy
National and Kapodistrian Univ of Athens
Athens, Greece

Csaba Szabo
University of Fribourg
Fribourg, Switzerland

ISSN 0171-2004 ISSN 1865-0325 (electronic)
Handbook of Experimental Pharmacology
ISBN 978-3-032-24431-4 ISBN 978-3-032-24432-1 (eBook)
https://doi.org/10.1007/978-3-032-24432-1

This Springer imprint is published by the registered company Springer Nature Switzerland AG
The registered company address is: Gewerbestrasse 11, 6330 Cham, Switzerland

Preface: Hydrogen Sulfide Biology at Thirty—Foundations in Place, Translation Ahead

The story of hydrogen sulfide (H_2S) in biology is a striking example of how scientific fields can lie dormant for decades before suddenly accelerating. For most of modern history, H_2S was viewed almost exclusively as an environmental toxin. From Ramazzini's eighteenth-century description of "sewer gas" toxicity to the twentieth-century toxicological literature defining its inhibition of mitochondrial respiration, H_2S was synonymous with poison. Yet, as outlined in a historical overview of the field (Szabo 2018), mammalian H_2S production had in fact been observed as early as the 1940s during the early studies of transsulfuration, in Vincent du Vigneaud's laboratory at Cornell University. These observations, however, failed to ignite sustained interest. They were noted, cataloged, and largely forgotten. The early hints that H_2S might play a biological role—detected in tissues and produced enzymatically—did not immediately challenge the dominant toxicological paradigm. The prevailing mindset was that endogenous H_2S, if present at all, was a biochemical curiosity or metabolic by-product. No unifying conceptual framework existed to interpret it as a signaling molecule.

That changed in 1996. In a landmark study, Hideo Kimura demonstrated that cystathionine-β-synthase (CBS) in the brain produces H_2S and proposed that this gas functions as a neuromodulator (Abe and Kimura 1996). His work suggested that H_2S enhances NMDA receptor–mediated responses and facilitates long-term potentiation. For the first time, H_2S was advanced as a physiological mediator. But even then, the field did not explode overnight. As with many transformative ideas, there was a lag phase. Skepticism persisted, and replication and extension were needed.

Momentum built in the late 1990s and early 2000s, particularly as vascular studies, demonstrated that H_2S induces vasorelaxation and interacts functionally with nitric oxide. The concept of "gasotransmitters," previously limited to nitric oxide and carbon monoxide, expanded to include H_2S. Enzymatic sources—CBS, cystathionine-γ-lyase (CSE), and 3-mercaptopyruvate sulfurtransferase (3-MST)—were characterized in increasing detail. Genetic and pharmacological tools emerged. The field crossed a threshold.

Three decades after Kimura's proposal, the scope and scale of H_2S biology are captured in the comprehensive *Physiological Reviews* article spanning nearly 250 pages and listing over 2000 references (Cirino et al. 2023). This synthesis documents H_2S as an endogenous regulator across virtually every organ system:

vascular tone, immune responses, neuronal transmission, cellular bioenergetics, endocrine regulation, metabolism, and aging. The enzymatic machinery of H_2S production and degradation is now mapped in detail. Mechanistic insights into protein persulfidation, redox interactions, and mitochondrial signaling have reshaped our understanding of sulfur biology.

The chapters assembled in this volume reflect the maturation of the field. H_2S is now studied in the context of vascular homeostasis, atherosclerosis, perivascular adipose tissue signaling, skeletal muscle physiology, organ transplantation, hormonal crosstalk, adipose tissue biology, and bioenergetics. Polysulfides and sulfane sulfur species have joined H_2S as part of a broader reactive sulfur signaling network. The intellectual foundations are firmly in place.

Yet translation remains incomplete.

From early on, the therapeutic potential of H_2S modulation was evident. If H_2S is deficient in certain cardiovascular and metabolic diseases, replacement strategies might restore homeostasis. Conversely, in settings of pathological overproduction—such as inflammation, Down syndrome, or cancer—enzyme inhibition might be beneficial. Preclinical models have yielded hundreds of studies supporting both concepts.

Several early translational efforts have moved into clinical development. IK-1001 (intravenous sodium sulfide) entered human trials. H_2S-releasing nonsteroidal anti-inflammatory drugs, including compounds like the donor SG1002 and H_2S-releasing nonsteroidal anti-inflammatory derivatives such as ATB-346, demonstrated improved gastrointestinal safety profiles in early-phase studies and suggested early efficacy in clinical trials. Mitochondria-targeted and slow-releasing H_2S donors have been synthesized and tested. Much progress has also been made in characterizing various inhibitors of CBS, CSE, or 3-MST. These efforts represent important milestones (Szabo and Papapetropoulos 2017). But, so far, none of the H_2S-based therapeutics have yet to complete the journey to fully established clinical therapies.

The reasons are familiar to translational science: dose optimization in the context of bell-shaped biological responses, pharmacokinetics of volatile and reactive species, biomarker development, patient stratification, and long-term safety assessment. Moreover, H_2S biology is deeply context-dependent; the same molecule may be protective in one tissue and pathogenic in another. Precision in modulation—not mere augmentation or suppression—will be required.

Hydrogen sulfide biology at thirty stands at a pivotal moment. The field has evolved from skepticism to encyclopedic mechanistic depth. Its enzymology, signaling pathways, and physiological roles are well characterized. The next frontier is clinical translation.

Athens, Greece Andreas Papapetropoulos
Fribourg, Switzerland Csaba Szabo

References

Abe K, Kimura H (1996) The possible role of hydrogen sulfide as an endogenous neuromodulator. J Neurosci 16:1066–1071

Cirino G, Szabo C, Papapetropoulos A (2023) Physiological roles of hydrogen sulfide in mammalian cells, tissues, and organs. Physiol Rev 103:31–276

Szabo C, Papapetropoulos A (2017) International Union of Basic and Clinical Pharmacology. CII: pharmacological modulation of H_2S levels: H_2S donors and H_2S biosynthesis inhibitors. Pharmacol Rev 69:497–564

Szabo C (2018) A timeline of hydrogen sulfide (H_2S) research: From environmental toxin to biological mediator. Biochem Pharmacol 149:5–19

Contents

Physiological Roles of Hydrogen Sulfide/Polysulfides and Regulation of their Production

Hideo Kimura

Contents

Abstract

Hydrogen sulfide (H_2S), produced enzymatically, acts as a signaling molecule involved in the regulation of neurotransmission, vascular tone, cytoprotection, inflammation, oxygen sensing, and energy metabolism. Hydrogen polysulfides (H_2S_n, $n \geq 2$) and other per- and polysulfides, such as cysteine persulfide, are enzymatically produced to regulate target protein activity. The chemical interaction of H_2S with NO also generates H_2S_n. S-sulfuration (S-sulfhydration) has been proposed as a mode of action of H_2S to regulate the activity of target molecules. H_2S_n S-sulfurate cysteine residues, while H_2S S-sulfurates oxidized cysteine residues such as those S-nitrosylated and S-sulfinated. Transient receptor potential ankyrin 1 (TRPA1) channels, which convert stimulations such as pain,

H. Kimura (✉)
Department of Pharmaceutical Engineering, Faculty of Engineering, Sanyo-Onoda City University, Sanyo-Onoda, Japan
e-mail: kimura@rs.socu.ac.jp

A. Papapetropoulos, C. Szabo (eds.), *Hydrogen Sulfide: A Toxicant with Healing Potential*, Handbook of Experimental Pharmacology 292,
https://doi.org/10.1007/164_2025_754

pressure, and temperature to the neuronal signal, were identified as a target of H_2S_n. Recently the channels were found to be involved in the regulation of the transcription/translation of sulfur metabolizing enzymes and the intracellular levels of sulfur containing amino acids, cysteine and methionine, and even those of neurotransmitter GABA. The physiological roles of these molecules will be discussed.

Keywords

Hydrogen sulfide · Polysulfides · S-sulfuration · Transmitters release · TRPA1 channel

Abbreviations

ATP	Adenosine triphosphate
CARS	Cysteinyl tRNA synthetase
CAT	Cysteine aminotransferase
CBS	Cystathionine β–synthase
CysSSH	Cysteine persulfide
CO	Carbon monoxide
CSE	Cystathionine γ–lyase
DAO	D-Amino acid oxidase
DTT	Dithiothreitol
GAPDH	Glyceraldehydes-3-phosphate dehydrogenase
GCL	Glutamate cysteine ligase
GSH	Glutathione
GSSH	Glutathione persulfide
HNO	Nitroxyl
H_2O_2	Hydrogen peroxide
HSNO	Thionitrous acid
H_2S	Hydrogen sulfide
H_2S_n	Hydrogen polysulfides
Keap1	Kelch ECH associating protein 1
LTP	Long-term potentiation
3MP	3-Mercaptopyruvate
3MST	3-Mercaptopyruvate sulfurtransferase
NaHS	Sodium hydrosulfide
NMDA	N-Methyl D,L-aspartate
NO	Nitric oxide
Nrf2	Nuclear factor erythroid 2-related factor 2
O_2	Oxygen
PTEN	Phosphatase and tensin homolog

SOD	Super oxide dismutase
SQR	Sulfur quinone oxydoreductase
$SSNO^-$	Nitrosopersulfide
TRPA1	Transient receptor potential ankyrin 1

Hydrogen sulfide (H_2S), produced enzymatically, acts as a signaling molecule involved in the regulation of neurotransmission, vascular tone, cytoprotection, inflammation, oxygen sensing, and energy metabolism. Hydrogen polysulfides (H_2S_n, $n \geqq 2$) and other per- and polysulfides, such as cysteine persulfide, are enzymatically produced to regulate target protein activity. The chemical interaction of H_2S with NO also generates H_2S_n. S-sulfuration (S-sulfhydration) has been proposed as a mode of action of H_2S to regulate the activity of target molecules. H_2S_n S-sulfurate cysteine residues, while H_2S S-sulfurates oxidized cysteine residues such as those S-nitrosylated and S-sulfinated. Transient receptor potential ankyrin 1 (TRPA1) channels, which convert stimulations such as pain, pressure, and temperature to the neuronal signal, were identified as a target of H_2S_n. Recently the channels were found to be involved in the regulation of the transcription/translation of sulfur metabolizing enzymes and the intracellular levels of sulfur containing amino acids, cysteine and methionine, and even those of neurotransmitter GABA. The physiological roles of these molecules will be discussed.

1 Introduction

Hydrogen sulfide (H_2S), initially known as a toxic environmental gas, has been identified in mammalian brains as a signaling molecule synthesized by enzymes such as cystathionine β-synthase (CBS), cystathionine γ-lyase (CSE), and 3-mercaptopyruvate sulfurtransferase (3MST) (Stipanuk and Beck 1982; Warenycia et al. 1989; Abe and Kimura 1996; Hosoki et al. 1997; Shibuya et al. 2009, 2013). S-sulfuration (S-sulfhydration) is a mechanism through which H_2S induces conformational changes in target proteins, thereby modifying their activity (Mustafa et al. 2009). Hydrogen polysulfides (H_2S_n, $n \geqq 2$) were identified to activate transient receptor potential ankyrin 1 (TRPA1) channels much more potently than H_2S by S-sulfurating two cysteine residues localized to the amino-terminus of the channels (Nagai et al. 2004, 2006; Oosumi et al. 2010; Kimura et al. 2013; Hatakeyama et al. 2015). Subsequently, various other proteins of which the activities were regulated through S-sulfuration by H_2S_n were identified; a tumour suppressor phosphatase tensin homologue (PTEN) (Greiner et al. 2013), a transcription factor complex kelch-like ECH-associated protein 1 (Keap1)/nuclear factor-like 2 (Nrf2) (Koike et al. 2013), protein kinase G1α (PKGα) (Stubbert et al. 2014), and GAPDH (Mustafa et al. 2009; Jarosz et al. 2015). Considering the oxidation state of sulfur in H_2S, H_2S_n, and cysteine residues, H_2S_n S-sulfurate cysteine residues of the target proteins. However, approximately 5–12% of cysteine residues in cells are oxidized

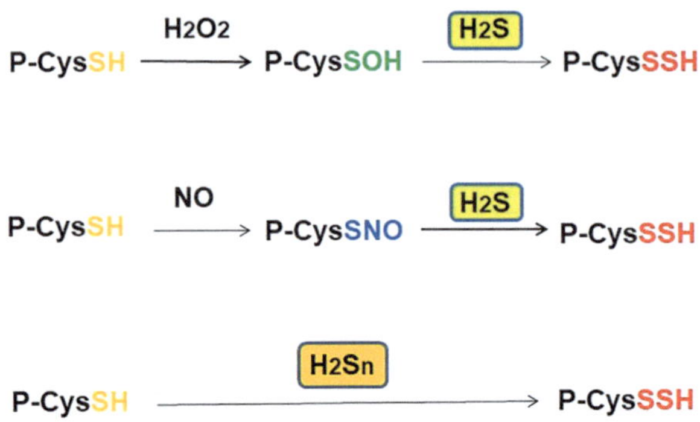

Fig. 1 S-sulfuration (S-sulfhydration) of cysteine residues by H_2S and H_2S_n. Cysteine residues are S-sulfenylated by H_2O_2 and S-nitrosylated by NO. These oxidized cysteine residues are S-sulfurated by H_2S. In contrast, cysteine residues are S-sulfurated by H_2S_n. (This figure is diverted from Kimura (2021))

(Go and Jones 2013) such as S-nitrosylated, S-sulfinated and generating a cysteine disulfide bond. These oxidized cysteine residues are S-sulfurated by H_2S. In addition, H_2S reduces cysteine disulfide bond of target proteins. For example, the first step of the metabolism of H_2S is the reduction of cysteine disulfide bond of sulfide quinone oxidoreductase (SQR) in mitochondria (Hildebrandt and Gieshaber 2008), and the activity of N-methyl D-aspartate (NMDA) receptors is enhanced by the reduction of cysteine disulfide bond localized to the hinge of the ligand binding domain of the receptors (Aizenman et al. 1989; Abe and Kimura 1996) (Fig. 4). The latter activity is minor because of the low levels of endogenous H_2S.

Several enzymes have been identified to produce per- and poly-sulfides including 3MST, SQR, haemoglobin, neuroglobin, catalase, super oxide dismutase, cysteinyl tRNA synthetase, lactoperoxidase, and myeloperoxidase (Nakamura et al. 1984; Searcy et al. 1995; Kimura et al. 2015; Vitvitsky et al. 2015; Bostelaar et al. 2016; Garai et al. 2017; Ruetz et al. 2017; Olson et al. 2017; Kimura et al. 2017; Landry et al. 2017; Akaike et al. 2017). Not only these producing enzymes but TRPA1 channels, which are the target of polysulfides, were found to change the endogenous levels of H_2S and polysulfides by regulating the levels of 3MST and cysteine, one of the substrates for cysteine aminotransferase, which generates 3-mercaptopyruvate, a substrate for 3MST (Furuie et al. 2023). Deficiency of TRPA1 channels in Drosophila causes a decrease in the levels of cysteine and methionine and an increase in those of GABA (Lee et al. 2016), while the application of resiniferatoxin, an ultra-potent capsaicin analogue antagonist of TRP vanilloid type 1 and TRPA1 channels, to porcine urinary bladder increased the transcription of rhodanese and decreased that of sulfur dioxygenase (ETHE1) (Lepiarczyk et al. 2023). Not only the producing enzymes but a variety of factors including proteins may be involved in regulating the production of H_2S and polysulfides (Fig. 2).

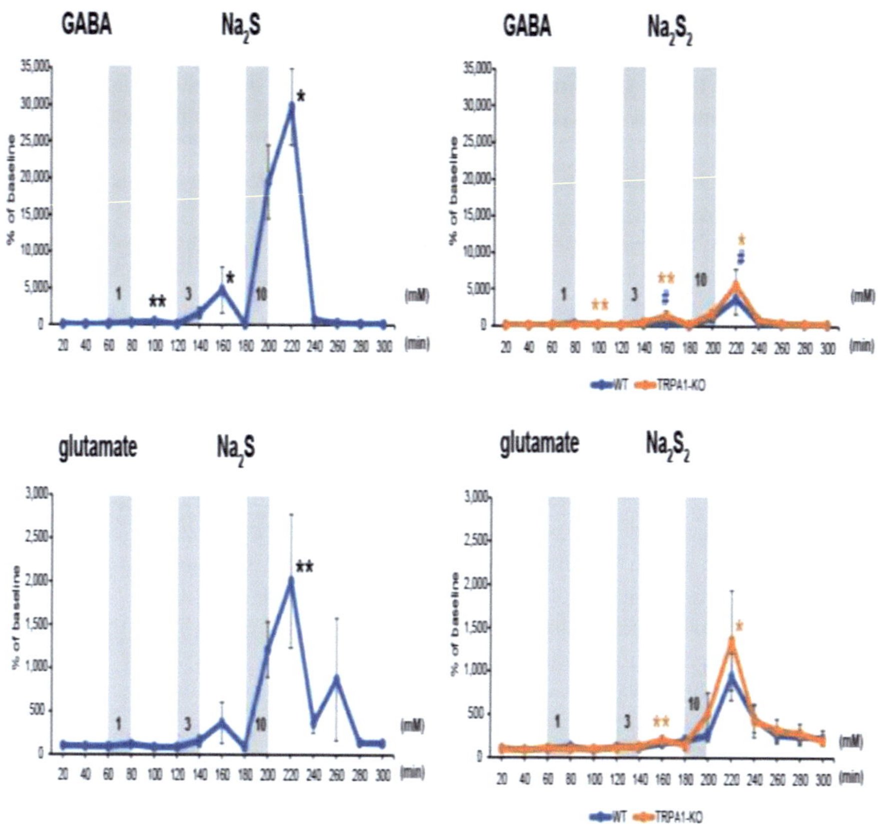

Fig. 2 The release of GABA and glutamate by H_2S and H_2S_2. In in vivo microdialysis at rat hippocampus, H_2S induced the release of GABA and glutamate much more greatly than H_2S_2. TRPA1 channels were not involved in the release of both neurotransmitters. (This figure is produced by modifying Furuie et al. (2023))

2 Endogenous H_2S

H_2S exhibits higher solubility in water (413 mg/100 ml at 20°C) compared to other gaseous signaling molecules such as nitric oxide (NO) (5.6 mg/100 ml, 20°C), carbon monoxide (CO) (2.76 mg/100 ml, 20°C), and even oxygen (O_2) (3.93 mg/100 ml, 25°C).

$$H_2S = H + + HS^- \text{ (pK1 7.04)}$$

$$HS^- = H + + S^{2-} \text{ (pK2 11.96)}.$$

Under physiological conditions (pH 7.4 at 37°C), approximately 20% of H_2S exists as a gas and the remaining as HS^- with a trace amount of S^{2-}. Considering the intracellular pH being approximately 7.0 in the cytosol, 8.0 in mitochondria, 4.7 in lysosomes, the ratio of H_2S/HS^- varies among subcellular compartments (Casey et al. 2010).

Endogenous H_2S levels in the brain were initially discovered while assessing tissue accumulation of H_2S following animal exposure (Werenycia et al. 1989). Since the methylene blue method was used to measure the levels of H_2S, the endogenous levels were overestimated ranging from 50 to 160 μM. This method works in the presence of high concentrations of hydrochloric acid, which even releases H_2S mostly from the iron-sulfur center of enzymes belonging to the respiratory chain localized to mitochondria where pH is approximately 8; under such physiological conditions, H_2S is not released from iron-sulfur center (Ishigami et al. 2009).

Endogenous levels of H_2S were measured by several methods. In the brain 14 nM was obtained where H_2S was released from homogenates under the acidic condition (pH 5.7) and measured by gas choromatography (Furne et al. 2008). Since brain homogenates, in which the intracellular and extracellular substances are mixed to cause the reaction never occurs under physiological conditions, and since homogenates have capacity to absorb exogenously applied H_2S (Ishigami et al. 2009), it is better to homogenize with substances which immediately react with thiol. Approximately 3 μM was measured when the tissue was homogenized with monobromobimane at pH 7 and the supernatant was applied to LC-MS/MS (Koike et al. 2017).

3 Sulfane Sulfur (Bound Sulfur)

Sulfane sulfur (bound sulfur), which releases H_2S under reducing conditions (e.g., in the presence of DTT), consists of S-sulfurated molecules such as H_2S_n and cysteine persulfide and constitutes an independent pool of acid-labile sulfur (Warenycia et al. 1990a; Ogasawara et al. 1993; Ishigami et al. 2009). The levels of sulfane sulfur (S-sulfurated molecules) differ between tissues; liver contains high, while low in the heart (Ishigami et al. 2009).

Exogenously applied H_2S is absorbed by tissue homogenates, and the capacity for absorption differs between tissues (Ishigami et al. 2009). Liver homogenates rapidly absorb H_2S upon exposure, while heart homogenates absorb it faster than those from brain tissues. Absorbed H_2S is recovered under reducing conditions, suggesting that H_2S is stored as sulfane sulfur in the form of S-sulfurated cysteines. These observations suggest that the liver contains a lot of oxidized cysteine residues such as those S-nitrosylated and S-sulfinated as well as cysteine disulfide bond, which are S-sulfurated by H_2S. Sulfane sulfur levels are low in the heart, while it efficiently absorbs H_2S, suggesting that heart contains reasonably high levels of oxidized cysteine residues (Ishigami et al. 2009).

4 Production of H_2S and Polysulfides

Cystathionine β-synthase (CBS), cystathionine γ-lyase (CSE), and 3-mercaptopyruvate sulfurtransferase (3MST) produce H_2S (Stipanuk and Beck 1982; Abe and Kimura 1996; Hosoki et al. 1997; Shibuya et al. 2009). 3MST metabolizes 3-mercaptopyruvate (3MP) produced by cysteine aminotransferase (CAT) from L-cysteine and α-ketoglutarate and 3MP by D-amino acid oxidase (DAO) from D-cysteine (Shibuya et al. 2009, 2013). A relatively high concentration of D-cysteine was recently identified in the embryonic brain, indicating its potential role as a signaling molecule (Semenza et al. 2021), suggesting that 3MST together with DAO pathway plays an important role in the developing brain to produce H_2S/H_2S_n.

3MST also produces H_2S_n and other per- and poly-sulfides including cysteine persulfide (Kimura et al. 2015, 2017). Sulfide quinone oxidoreductase (SQR) metabolizes H_2S to produce mainly GSSH (Landry et al. 2017). Cupper/zinc super oxide dismutase (SOD) can produce H_2S_n by oxidizing H_2S with O_2 and H_2O_2 as electron acceptor (Olson et al. 2018; Searcy et al. 1995). SOD oxidation prefers dissolved H_2S to HS^-, which rather inhibits the production of H_2S_n (Olson et al. 2018). Catalase oxidizes H_2S to generate H_2S_n in the presence of O_2 (Olson et al. 2017). Lactoperoxidase and myeloperoxidase can oxidize H_2S to generate H_2S_n in the presence of H_2O_2 (Garai et al. 2017; Nakamura et al. 1984). Haemoglobin, myoglobin neuroglobin, can also oxidize H_2S to produce H_2S_n. Centrally located iron (Fe^{3+}) plays a role to oxidize H_2S. H_2S_n is the main product when sulfide levels are high, while thiosulfate is the main product when sulfide levels are low (Vitvitsky et al. 2015). Cysteine t-RNA synthase (CARS) S-sulfurates cysteine to generate cysteine persulfide (Akaike et al. 2017).

Cavallini et al. demonstrated the production of cysteine persulfide by CSE from cystine, and CBS was also reported to have a similar activity to CSE (Cavallini et al. 1960; Ida et al. 2014). CBS and CSE are cytoplasmic enzymes predominantly exposed to cysteine rather than cystine. The endogenous concentrations of cystine in the liver and lung are approximately 0.2. and 0.05 μM, respectively, and those in the heart and brain are under detectable levels, suggesting that it is difficult to produce cysteine persulfide (Ida et al. 2014; Yadav et al. 2016). The cells expressing 3MST and CAT increased the levels of sulfane sulfur, which is detected as the release of H_2S when cells are homogenized in the presence of DTT, while cells expressing CBS do not increased the release of H_2S compared to a control, suggesting no production of sulfane sulfur including cysteine persulfide by CBS (Shibuya et al. 2009). Brains of 3MST knockout mice and rat showed significantly lower levels of sulfane sulfur compared to those of the wild type (Kimura et al. 2017; Furuie et al. 2023). In contrast, brains of CBS knockout mice have no significantly different levels of sulfane sulfur compared to those of the wild type (Kimura et al. 2017). Therefore, it is controversial for CSE and especially for CBS to produce cysteine persulfide in cells (Yadav et al. 2016).

5 S-sulfuration (S-sulfhydration) of Cysteine Residues of Target Molecules

Although CSE may not produce cysteine persulfide under physiological conditions as described previously, while CSE is able to produce cysteine persulfide in vitro in the presence of exogenously applied cystine (Cavallini et al. 1960; Ida et al. 2014; Yadav et al. 2016). Kato et al. observed that rat liver serine dehydratase was inhibited in the presence of CSE together with cystine (by generating cysteine persulfide), and it was thought that the formation of hydrodisulfide (S-sulfurated) suppressed the activity of the enzyme (Kato et al. 1966). Since then many enzymes were found to be inhibited by S-sulfuration, including tyrosine aminotransferase (Hargrove and Wichman 1987), ornithine decarboxylase (Murakami et al. 1984), and adenylate kinase (Conner and Russell 1983). This inhibition is reversible upon treatment with GSH and DTT, which reduce S-sulfurated cysteine residues back to thiol groups.

We found that H_2S activates TRP channels, which transmits the signal of pain, temperature, and pressure (Nagai et al. 2004). We later recognized H_2S_n are the entities to activate TRPA1 channels (Nagai et al. 2006; Oosumi et al. 2010; Kimura et al. 2013). Snyder and colleagues proposed that H_2S S-sulfhydrates (S-sulfurate) (oxidized) cysteine residues of the target molecules to induce conformational changes to modify their activity (Mustafa et al. 2009). Two cysteine residues localized to the amino terminus of TRPA1 channels can be S-sulfurated by H_2S_n, or one of them is able to S-sulfurated which reacts with the remaining residue to generate the cysteine disulfide bond, leading to the conformational change to activate the channels (Hatakeyama et al. 2015). S-sulfuration is highly unstable. When cultures of astrocytes expressing TRPA1 channels are superfused with the medium containing H_2S_n, the channels are activated to induce Ca^{2+} influx immediately after the changes of the medium. However, the activated state lasts until the medium is changed to that without H_2S_n, and it quickly returns to the quiescent state (Kimura et al. 2013). The intracellular GSH may reduce S-sulfurated cysteine residues to cease the channel activation.

Following TRPA1 channels, tumour suppressor phosphatase and tensin homologue (PTEN), protein kinase G1α (PKG1α), and glutaraldehyde 3-phosphate dehydrogenase (GAPDH) were found to be S-sulfurated by polysulfides rather than H_2S (Greiner et al. 2013; Stubbert et al. 2014; Jarosz et al. 2015). Considering these observations, the effects of H_2S previously found were thought to be mediated by polysulfides. However, recently we found that the release of GABA and glutamate was induced by H_2S much more potently than H_2S_2 in in vivo microdialysis (Furuie et al. 2023). The reduction of cysteine disulphide bond or S-sulfuration of oxidized cysteine residues by H_2S must be involved in the regulation of neurotransmitter release machinery (Fig. 3).

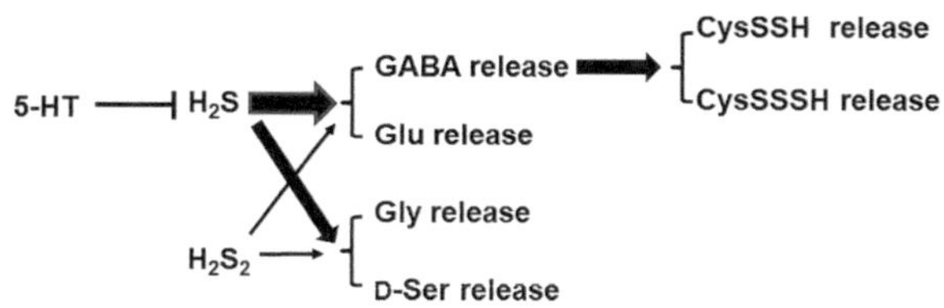

Fig. 3 H_2S and polysulfides regulate the release of neurotransmitters and vice versa. H_2S induces the release of GABA and glutamate much more greatly than H_2S_2. H_2S and H_2S_2 induce the release of glycine and D-serine in a similar fashion. GABA induces the release of cysteine-persulfide and – trisulfide. Serotonin suppresses the release of H_2S

6 Release of Neurotransmitters Induced by H_2S and H_2S_n

The application of high concentrations of H_2S to animals decreases the levels of neurotransmitters in the brain. Exposure of rats to high levels of H_2S during perinatal development significantly decreases neurotransmitter levels (Hannah et al. 1989). Sublethal doses of NaHS chronically applied to rodents suppressed neurotransmitter levels (Warenycia et al. 1990b). In contrast, patch-clamp electrophysiology in rat brain stem slices showed that 10 μM NaHS significantly increased the amplitude of evoked solitary tract excitatory postsynaptic currents (EPSCs), frequency of miniature EPSCs, and presynaptic terminal calcium fluorescence in the nucleus tractus solitarri (Austgen et al. 2011). They also showed that the CBS inhibitor aminooxyacetate significantly reduced the amplitude of EPSCs and presynaptic terminal calcium fluorescence without altering postsynaptic properties. NaHS (200 μM) selectively potentiated cholinergic fast excitatory postsynaptic potentials (EPSPs) evoked by splanchnic nerve stimulation but not those evoked by colonic nerve stimulation (Sha et al. 2013). They also showed that the CSE blocker DL-propargylglycine significantly reduced the amplitude of EPSPs evoked by splanchnic nerve stimulation but not those colonic, and similar observations were obtained in CSE knockout mice. NaHS (300 μM) reversibly increased the frequency of miniature end-plate currents associated with reversible increase in endplate current amplitude, which was abolished by activation of ryanodine receptors, suggesting that acetylcholine release induced by H_2S is related to an increase in the intracellular Ca^{2+} concentrations due to activation of ryanodine receptors (Gerasimova et al. 2015).

H_2S and H_2S_2 facilitate the release of neurotransmitters and gliotransmitters such as GABA, glutamate, glycine, D-serine, L-serine, and glutamine in in vivo microdialysis of rat hippocampus (Furuie et al. 2023). The release of GABA and glutamate was significantly increased by H_2S much more potently than by H_2S_2. The release of glycine, D-serine, L-serine, and glutamine was facilitated by H_2S and H_2S_2 with a similar efficiency, and the facilitation efficiency is much less than that on GABA and glutamate. The mechanism of the release of GABA and glutamate may be different from that of glycine, D-serine, L-serine, and glutamine. The facilitation of GABA release was robust enough to be detectable even in suspended brain cells, which are less sensitive compared to the microdialysis (Furuie et al. 2023).

The release of H_2S and polysulfides is regulated in turn by neurotransmitters. Serotonin suppresses the release of H_2S, while it facilitates that of H_2S_3 (Furuie et al. 2023). Acetylcholine and norepinephrine facilitate the release of cysteine persulfide. H_2S/polysulfides and neurotransmitters thus reciprocally regulate their release.

Abnormalities in the balance between glutamatergic and GABAergic activity have been implicated in the pathophysiology of schizophrenia. Serine racemase, an enzyme that converts L-serine to D-serine, is associated with susceptibility to schizophrenia in humans and a model animal, and the activation of $GABA_A$ receptors regulates the extracellular levels of D-serine (Labrie et al. 2009; Umino et al. 2017). H_2S may maintain a balance between glutamatergic and GABAergic activities. H_2S facilitates the release of GABA, which activates $GABA_A$ receptors to enhance the release of D-serine. H_2S also increase the release of glutamate together with D-serine to activate NMDA receptors. 3MST-KO rat brains, which have H_2S and polysulfides much less compared to those in the wild-type, showed behavioural hyperactivity observed in schizophrenia induced by MK-801, non-selective NMDA receptor antagonist, significantly greater than the wild type (Wolf 1998; Belforte et al. 2010; Furuie et al. 2023).

The physiological role of H_2S was identified as its facilitation of hippocampal long-term potentiation, a synaptic model of memory formation by enhancing the activity of NMDA receptors (Abe and Kimura 1996). Dithiothreitol (DTT) enhances the activity of NMDA receptors by reducing cysteine disulphide bond localized to the hinge of ligand binding domain of NMDA receptors (Aizenman et al. 1989). H_2S can enhance the activity of NMDA receptors by reducing the cysteine disulphide bond in NMDA receptors. However, H_2S with one-tenth concentration of DTT much more potently facilitates the induction of LTP than DTT, suggesting the existence of an alternative mechanism.

H_2S facilitates the release of GABA, glutamate, and D-serine as well as glycine (Furuie et al. 2023; Umino et al. 2017). Glutamate activates NMDA receptors, and D-serine and glycine enhance them. $GABA_A$ receptor activation also induces the release of D-serine (Umino et al. 2017). It may be an alternative mechanism for the enhancement of LTP induction by H_2S.

It was previously reported that the activation of TRPA1 channels localized to astrocytes facilitates the induction of LTP by enhancing the release of D-serine (Shigetomi et al. 2013), and TRPA1 channels are activated by H_2S_n (Kimura et al. 2013). However, in in vivo microdialysis, there observed no significant difference in the release of D-serine between TRPA1-KO rats and the wild type, suggesting that TRPA1 channels are not involved in the release of D-serine (Furuie et al. 2023). However, the induction of LTP was suppressed in TRPA1-KO rat brain that agreed with the previous observation by Shigetomi et al. (2013) that TRPA1 channels are required for the induction of LTP.

LTP induction was suppressed in 3MST-KO rat brain as predicted from the fact that the levels of H_2S and polysulfides are decreased, leading to the less enhancement of glutamate, GABA, and D-serine release (Furuie et al. 2023) (Fig. 4).

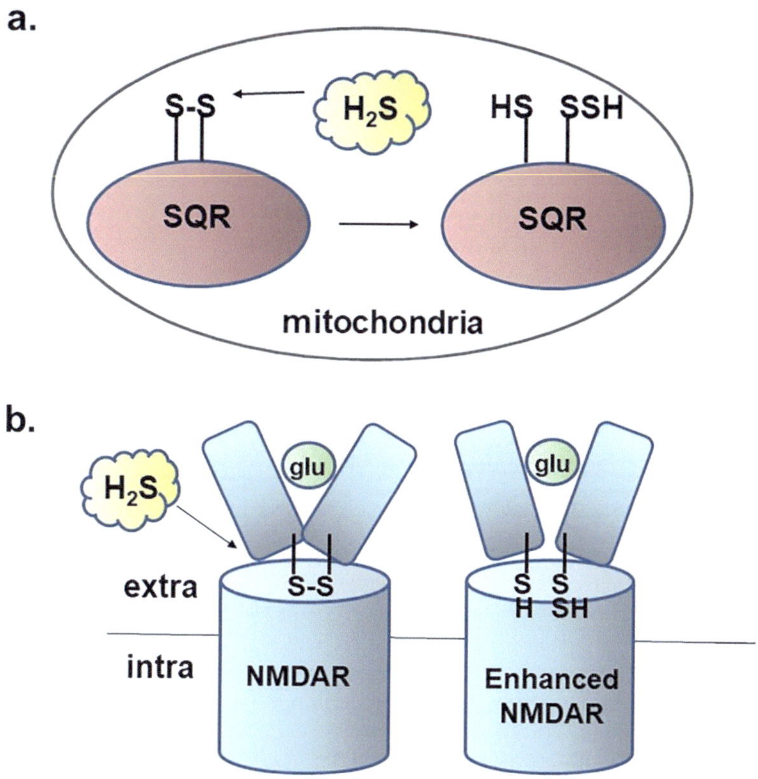

Fig. 4 H_2S reduces a cysteine disulfide bond. (**a**) The first step of the metabolism of H_2S is the reduction of the cysteine disulfide bond of sulfide quinone oxidoreductase (SQR) in mitochondria. Note that it was recently identified as the cysteine trisulfide bond (Landry et al., 2019). (**b**) H_2S may reduce the cysteine disulfide bond localized to the hinge of the ligand binding domain of N-methyl D-aspartate receptors to enhance the activity of the receptors. (This figure is diverted from Kimura (2024))

7 TRPA1 Channels Regulate the Metabolism of Sulfur-Containing Substances

TRPA1 channels are not only the target of H_2S_n but also involved in the metabolism of sulfur containing amino acids and sulfur metabolizing enzymes as well as the levels of GABA (Lee et al. 2016; Furuie et al. 2023). TRPA1 channel deficient Drosophila neurons showed decreased levels of methionine and cysteine, 15 and 50%, respectively, of the wild type (Lee et al. 2016). Approximately 40% of decrease in cysteine level was observed in TRPA1-KO rat brain (Furuie et al. 2023). Resiniferatoxin, an ultrapotent capsaicin analogue antagonist of TRP vanilloid type 1 (TRPV1) and TRPA1 channels, changed the transcription of sulfur metabolizing enzymes: upregulated rhodanese and downregulated sulfur

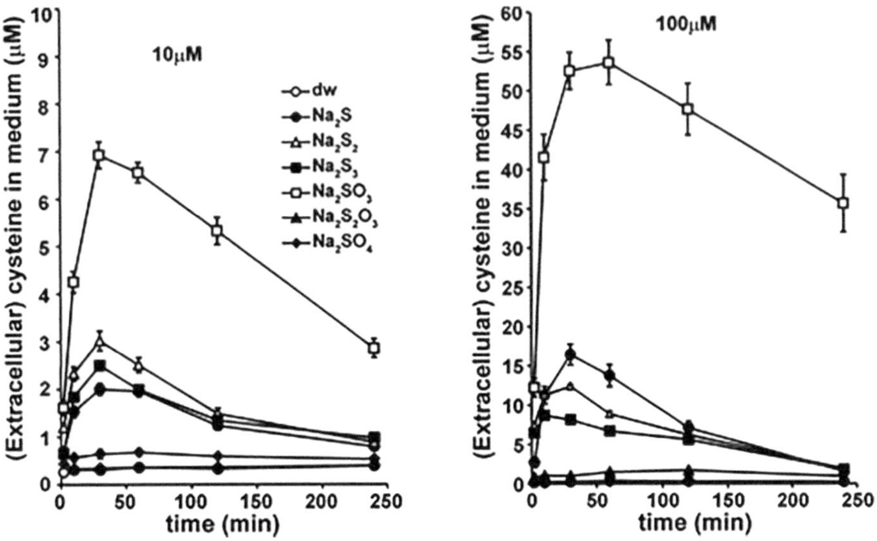

Fig. 5 Sulfite reacts with cystine to generate cysteine and S-cysteinesulfonate. Na_2SO_3 efficiently converted cystine to cysteine in the medium, whereas Na_2S_2 and Na_2S_3 resulted in approximately one-third of the production induced by Na_2SO_3. (This figure is diverted from Kimura et al. (2019))

dioxygenase (ETHE1) together with molybdenum cofactor synthesis 2 (Lepiarczyk et al. 2023; Pecze et al. 2009).

In TRPA1-KO rat brain, the levels of 3MST were also decreased (Furuie et al. 2023). 3MP, a substrate of 3MST, is produced from cysteine and α-ketoglutarate by CAT. Therefore, the decrease in cysteine and 3MST levels results in the decreased production of H_2S, H_2S_2, H_2S_3, and cysteine persulfide (Furuie et al. 2023). These observations suggest that not only H_2S, H_2S_n, and other polysulfides producing enzymes but also TRPA1 channels influence the levels of these signalling molecules.

GABA levels are also changed in TRPA1 deficient Drosophila neurons, TRPA1-KO rat brain, and 3MST-TRPA1 double KO rat brain, approximately 44, 22, and 40% increase, respectively (Lee et al. 2016; Furuie et al. 2023), suggesting that TRPA1 channels may influence the production and metabolism to decrease the levels of GABA for the efficient transmission of signals mediated by the channels. It is possible that the metabolism of some other proteins may also be regulated by TRPA1 channels (Fig. 5).

8 Cytoprotective Effect of H_2S, H_2S_n, and H_2SO_3 Against Oxidative Stress (Oxytosis and Ferroptosis)

In contrast to excitotoxicity mediated by the activation of ionotropic glutamate receptors (Choi 1988), millimolar concentrations of glutamate induce cell death by suppressing cystine/glutamate antiporter leading to decreasing the import of cystine, which is reduced to cysteine in the cell and used for the production of glutathione (GSH), the intracellular major antioxidant (Murphy et al. 1990). Ionotropic glutamate receptors are yet expressed in the embryonic brains, and millimolar concentrations of glutamate induce oxidative cell death (Murphy et al. 1990). It was later called oxytosis, which has been proposed to be a similar or identical to ferroptosis (Tan et al. 2001; Maher et al. 2020). H_2S protects embryonic neurons from oxidative stress by enhancing the activity of cystine/glutamate antiporter and glutamate cysteine ligase (GCL) to increase the production of GSH. H_2S_n S-sulfurate a cysteine residue of Keap1 to release Nrf2 to the nucleus where Nrf2 upregulates the transcription of antioxidant genes including GCL, resulting in the increased production of GSH (Koike et al. 2013).

More highly oxidized sulfite (H_2SO_3) provides a stronger cytoprotective effect on neurons against oxidative stress (oxytosis) compared to H_2S and H_2S_n (Kimura et al. 2019). The endogenous concentration of free H_2SO_3 is approximately 2 μM in the rat brain and comprised approximately 6% of the total H_2SO_3 including the bound form. One μM H_2SO_3 completely protected the neurons from oxytosis by increasing the levels of GSH, while higher concentration (10 μM) was required for H_2S and H_2S_3 (Kimura et al. 2019). H_2SO_3 reacts with cystine to produce cysteine and S-cysteinsulfonate much more greatly than H_2S, H_2S_2, and H_2S_3 (1) (Clarke et al. 1932; Kimura et al. 2019).

$$\text{Sulphite} + \text{cystine} = \text{cysteine} + \text{S} - \text{cysteinesulfonate} \tag{1}$$

It is possible that cysteine produced by the reaction between cystine and H_2SO_3 is transported into cells by cysteine transporter more efficiently than cystine by cystine/glutamate antiporter and contributes to produce GSH (Kimura et al. 2005, 2010). Alternatively, H_2SO_3 may cleave cysteine disulphide bond of proteins, which are involved in the cytoprotection, to activate them.

9 Perspective

Since the identification of H_2S_n and other S-sulfurated molecules as chemical entities to S-sulfurate cysteine residues of target molecules to modify their activity, most of the effects of H_2S previously found have been thought to be mediated by polysulfides. Initially, even the ability of H_2S to reduce cysteine disulfide bonds in target proteins was met with skepticism (Furne et al. 2008). However, H_2S reduces the cysteine disulphide bond of SQR at the first step of its metabolism in mitochondria (Hildebrandt and Grieshaber 2008) that was recentely identified as cysteine trisulfide bond (Landry et al., 2019). The release of GABA and glutamate induced by H_2S was much greater than that by H_2S_2 in in vivo microdialysis as described previously (Furuie et al. 2023). Most of cysteine residues inside cells are thiol, but some of them are oxidized such as nitrosylated or sulfinated, and these oxidized cysteine residues can be S-sulfurated by H_2S (Go and Jones 2013). Identifying the target molecules of H_2S that regulate the release of GABA and glutamate is particularly intriguing, including transporters and molecules involved in thexocytosis of synaptic vesicles.

The neuroprotective effect of H_2S against oxidative stress induced by high concentrations of glutamate is similarly potent to that of H_2S_n, and sulphite is more potent (Kimura et al. 2019). Clarke demonstrated that sulphite, whose sulfur is more oxidized than that of H_2S_n, reacts with cystine to produce cysteine and S-cysteinesulfonate (1932), and the producing potency of cysteine from cystine by sulphite is much greater than that by H_2S and H_2S_n (Kimura et al. 2019), suggesting that sulphite may cleave cysteine disulphide bond more effectively than H_2S. It is interesting that sulphite, which is more oxidized compared to H_2S and H_2S_n, may cleave the less oxidized cysteine disulphide bond.

TRPA1 channel, which responds to the changes in the temperature and pressure and transmits pain, is a target of H_2S_n (Kimura et al. 2013). It was recently identified to be involved in the regulation of the levels of sulfur containing amino acids such as methionine and cysteine and those of sulfur metabolizing enzymes including 3MST, and even those of GABA, resulting in the regulation of the levels of H_2S and H_2S_n (Lee et al. 2016; Furuie et al. 2023; Leparczyk et al. 2023). TRPA1 channels are activated by H_2S_n, while they increase the levels of H_2S_n as well as H_2S, inducing the positive feedback, leading to the increase in the transmission of the signal induced by the channels. In contrast, the channels suppress the levels of GABA, of which the release is increased by H_2S and to the lesser extent by H_2S_n, resulting in the decreased effect of inhibitory neurotransmission by TRPA1 channels (Furuie et al. 2023). It is possible that there may be some other proteins which modify the production of H_2S and polysulfides in addition to their producing enzymes.

Serotonin suppresses the release of H_2S (Furuie et al. 2023). By identifying the regulation of H_2S/H_2S_n by neurotransmitters and vice versa may help understand the regulation of neurotransmission by these molecules and their therapeutic applications to neuronal and psychiatric diseases.

Acknowledgements This work was supported by Grants-in-Aid for Scientific Research from the Ministry of Education, Sciences, Sports, and Technology, Japan (23 K06016), and Smoking Research Foundation to HK.

Author Disclosure Statement There is no conflict of interest.

References

Abe K, Kimura H (1996) The possible role of hydrogen sulfide as an endogenous neuromodulator. The Journal of Neuroscience 16:1066–1071

Aizenman E, Lipton DA, Loring RH (1989) Selective modulation of NMDA responses by reduction and oxidation. Neuron 2:1257–1263

Akaike T, Ida T, Wei F-Y, Nishida M, Kumagai Y, Alam MM et al (2017) Cysteinyl-tRNA synthetase governs cysteine polysulfidation and mitochondrial bioenergetics. Nature Communications 8:1177

Austgen JR, Hermann GE, Dantzler HA, Rogers RC, Kline DD (2011) Hydrogen sulfide augments synaptic neurotransmission in the nucleus of the solitary tract. J Neurophysiol. 106:1822–1832

Belforte JE et al (2010) Postnatal NMDA receptor ablation in corticolimbic interneurons confers schizophrenia-like phenotypes. Nat. Neurosci. 13:76–83

Bostelaar T, Vitvitsky V, Kumutima J, Lewis BE, Yadav PK, Brunold TC et al (2016) Hydrogen sulfide oxidation by myoglobin. Journal of the American Chemical Society 138:8476–8488

Casey JR, Grinstein S, Orlowski J (2010) Sensors and regulators of intracellular pH. Nat Rev Mol Cell Biol. 11:50–61. https://doi.org/10.1038/nrm2820

Cavallini D, Marco CD, Mondavi B, Mori BG (1960) The cleavage of cystine by cystathionase and the transulfuration of hypotaurine. Enzymologia 22:161–173

Choi DW (1988) Glutamate neurotoxicity and diseases of the nervous system. Neuron 1:623–634

Clarke HT (1932) The action of sulfite upon cystine. J Biol Chem 97:235–248

Conner J, Russell PJ (1983) Elemental sulfur: A novel inhibitor of adenylate kinase. Biochem. Biophys. Res. Commun. 113:348–352

Furne J, Saeed A, Levitt MD (2008) Whole tissue hydrogen sulfide concentrations are orders of magnitude lower than presently accepted values. Am. J. Physiol. Regul. Integr. Comp. Physiol. 295:R1479–R1498

Furuie H, Kimura Y, Akaishi T, Yamada M, Miyasaka Y, Saitoh A, Shibuya N, Watanabe A, Kusunose N, Mashimo T et al (**2023**) Hydrogen sulfide and polysulfides induce GABA/glutamate/D-serine release, facilitate hippocampal LTP, and regulate behavioral hyperactivity. Sci. Rep. 13:17663

Garai D, Ríos-González BB, Furtmüller PG, Fukuto JM, Xian M, López-Garriga J et al (2017) Mechanisms of myeloperoxidase catalyzed oxidation of H_2S by H_2O_2 or O_2 to produce potent protein Cys–polysulfide-inducing species. Free Radical Biology and Medicine 113:551–563

Gerasimova E, Lebedeva J, Yakovlev A, Zefirov A, Giniatullin R, Sitdikova G (2015) Mechanisms of hydrogen sulfide (H_2S) action on synaptic transmission at the mouse neuromuscular junction. Neuroscience 303:577–585

Go YM, Jones DP (2013) The redox proteome. J Biol Chem 288:26512–26520

Greiner R, Palinkas Z, Basell K, Becher D, Antelmann H, Nagy P, Dick TP (2013) Polysulfides link H_2S to protein thiol oxidation. *Antioxid Redox Signal* 19:1749–1765

Hannah RS, Hayden LJ, Roth SH (1989) Hydrogen sulfide exposure alters the amino acid content in developing rat CNS. Neurosci. Lett. 99:323–327

Hargrove JL, Wichman RD (1987) A cystine-dependent inactivator of tyrosine aminotransferase co-purifies with gamma-cystathionase (cystine desulfurase). The Journal of Biological Chemistry 262:7351–7357

Hatakeyama Y, Takahashi K, Tominaga M, Kimura H, Ohta T (2015) Polysulfide evokes acute pain through the activation of nociceptive TRPA1 in mouse sensory neurons. Molecular Pain 11:24

Hildebrandt TM, Grieshaber MK (2008) Three enzymatic activities catalyze the oxidation of sulfide to thiosulfate in mammalian and invertebrate mitochondria. FEBS J. 275:3352–3361. https://doi.org/10.1111/j.1742-4658.2008.06482.x

Hosoki R, Matsuki N, Kimura H (1997) The possible role of hydrogen sulfide as an endogenous smooth muscle relaxant in synergy with nitric oxide. Biochemical and Biophysical Research Communications 237:527–531

Ida T, Sawa T, Ihara H, Tsuchiya Y, Watanabe Y, Kumagai Y, Suematsu M, Motohashi H, Fujii S, Matsunaga T et al (2014) Reactive cysteine persulfides and S-polythiolation regulate oxidative stress and redox signaling. Proc. Natl. Acad. Sci. USA 111:7606–7611

Ishigami M, Hiraki K, Umemura K, Ogasawara Y, Ishii K, Kimura H (2009) A source of hydrogen sulfide and a mechanism of its release in the brain. Antioxid. Redox Signal. 11:205–214

Jarosz AP, Wei W, Gauld JW, Auld J, Özcan F, Aslan M, Mutus B (2015) Glyceraldehyde 3-phosphate dehydrogenase (GAPDH) is inactivated by S-sulfuration in vitro. Free Radic Biol Med. 89:512–521. https://doi.org/10.1016/j.freeradbiomed.2015.09.007

Kato A, Ogura M, Suda M (1966) Control mechanism in the rat liver enzyme system converting L-methionine to L-cystine. 3. Noncompetitive inhibition of cystathionine synthetase-serine dehydratase by elemental sulfur and competitive inhibition of cystathionase-homoserine dehydratase by L-cysteine and L-cystine. J. Biochem. 59:40–48

Kimura H (2021) Hydrogen sulfide (H_2S) and polysulfides (H_2S_n) signaling: The first 25 years. biomolecules 11:896

Kimura H (2024) Hydrogen sulfide (H_2S)/polyfulfides (H_2S_n) signaling and TRPA1 channels modification on sulfur metabolism. biomolecules 14:129

Kimura H, Nagai Y, Umemura K, Kimura Y (2005) Physiological roles of hydrogen sulfide: synaptic modulation, neuroprotection, and smooth muscle relaxation. Antioxid Redox Signal. 7:795–803

Kimura Y, Goto Y, Kimura H (2010) Hydrogen sulfide increases glutathione production and suppresses oxidative stress in mitochondria. Antioxid Redox Signal. 12:1–13

Kimura Y, Mikami Y, Osumi K, Tsugane M, Oka J-I, Kimura H (2013) Polysulfides are possible H2S-derived signaling molecules in rat brain. The FASEB Journal 27:2451–2457. https://doi.org/10.1096/fj.12-226415

Kimura Y, Toyofuku Y, Koike S, Shibuya N, Nagahara N, Lefer D et al (2015) Identification of H2S3 and H2S produced by 3- mercaptopyruvate sulfurtransferase in the brain. Scientific Reports 5:14774

Kimura Y, Koike S, Shibuya N, Lefer D, Ogasawara Y, Kimura H (2017) 3-Mercaptopyruvate sulfurtransferase produces potential redox regulators cysteine- and glutathione-persulfide (Cys-SSH and GSSH) together with signaling molecules H2S2, H2S3 and H2S. Scientific Reports 7:10459. https://doi.org/10.1038/s41598-017-11004-7

Kimura Y, Shibuya N, Kimura H (2019) Sulfite protects neurons from oxidative stress. Br. J. Pharm. 176:571–582

Koike S, Ogasawara Y, Shibuya N, Kimura H, Ishii K (2013) Polysulfide exerts a protective effect against cytotoxicity caused by t-buthylhydroperoxide through Nrf2 signaling in neuroblastoma cells. FEBS Letters 587:3548–3555

Koike S, Kawamura K, Kimura Y, Shibuya N, Kiimura H, Ogasawara Y (2017) Analysis of endogenous H2S and H2Sn in mouse brain by high performance liquid chromatography with fluorescence and tandem mass spectrometric detection. Free Radic. Biol. Med. 113:355–362

Labrie V et al (2009) Serine racemase is associated with schizophrenia susceptibility in humans and in a mouse model. Hum. Mol. Genet. 18:3227–3243

Landry AP, Ballou DP, Banerjee R (2017) H2S oxidation by nanodisc-embedded human sulfide quinone oxidoreductase. The Journal of Biological Chemistry 292:11641–11649

Landry AP, Moon S, Kim H, Yadav PK, Guha A, Cho US, Banerjee R (2019) A catalytic trisulfide in human sulfide quinone oxidoreductase catalyzes coenzyme a persulfide synthesis and inhibits butyrate oxidation. Cell Chem Biol 26:1515–1525. e4. https://doi.org/10.1016/j.chembiol.2019.09.010

Lee JE, Kim Y, Kim KH, Lee DY, Lee Y (2016) Contribution of Drosophila TRPA1 to Metabolism. PLoS ONE 11:e0152935

Lepiarczyk E, Paukszto Ł, Wiszpolska M, Łopie'nska-Biernat E, Bossowska A, Majewski MK, Majewska M (2023) Molecular influence of resiniferatoxin on the urinary bladder wall based on differential gene expression profiling. Cells 12:462

Maher P, Currais A, Schubert D (2020) Using the oxytosis/ferroptosis pathway to understand and treat age-associated neurodegenerative diseases. Cell Chem Biol. 27:1456–1471

Murakami Y, Kameji T, Hayashi S (1984) Cysteine-dependent inactivation of hepatic ornithine decarboxylase. The Biochemical Journal 217:573–580

Murphy TH, Schnaar RL, Coyle JT (1990) Immature cortical neurons are uniquely sensitive to glutamate toxicity by inhibition of cystine uptake. FASEB J. **4**:1624–1633

Mustafa AK, Gadalla MM, Sen N, Kim S, Mu W, Gazi SK et al (2009) H2S signals through protein S-sulfhydration. Science Signaling 2:ra72

Nagai Y, Tsugane M, Oka J, Kimura H (2004) Hydrogen sulfide induces calcium waves in astrocytes. The FASEB Journal 18:557–559. https://doi.org/10.1096/fj.03-1052fje

Nagai Y, Tsugane M, Oka J-I, Kimura H (2006) Polysulfides induce calcium waves in rat hippocampal astrocytes. Journal of Pharmacological Sciences 100:200

Nakamura S, Nakamura M, Yamazaki I, Morrison M (1984) Reactions of ferryl lactoperoxidase (compound II) with sulfide and sulfhydryl compounds. The Journal of Biological Chemistry 259:7080–7085

Ogasawara Y, Ishii K, Togawa T, Tanabe S (1993) Determination of bound sulfur in serum by gas dialysis/high-performance liquid chromatography. Anal. Biochem. 215:73–81

Olson KR, Gao Y, DeLeon ER, Arif M, Arif F, Arora N, Straub KD (2017) Catalase as a sulfide–sulfur oxido-reductase: An ancient (and modern?) regulator of reactive sulfur species (RSS). Redox Biology 12:325–339

Olson KR, Gao Y, Arif F, Arora K, Patel S, DeLeon ER, Straub KD (2018) Metabolism of hydrogen sulfide (H2S) and productionof reactive sulfur species (RSS) by superoxide dismutase. Redox Biol. 15:74–85

Oosumi K, Tsugane M, Ishigami M, Nagai Y, Iwai T, Oka J-I, Kimura H (2010) Polysulfide activates TRP channels and increases intracellular Ca2+ in astrocytes. Neuroscience Research 685:e109–e222

Pecze L, Pelsoczi P, Kecskés M, Winter Z, Papp A, Kaszás K, Letoha T, Vizler C, Oláh Z (2009) Resiniferatoxin mediated ablation of TRPV1+ neurons removes TRPA1 as well. Can. J. Neurol Sci. 36:234–241

Ruetz M, Kumutima J, Lewis BE, Filipovic MR, Lehnert N, Stemmler TL, Banerjee R (2017) A distal ligand mutes the interaction of hydrogen sulfide with human neuroglobin. The Journal of Biological Chemistry 292:6512–6528

Searcy DG, Whitehead JP, Maroney MJ (1995) Interaction of Cu, Zn superoxide dismutase with hydrogen sulfide. Archives of Biochemistry and Biophysics 318:251–263. https://doi.org/10.1006/abbi.1995.1228

Semenza ER, Harraz MM, Abramson E, Malla AP, Vasavda C, Gadalla MM, Kornberg MD, Snyder SH, Roychaudhuri R (**2021**) D-cysteine is an endogenous regulator of neural progenitor cell dynamics in the mammalian brain. Proc. Natl. Acad. Sci. USA 118:e2110610118

Sha L, Linden DR, Farrugia G, Szurszewski JH (2013) Hydrogen sulfide selectively potentiates central preganglionic fast nicotinic synaptic input in mouse superior mesenteric ganglion. J Neurosci. 33:12638–12646

Shibuya N, Tanaka M, Yoshida M, Ogasawara Y, Togawa T, Ishii K, Kimura H (2009) 3-Mercaptopyruvate sulfurtransferase produces hydrogen sulfide and bound sulfane sulfur in the brain. Antioxidants and Redox Signaling 11:703–714

Shibuya N, Koike S, Tanaka M, Ishigami-Yuasa M, Kimura Y, Ogasawara Y et al (2013) A novel pathway for the production of hydrogen sulfide from D-cysteine in mammalian cells. Nature Communications 4:1366

Shigetomi E, Jackson-Weaver O, Huckstepp RT, O'Dell TJ, Khakh BS (2013) TRPA1 channels are regulators of astrocyte basalcalcium levels and long-term potentiation via constitutive D-serine release. J. Neurosci. 33:10143–10153

Stipanuk MH, Beck PW (1982) Characterization of the enzymic capacity for cysteine desulphhydration in liver and kidney of the rat. The Biochemical Journal 206:267–277. https://doi.org/10.1042/bj2060267

Stubbert D, Prysyazhna O, Rudyk O, Scotcher J, Burgoyne JR, Eaton P (2014) Protein kinase G Iα oxidation paradoxically underlies blood pressure lowering by the reductant hydrogen sulfide. Hypertension 64:1344–1351

Tan S, Schubert D, Maher P (2001) Oxytosis: A novel form of programmed cell death. *Cur. Top. Med. Chem.* **1**:497–506

Umino A et al (2017) Evidence for tonic control by the GABAA receptor of extracellular d-serine concentrations in the medial prefrontal cortex of rodents. Front. Mol. Neurosci. 10:240

Vitvitsky V, Yadav PK, Kurthen A, Banerjee R (2015) Sulfide oxidation by a noncanonical pathway in red blood cells generates thiosulfate and polysulfides. Journal of Biological Chemistry 290:8310–8320

Warenycia MW, Goodwin LR, Benishin CG, Reiffenstein RJ, Grancom DM, Taylor JD, Dieken FP (1989) Acute hydrogen sulfide poisoning. Demonstration of selective uptake of sulfide by the brainstem by measurement of brain sulfide levels. Biochem. Pharmacol. 38:973–981

Warenycia MW, Goodwin LR, Francom DM, Dieken FP, Kombian SB, Reiffenstein RJ (1990a) Dithiothreitol liberates non-acid labile sulfide from brain tissue of H2S-poisoned animals. Arch. Toxicol. 64:650–655

Warenycia MW, Kombian SB, Reiffenstein RJ (1990b) Stress-induced increases in brainstem amino acid levels are prevented bychronic sodium hydrosulfide treatment. Neurotoxicol 11: 93–97

Wolf ME (1998) The role of excitatory amino acids in behavioral sensitization to psychomotor stimulants. Prog. Neurobiol. 54:679–720

Yadav PK, Martinov M, Vitvitsky V, Seravalli J, Wedmann R, Filipovic MR, Banerjee R (2016) Biosynthesis and Reactivity of Cysteine Persulfides in Signaling. J. Am. Chem. Soc. 138:289–299

Hydrogen Sulfide Consumption and Cell Bioenergetics in Human (Patho)physiology

Maria Petrosino and Karim Zuhra

Contents

Abstract

Hydrogen sulfide (H_2S), once considered merely a toxic gas, is now recognized as a key endogenous signaling molecule with profound effects on vascular tone, inflammation, and cytoprotection. Central to its physiological roles is a tightly regulated balance between synthesis and degradation. While much attention has been given to H_2S biosynthesis and its signaling functions, its catabolism – particularly through the mitochondrial sulfide oxidizing pathway – has received comparatively less pharmacological exploration. This pathway not only serves as

M. Petrosino · K. Zuhra (✉)
Section of Pharmacology, Department of Oncology, Microbiology and Immunology, Faculty of Science and Medicine, University of Fribourg, Fribourg, Switzerland
e-mail: karim.zuhra@unifr.ch

A. Papapetropoulos, C. Szabo (eds.), *Hydrogen Sulfide: A Toxicant with Healing Potential*, Handbook of Experimental Pharmacology 292,
https://doi.org/10.1007/164_2025_776

a critical detoxification mechanism but also links H_2S oxidation directly to cellular bioenergetics by contributing to mitochondrial ATP production. Such coupling underscores a unique intersection between gasotransmitter regulation and energy metabolism. This chapter highlights the bioenergetic significance of H_2S degradation, emphasizing how modulation of its mitochondrial catabolic machinery could serve as a novel therapeutic strategy. By modulating H_2S clearance, especially in pathologies marked by disrupted sulfur homeostasis and mitochondrial dysfunction, targeted pharmacological intervention may restore metabolic balance and cellular energy efficiency.

Keywords

Bioenergetics · Catabolism · Hydrogen sulfide · Sulfide oxidation pathway

1 Introduction

Hydrogen sulfide (H_2S), historically regarded as a noxious gas, has emerged as a critical endogenous signaling molecule involved in diverse physiological processes, including vasodilation, neuromodulation, inflammation, and cytoprotection (Cirino et al. 2023). In human physiology, the synthesis and degradation of H_2S is tightly regulated to maintain sufficient bioavailability for these protective and regulatory roles, while preventing cytotoxic and pathological accumulation. This balance is achieved through a dynamic interplay between biosynthetic and degradative enzymatic pathways, with steady-state levels typically in the nanomolar range in most tissues (Furne et al. 2008; Kabil and Banerjee 2014).

Although the biology of H_2S metabolism is now well characterized, particularly in terms of its enzymatic sources and signaling functions, the degradative arm of this system remains relatively underexplored from a pharmacological perspective. At the core of this catabolic route lies the sulfide-oxidizing pathway, a mitochondrial enzymatic machinery that not only detoxifies H_2S but also couples its oxidation to ATP production, an elegant convergence of gasotransmitter regulation and cellular energetics.

This chapter explores the enzymatic pathways responsible for H_2S degradation, with a particular emphasis on their emerging relevance to pharmacology. We examine how modulation of the sulfide-oxidizing pathway and related catabolic enzymes may present novel therapeutic opportunities in conditions where H_2S homeostasis is disrupted, such as cardiovascular, neurodegenerative, and metabolic diseases. By shifting focus to the less-charted degradative side of H_2S biology, we aim to provide a framework for future pharmacological intervention targeting sulfur metabolism.

2 Mitochondrial H_2S Consumption Pathway: Structure and Function of H_2S Consumption Enzymes

The mitochondrial H_2S consumption pathway is a key biochemical system that regulates the intracellular levels of H_2S. This pathway serves to oxidize H_2S into less reactive species, primarily in the mitochondrial matrix, preventing its accumulation to toxic levels while enabling its role in cell signaling. The primary enzymes involved include sulfide:quinone oxidoreductase (SQR), persulfide dioxygenase (ETHE1), thiosulfate sulfurtransferase (TST, also known as rhodanese), and sulfite oxidase (SOx) (Fig. 1). SQR is the first and central enzyme in the mitochondrial H_2S oxidation pathway (Jackson et al. 2012). It is an inner mitochondrial membrane-bound flavoprotein (FAD) that catalyzes the two-electron oxidation of H_2S to a persulfide intermediate while transferring electrons to coenzyme Q (ubiquinone, CoQ) in the electron transport chain (ETC) (Fig. 1a). The resulting persulfide is then transferred to glutathione (GSH) or other acceptor molecules. In the first step of the reaction, SQR is catalytically promiscuous: the enzyme can recognize as main acceptors sulfite (kcat/KM values of $2.0 \times 10^6\ M^{-1}\ s^{-1}$) (Jackson et al. 2012; Landry et al. 2017) thus producing thiosulfate, or GSH (kcat/KM value of $1.1 \times 10^4\ M^{-1}s^{-}$) (Landry et al. 2017) thus producing glutathione persulfide (GSSH). Although sulfite has higher catalytic efficiency with SQR than GSH, the latter is favored as the physiological sulfane sulfur acceptor due to its much higher intracellular concentration. Kinetic data show that under typical cellular conditions, GSH supports a significantly higher turnover rate and is likely preferred over sulfite by at least 5: 1 – and potentially much more – despite the superior enzymatic efficiency boosted by sulfite (Landry et al. 2017). In the second half of the reaction, electrons are transferred from H_2S to the FAD cofactor, and subsequently to CoQ. These H_2S-derived electrons enter the mitochondrial electron transport chain at complex III, thereby contributing to oxidative phosphorylation (OXPHOS) and ATP production. The structure of human SQR was determined for the first time by Jackson et al. in 2019 (PDB: ID 6M06) (Jackson et al. 2019). It consists of two tandem Rossmann fold domains and a C-terminal domain. The first Rossmann fold starts near the N-terminus and binds the ADP portion of FAD, the second one is closer to the isoalloxazine ring of FAD with the two active-site cysteine residues (Cys201, Cys379) located just above the re-face of the flavin ring. The C-terminal domain contains two helices (denoted the C-terminal and penultimate C-terminal helices) which lie in the plane of the mitochondrial membrane, with their hydrophobic faces communicating with the membrane interior. Indeed, the C-terminal helix is strongly amphipathic and protrudes from the globular body of the protein and the penultimate helix, which is located just upstream of the C-terminal helix, also contributes to a strongly hydrophobic outward-facing surface. This non-polar environment creates a cavity that represents the binding site for CoQ (Fig. 1b).

In the mitochondrial matrix, GSSH can be further processed either by ETHE1 or by TST (Fig. 1a) depending on metabolic context and enzyme availability (Filipovic et al. 2018). ETHE1 is a non-heme iron-dependent sulfur dioxygenase that catalyzes the oxidation of GSSH to sulfite, thereby facilitating the clearance of reactive sulfur

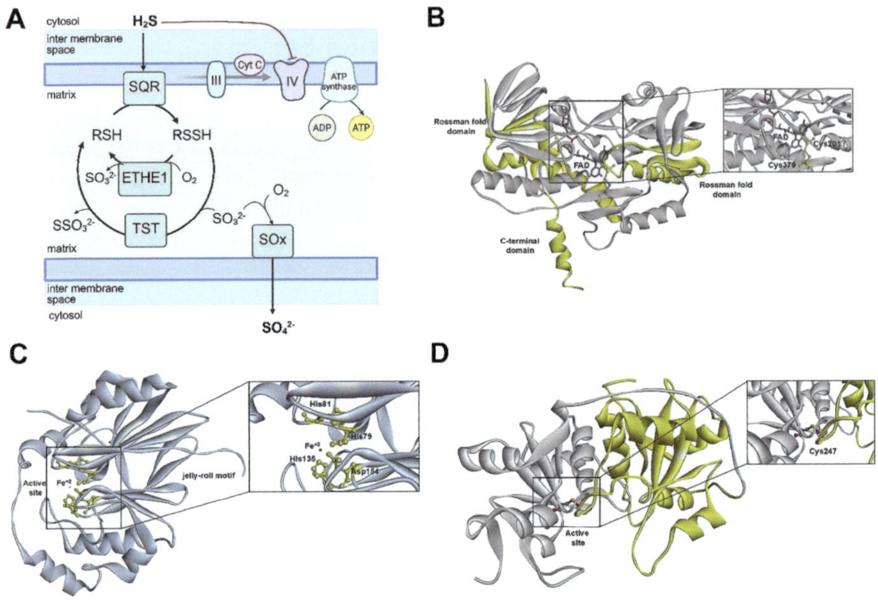

Fig. 1 Mitochondrial H_2S catabolism and its interaction with the respiratory chain. (**a**) Schematic illustration of the mitochondrial pathway of H_2S catabolism and its integration with the electron transport chain (ETC). H_2S, originating from the cytosol, is oxidized by sulfide:quinone oxidoreductase (SQR), located in the inner mitochondrial membrane. This oxidation reduces coenzyme Q (ubiquinone, Q_{ox} to ubiquinol, Q_{red}) and forms glutathione persulfide (GSSH). GSSH undergoes further metabolism in the mitochondrial matrix via a cascade involving multiple enzymes: ETHE1 (persulfide dioxygenase) converts GSSH and O_2 into sulfite (SO_3^{2-}); TST (thiosulfate sulfurtransferase) facilitates conversion of GSSH into thiosulfate (SSO_3^{2-}); and SOx (sulfite oxidase) oxidizes sulfite into sulfate (SO_4^{2-}). Glutathione (GSH) is regenerated in the process. Additionally, H_2S can act as an inhibitor of Complex IV (cytochrome c oxidase) within the ETC, potentially impacting mitochondrial respiration. (**b**) Crystal structure of sulfide:quinone oxidoreductase (SQR), PDB: 6MP5. The structure highlights the two Rossman fold domain at the N-terminus, the two cysteine residues (Cys201 and Cys379) that interact with the isoalloxazine ring of FAD and the C-terminal domain composed of the C-terminal and the penultimate C-terminal helices. (**c**) Crystal structure of persulfide dioxygenase (ETHE1), PDB: 4CHL.The structure shows the characteristic β-barrel architecture of a cupin-folded protein. Two conserved β-sheets, that form a jelly-roll motif, contribute to stabilize the active site which coordinates a single Fe^{2+} ion via a 2-His-1-carboxylate facial triad, consisting of His79, His81, His135, and Asp154 residues. (**d**) Crystal structure of thiosulfate sulfurtransferase (TST), PDB: 2ORA. The structure shows two similar domains arranged in tandem, each adopting an α/β fold and the conserved catalytic cysteine residue (Cys247 in humans), which is essential for sulfur transfer activity

species (Fig. 1a). Unlike many metalloenzymes, ETHE1 does not require an external cofactor, instead it shows a mononuclear iron center to activate molecular oxygen for oxidative catalysis (Henriques et al. 2014). The mechanism of ETHE1 remains partially unresolved, but current evidence supports a model in which the iron center activates O_2 to form a reactive iron–oxygen species that initiates the oxidation of the persulfide substrate. The enzyme specifically recognizes and oxidizes persulfide-bound glutathione, thereby preventing the accumulation of potentially toxic sulfane sulfur species. Structurally, ETHE1 adopts a cupin fold, a characteristic β-barrel architecture composed of two conserved β-sheets that form a jelly-roll motif (Fig. 1c). This structural motif stabilizes the active site and coordinates a single Fe^{2+} ion via a 2-His-1-carboxylate facial triad, a common metal-binding motif in non-heme iron enzymes (Pettinati et al. 2015; Goudarzi et al. 2018). ETHE1 function is not only essential for sulfur metabolism but also for broader mitochondrial redox balance. By oxidizing GSSH to sulfite, ETHE1 prevents the backflow of reactive sulfur species into signaling pathways and avoids excessive H_2S buildup, which can inhibit cytochrome c oxidase (CcOX, also known as complex IV) and impair oxidative phosphorylation. Disruption of ETHE1 function – through genetic mutations that alter its folding, metal-binding site, or catalytic activity – leads to ethylmalonic encephalopathy (EE), a rare but severe metabolic disorder characterized by elevated tissue sulfide levels, lactic acidosis, neurological dysfunction, and early-onset gastrointestinal symptoms. Patients with ETHE1 deficiency often exhibit signs of mitochondrial failure and typically leads to premature death due to uncontrolled sulfide accumulation and bioenergetic collapse (Burlina et al. 1991; Tiranti et al. 2009).

TST is a mitochondrial matrix enzyme that plays a critical role in both sulfur metabolism and cyanide detoxification (Suman and Gretarsdottir 2019). It is a member of the rhodanese superfamily, characterized by its sulfurtransferase activity and a highly conserved catalytic mechanism. TST contributes to the terminal steps of mitochondrial H_2S oxidation by transferring sulfane sulfur intermediate (e.g., GSSH) to various acceptors (such as thiosulfate or sulfite) (Fig. 1a), thereby preventing the accumulation of reactive sulfur species and contributing to redox homeostasis (Luo et al. 2025). A peculiar sulfur acceptor is cyanide, for which TST plays a central role in its enzymatic detoxification. In this reaction, TST catalyzes the transfer of a sulfane sulfur atom from thiosulfate to cyanide (kcat/KM $31 \times 10^3\ M^{-1}\ s^{-1}$) converting it into thiocyanate, a considerably less toxic compound that is readily excreted in the urine (Libiad et al. 2015). Clinically, thiosulfate administration is widely recognized as a first-line therapeutic approach in cyanide poisoning, leveraging the endogenous activity of TST to accelerate cyanide clearance and reduce systemic toxicity (Zhang et al. 2021). More recently, the discovery that cyanide can be endogenously produced in mammalian cells has renewed interest in TST, not only as a detoxifying enzyme but also as a key regulator of intracellular cyanide levels under physiological conditions (Zuhra et al. 2025; Zuhra and Szabo 2022). This expanding view positions TST as a potentially important modulator of cyanide signaling and sulfur metabolism beyond its traditionally understood role in toxicology. Structurally, TST is composed of two similar domains arranged in tandem, each adopting an α/β fold, a hallmark of the rhodanese family. While both domains share structural homology, only the C-terminal domain harbors the active site, which contains a conserved catalytic cysteine residue (Cys247 in humans) essential for sulfur transfer activity (Fig. 1d). This cysteine undergoes reversible persulfidation, forming a Cys-SSH intermediate during the catalytic cycle (Filipovic et al. 2018). The enzyme follows a classic ping-pong mechanism, where it first reacts with a sulfur donor (e.g., thiosulfate), transferring the sulfur atom to the catalytic cysteine to form the persulfide intermediate. Subsequently, sulfur is transferred from this intermediate to a sulfur acceptor such as cyanide, resulting in the formation of thiocyanate (Westley 1973). This two-step mechanism, comprising sulfur acquisition and subsequent transfer, is tightly coordinated by the structural integrity of the active site and surrounding residues. Key features that facilitate this process include the positioning of nucleophilic side chains, the hydrogen bonding network that stabilizes the transition state, and the flexible loop regions that accommodate different sulfur donors and acceptors. Disruption of this architecture through mutation of the catalytic cysteine or destabilization of the domain interface can compromise enzymatic activity and lead to the accumulation of persulfides or cyanide, both of which are toxic to mitochondrial function (Billaut-Laden et al. 2006).

The terminal enzyme of the mitochondrial sulfide oxidation pathway, SOx, catalyzes the oxidation of sulfite to sulfate (Fig. 1) (Velayutham et al. 2016). This reaction is essential for the complete detoxification of H_2S and for preventing the accumulation of reactive sulfur intermediates.

SOx is localized to the intermembrane space of mitochondria and plays a pivotal role in linking sulfur metabolism to the mitochondrial respiratory chain. Structurally, it functions as a homodimer, with each monomer composed of two distinct redox-active domains: a molybdenum cofactor (MoCo)-binding domain at the C-terminus, which harbors the catalytic site, and a heme-binding cytochrome b5-like domain at the N-terminus, which participates in electron transfer (Johnson-Winters et al. 2010). These two domains are connected by a flexible polypeptide linker which permits dynamic positioning and optimal orientation of the redox centers during catalysis. This flexible architecture is critical for efficient intra-subunit electron shuttling. At the active site within the MoCo domain, sulfite is oxidized to sulfate in a reaction that involves a two-electron transfer. The molybdenum ion, typically in the Mo (VI) oxidation state, is reduced to Mo(IV) as it accepts electrons from sulfite during its conversion to sulfate. These electrons are then relayed from the molybdenum center to the intrinsic heme group located in the cytochrome b_5-like domain. From there, electrons are transferred to cytochrome c in the mitochondrial electron transport chain, thereby coupling sulfur metabolism with cellular energy production (Johnson-Winters et al. 2010). The spatial arrangement and electronic communication between MoCo and heme are finely tuned by the interdomain linker, which governs not only the distance between centers but also the flexibility required for conformational changes during electron transfer. Mutations affecting SOx structure, especially in the MoCo or heme domains, can severely impair enzymatic function, leading to SOx deficiency (or sulfocysteinuria), a rare but severe metabolic disorder characterized by elevated sulfite and thiosulfate levels in tissues and urine, progressive neurological deterioration, seizures, and early mortality (Johnson et al. 2002).

3 Cellular Bioenergetics and the Multifaceted Role of H_2S in Human Physiology

The key players in fueling the metabolism of cellular bioenergetic homeostasis include glucose, fatty acids, and amino acids. These nutrients serve as primary energy sources, entering various metabolic pathways to be broken down into simpler molecules that feed into glycolysis, the Krebs cycle, and OXPHOS, ultimately supporting ATP production and maintaining energy balance within the cell. *Glycolysis* is the first step, occurring in the cytoplasm, where glucose is broken down into pyruvate, producing overall 2 molecules of ATP and 2 molecules of NADH per glucose molecule. The *Krebs cycle*, which takes place in the mitochondria, further processes pyruvate, generating overall 6 molecules of NADH and 1 molecule of $FADH_2$, which carry electrons to the next stage. *OXPHOS*, also in the mitochondria, uses these electrons to drive the production of a large amount of ATP (30–32 molecules per molecule of glucose) through the ETC, with oxygen (O_2) as the final electron acceptor. Modulation of ATP synthesis to meet fluctuations of cellular energetic needs is still matter of investigation and are achieved through (i) negative feedback of ATP; (ii) reducing equivalents delivery to the ETC; (iii) O_2 availability; (iv) calcium homeostasis; (v) direct control of ETC complexes activity via post-translational modification or small molecules modulators such as gasotransmitters (Balaban et al. 1990; Brown 1992, 2001; Cárdenas et al. 2010; Stram and Payne 2016; Giuffrè et al. 2018). Small gaseous molecules such as nitric oxide (NO), carbon monoxide (CO), and hydrogen sulfide (H_2S) are part of the family of gasotransmitters and, despite their simple structure, they play a remarkably complex and dual role in human physiology, being cytoprotective at low concentrations and a lethal poison at higher concentrations. A common feature of gasotransmitters is the ability to negatively modulate cellular bioenergetics by binding to ferric heme a_3 of CcOX thus interfering with O_2 binding (Wang 2014). However, unlike NO and CO, the role of H_2S in cellular bioenergetics is uniquely marked by its dose-dependent effects. At low concentrations, H_2S acts as a stimulator of ATP synthesis rather than an inhibitor, thus enhancing mitochondrial function and energy production (Szabo et al. 2014). This duality, often represented as a bell-shaped dose-response curve, underscores the fine balance that cells must maintain to harness the benefits of H_2S while avoiding its toxic consequences.

3.1 Molecular Mechanisms Underlying H_2S Toxicity and Mitochondrial Respiration Inhibition

The cytotoxic effects of H_2S at elevated concentrations are well-documented, yet the molecular mechanisms that mediate this toxicity remain an area of active investigation. One of the key mechanisms by which H_2S exerts its harmful effects is through the inhibition of mitochondrial respiration. Specifically, H_2S can bind reversibly to the heme moiety of CcOX, the terminal enzyme of the ETC, playing a critical role in the reduction of O_2 to water, a process that drives the generation of the proton gradient necessary for ATP synthesis. By binding to CcOX, H_2S effectively impairs this enzyme function, disrupting the flow of electrons through the ETC and, consequently, diminishing ATP production (Vicente et al. 2016). The inhibition of CcOX by H_2S is characterized by a relatively low inhibition constant (Ki), with studies using isolated enzymes reporting Ki values as low as 0.2 μM at physiological pH (7.4) (Petersen 1997). This high affinity suggests that even low concentrations of H_2S can significantly impact mitochondrial function. However, in more complex systems, such as isolated mitochondria or cultured cells, the concentration of H_2S required to observe inhibition of the ETC is typically much higher, often in the range of tens of micromolar (Leschelle et al. 2005). This discrepancy may be due to several factors, including the presence of protective mechanisms within cells, the compartmentalization of H_2S within subcellular organelles, and the dynamic interactions between H_2S and other cellular components.

3.2 The Sulfide Oxidation Pathway: A Cellular Defense Mechanism

To counteract the potential toxicity of H_2S, mammalian cells have evolved a specialized metabolic pathway known as the sulfide oxidation pathway. This pathway not only serves to detoxify H_2S but also links sulfide metabolism with ATP production, thereby integrating waste management with energy generation. The sulfide oxidation pathway is centered around the enzyme SQR, which catalyzes the first step in the oxidation of H_2S. SQR facilitates the transfer of a sulfur atom from H_2S to an acceptor molecule, such as GSH, while simultaneously transferring the electrons derived from this reaction to CoQ in the mitochondrial ETC (Fig. 1). This electron transfer contributes to the proton gradient used to generate ATP via OXPHOS (Libiad et al. 2014). The activity of the sulfide oxidation pathway can be studied either by measuring O_2 consumption rate (OCR) in whole cells (integrative approach) or by measuring the reduction of CoQ in tissue/cell mitochondrial extract (reductionist approach) in response to increasing concentrations of H_2S (Fig. 2a, b). In experimental settings, H_2S is often administered as sodium sulfide (Na_2S) to mimic physiological conditions. In the integrative approach, where the ETC remains coupled, proper ETC function is maintained, allowing assessment of the potent inhibitory effects of H_2S on CcOX, Na_2S it is typically supplied in a controlled manner, often through continuous infusion, to prevent toxic spikes in H_2S levels that could lead to cellular dysfunction. The OCR driven by sulfide can be halted by inhibitors of complex III, such as Antimycin A, confirming that the electrons from sulfide enter the ETC at the level of CoQ (Goubern et al. 2007). The mitochondrial sulfide-oxidizing capability is typically evaluated at 37 °C using a high-resolution respirometer (Oxygraph-2 k, Oroboros Instruments, Innsbruck, Austria), which features dual 1.5 mL chambers and a micropump (TIP-2 k) for precise sulfide injections into the chambers. This procedure involves injecting sulfide into a cell suspension at a gradually increasing flow rate (controlled by the pump), with the mitochondrial sulfide-detoxifying function being gauged by the resulting increase in

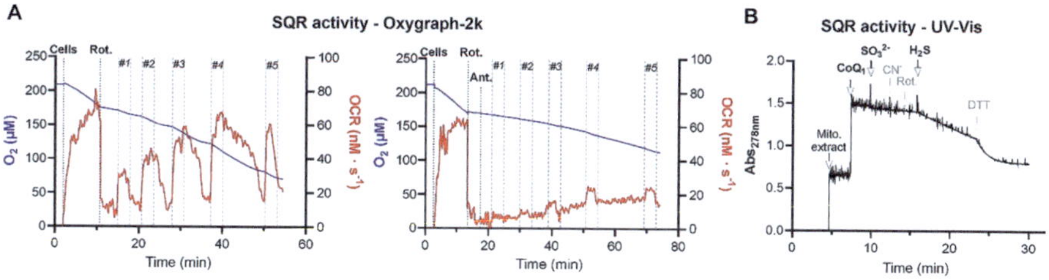

Fig. 2 SQR activity using two different approaches. (**a**) Integrative approach: the effect of sulfide on oxygen consumption in SW480 colorectal cancer cells by measuring oxygen consumption rates (OCR). Cells were first treated with rotenone (Rot.) alone (left panel) or with supplementation of Antimycin A (Ant.) and then subjected to stepwise injections of sulfide (10–160 nL·s^{-1}; injections #1–5) (adapted from Malagrinò et al. 2019). (**b**) Reductionist approach: SQR activity in mitochondrial extracts, monitored spectroscopically by tracking the reduction of CoQ1 in the presence of the SQR substrates sulfite and H_2S. Cyanide was used to inhibit complex IV, while rotenone was added to block H_2S-driven reverse electron flow through complex I. (Adapted from Zuhra et al. 2019)

OCR (Abou-Hamdan et al. 2015). As the sulfide injection rate rises, the concentration of sulfide in the solution –and consequently the sulfide-driven cellular OCR – also increases until the point where the injected sulfide concentration inhibits CcOX. SQR-mediated sulfide detoxification catalyzes electron transfer to O_2 via the quinol c reductase (complex III)/cytochrome c/CcOX pathway, as well as reverse electron transfer through complex I (Lagoutte et al. 2010). To avoid the misestimation of mitochondrial sulfide-oxidizing activity caused by electrons from SQR-mediated sulfide oxidation being redirected away from O_2 reduction, measurements are recommended to be conducted in the presence of rotenone, a known complex I inhibitor. Following the depression in the OCR induced by rotenone, increasing flow rates of sulfide account for a concomitant increase on OCR. This, however, cannot be reproduced in the presence of antimycin A. By subtracting these results from those obtained in the absence of antimycin A, the actual mitochondrial oxygen consumption due to sulfide oxidation can be determined, leading to an estimate of the H_2S-oxidizing activity, based on the reported consumption of approximately 1.33 molecules of H_2S per molecule of O_2 in the mitochondrial sulfide-oxidizing pathway (Lagoutte et al. 2010). The capacity for sulfide oxidation varies considerably across different cell types, reflecting their distinct metabolic demands and exposure to H_2S. For example, cells in the nervous system exhibit minimal sulfide oxidizing activity, thus making this tissue particularly vulnerable to fluctuations of H_2S levels (Lagoutte et al. 2010; Linden et al. 2012; Fagerberg et al. 2014). In contrast, colonocytes display remarkably high sulfide-oxidizing activity. This is particularly important because these cells are chronically exposed to high concentrations of H_2S, often reaching millimolar levels, produced by the gut microbiota during the fermentation of sulfur-containing amino acids. The efficient oxidation and detoxification of H_2S by colonocytes are crucial for maintaining the integrity of the colonic epithelium and preventing inflammatory conditions such as colitis (Goubern et al. 2007). The enzymes involved in the sulfide oxidation pathway, including SQR, are highly expressed and precisely localized within colonic tissues, ensuring rapid and effective detoxification of H_2S (Libiad et al. 2019). The oxidation of H_2S by colonocytes is coupled with ATP production, consuming approximately 0.75 molecules of O_2 per molecule of H_2S oxidized (0.25 by CcOX and 0.5 by ETHE1) (Lagoutte et al. 2010). Although this process is less energy-efficient than other metabolic pathways, its primary role is the safe disposal of potentially toxic H_2S, rather than maximal energy production. The additional consumption of O_2 by ETHE1 further reduces the efficiency of this pathway but is a necessary trade-off to prevent the accumulation of harmful levels of H_2S in the colon.

3.3 H_2S as an Emergency Substrate and Its Role in Disease

Beyond its role in normal physiology, H_2S has been proposed to act as an "emergency" substrate for the mitochondrial ETC, particularly under conditions where traditional electron donors, such as those derived from the Krebs cycle, are insufficient to meet cellular energy demands. This scenario may occur during periods of metabolic stress or nutrient deprivation, where the availability of glucose and fatty acids is limited. In such cases, H_2S can serve as an alternative electron donor, helping to maintain ATP production and cellular viability. The ability of cells to utilize H_2S in this manner is particularly relevant in tissues with high metabolic demands, such as the heart and skeletal muscles, as well as in pathological conditions where traditional metabolic pathways are compromised (Módis et al. 2014). A critical aspect of sulfide biology is that its catabolism is sensitive to O_2 availability (Olson 2015). Therefore, under O_2 deprivation (hypoxia), H_2S degradation slows down, resulting in higher H_2S mitochondrial concentrations (Arndt et al. 2017). In such situations, H_2S may orchestrate a temporary "shutoff" mechanism by negatively modulating O_2 consumption, thus allowing the body to redistribute O_2 availability. H_2S also promotes vasodilation, aiding O_2 delivery to hypoxic tissues. In cancer, H_2S signaling supports tumor cell survival under hypoxia by enhancing mitochondrial and cytosolic production via cystathionine β-synthase (CBS) and cystathionine γ-lyase CSE, promoting ATP synthesis and resistance to oxidative stress (Szabo 2016; Giuffrè et al. 2020; Teng et al. 2013; Wang et al. 2014). The role of H_2S in supporting cellular bioenergetics in cancer cells is linked to its ability to induce a metabolic shift under hypoxic conditions. Specifically, H_2S promotes a transition from OXPHOS (which depends on O_2) to glycolysis, an anaerobic process. This shift enables cells to generate ATP even in O_2-limited environments. In this context, H_2S catabolism through the SQR pathway plays a critical role by regulating H_2S levels and preventing cytotoxicity while still feeding ETC, thus maintaining tonic OXPHOS (Malagrinò et al. 2019). This concept has been emphasized in rodent models of cerebral ischemia. Although brain tissue expresses low levels of SQR – which accounts for almost undetectable enzymatic activity – it has been shown that mice chronically exposed to controlled low doses of H_2S (H_2S preconditioning) displayed improved outcomes in resolving transient focal cerebral ischemia and post-ischemic cerebral edema, compared to mice that only received saline (Gheibi et al. 2014). Notably, the enzyme SQR appears to play a crucial role. Preconditioning with H_2S has been linked to higher levels of SQR expression in brain tissue, whereas SQR knockdown reduced the cytoprotective effect of H_2S preconditioning, thus suggesting that SQR may contribute to the increased resistance to the hypoxic conditions associated with cerebral ischemia (Marutani et al. 2021). The sulfide oxidizing pathway has therefore implications in cellular protection that may go well beyond the mere regulation of bioenergetics. The catalytic activity of SQR, besides injecting reducing equivalents in the ETC, is among the most physiologically relevant sources of sulfane sulfurs (including glutathione persulfide) which are believed to be the actual mediator of the majority of the cytoprotective properties commonly attributed to H_2S (Libiad et al. 2019).

3.4 Effect of Post-translational Modification on Cellular Bioenergetics: The Role of Persulfidation

The role of post-translational modification (PTM) on the regulation of relevant metabolic pathways has been a topic of extensive research and is now considered a milestone of cellular signal transduction. Following the discovery of protein phosphorylation, with the implementation of the sensitivity of the detection methods, protocols for sample preparations, and more recently, computational tools, new protein modifications and their functions were discovered (Stram and Payne 2016; Keenan et al. 2021). Cysteine, with its electron-rich thiolate group, stands out as the most reactive nucleophile among amino acids, facilitating a variety of chemical modifications. In recent years, protein *S*-persulfidation (*S*-sulfhydration), emerged as a major PTM, conceivably as prominent as phosphorylation (Mustafa et al. 2009; Paul and Snyder 2012). The role of *S*-persulfidation has been so far understudied due to the intrinsic reactivity of persulfides and their similarity to cysteine (Zivanovic et al. 2019). It is now believed that much of the biological role of H_2S under physiological conditions is actually accomplished via *S*-persulfidation of target protein thus contributing to the regulation of metabolic pathways. H_2S through the *S*-persulfidation signaling is now attracting as much interest as its “elder brother” NO with *S*-nitrosylation. Together, these PTMs signaling routes may define a potential new paradigm in signal transduction (Paul and Snyder 2012; Martínez-Ruiz and Lamas 2004), but despite some common features between the two PTMs, their biological role may differ, being synergistic or opposite, depending on the context. Like *S*-nitrosylation, *S*-persulfidation is reversible through the thioredoxin system, supporting its role as a PTM in living organisms (Krishnan et al. 2011; Benhar et al. 2009). For instance, the activity of glyceraldehyde-3-phosphate dehydrogenase (GAPDH), a key enzyme in glycolysis, relies heavily on cysteine residue Cys152; *S*-nitrosylation of this site impairs its glycolytic function but promotes the nuclear translocation of a GAPDH isoform involved in transcription (Hara et al. 2005; Sen et al. 2008). In contrast, S-sulfhydration of Cys150 in rodent GAPDH (Cys152 in humans) results in a seven-fold increase in glycolytic activity. Consequently, in the liver of mice lacking CSE, GAPDH catalytic activity is reduced by 35%, although its protein levels remain unchanged (Mustafa et al. 2009). Within mitochondria, exposure to the NO donor *S*-nitrosoglutathione (GSNO) has been shown to induce S-nitrosylation at more than 80 cysteine sites distributed across approximately 60 proteins, one of which is cysteine 294 within the α-subunit of ATP synthase (ATP5A1) (Sun et al. 2007; Murray et al. 2011). Structural analysis from X-ray crystallography of Bovine Mitochondrial F1-ATPase (PDB: 1E1Q) (Gibbons et al. 2000) revealed that Cys294 is positioned on the surface of the ATP5A1, surrounded by multiple basic amino acid residues. This local microenvironment promotes its deprotonation under physiological pH, rendering the thiol group more reactive toward oxidative modifications. Consequently, Cys294 is thought to function as a redox-sensitive site, enabling ATP synthase to sense and adapt to fluctuations in the mitochondrial redox state, thereby modulating its catalytic activity. Further work by Sun et al. (2007) showed that preconditioning mouse hearts with GSNO promoted S-nitrosylation of the ATP5A1, which was associated with reduced ATP hydrolytic activity. The authors suggested that this inhibition helps conserve intracellular ATP during myocardial ischemia by switching ATP synthase from ATP hydrolysis to ATP preservation. Such a shift reduces mitochondrial membrane potential and, in turn, diminishes the driving force for calcium accumulation within the mitochondrial matrix. Collectively, these changes in energy utilization were proposed to contribute to a protective mechanism during ischemic stress. In contrast, the same ATP synthase α-subunit (ATP5A1) is also a target of *S*-persulfidation. Módis et al. (2016) reported that H_2S induces *S*-persulfidation of ATP5A1 in cells, which enhances its enzymatic activity. This modification occurs at cysteine residues Cys244 and Cys294, as confirmed by site-directed mutagenesis experiments. In mice lacking CSE, *S*-persulfidation of ATP5A1 was significantly reduced, indicating a physiological role for H_2S in this process. Furthermore, in a burn injury model, elevated H_2S levels correlated with increased *S*-persulfidation of ATP5A1, which subsequently declined as H_2S levels dropped. Overall, these findings suggest that *S*-persulfidation of ATP synthase, regulated by H_2S, supports the maintenance of mitochondrial function under both physiological and pathological conditions.

H_2S-metabolizing enzymes regulation may be emblematic of the crosstalk among gasotransmitters. S-nitrosylation of critical cysteine residues, such as the catalytic thiol in the TST active center (Cys247), has been reported to inhibit enzymatic activity, providing a redox-sensitive mechanism for downregulating H_2S catabolism under nitric oxide-rich conditions (Kwiecień et al. 2003). Intriguingly, *S*-nitrosylation negatively modulates the enzymatic activity of CSE (Fernandes et al. 2021) conceivably influencing *S*-persulfidation of downstream proteins, as part of a widely described crosstalk between gasotransmitters (Giuffrè et al. 2018). In H_2S-metabolizing enzymes, S-sulfhydration is not only a reversible regulatory modification, but also an essential catalytic intermediate, enabling efficient sulfur transfer (Landry et al. 2018). In SQR, for example, catalysis proceeds via the formation of a trisulfide intermediate on a conserved active-site cysteine (Cys379) which facilitates electron transfer to CoQ during sulfide oxidation (Landry et al. 2019; Filipovic et al. 2018). Similarly, in TST, the formation of a Cys-SSH intermediate is central to its two-step sulfur transfer mechanism (Vignane and Filipovic 2023). Beyond its role in catalysis, emerging evidence suggests that non-catalytic persulfidation of adjacent or regulatory cysteines may modulate enzyme conformation, protect from irreversible oxidation, or affect interactions with redox partners.

Beyond directly modulating mitochondrial activity, H_2S also acts indirectly by enhancing oxygen delivery via vascular regulation, scavenging ROS to protect mitochondria, and supporting glucose metabolism by influencing insulin release and energy balance. In the endocrine system, H_2S plays a role in glucose metabolism by reducing insulin release from β-cells, via several mechanisms, including activation of K_{ATP} channels (Chen et al. 2017; Zhang et al. 2020). It also seems to have a role in fat metabolism, influencing lipid accumulation and adipogenesis (Katsouda et al. 2022). In the central nervous system H_2S seems to be protective against oxidative stress, apoptosis, and degeneration, directly scavenging ROS, reactive nitrogen species (RNS) (Popov 2013). H_2S also promotes the activation of the Nrf2-mediated antioxidant response, which helps in protecting cells from oxidative damage, via *S*-persulfidation of Keap1 (Cys151) (Yang et al. 2013; Panieri et al. 2020). Finally, recent studies suggested that H_2S via protein *S*-persulfidation may also have a role in protecting from lifespan stress stimuli thus promoting resistance to oxidative stress and mitochondrial function (Zivanovic et al. 2019; Vo et al. 2022; Wang 2012). Altogether, this dual role of S-sulfhydration, as both a metabolic intermediate and regulatory signal, positions it as a key modulator of mitochondrial H_2S oxidation capacity.

4 Pathological Aspects of Hydrogen Sulfide

4.1 The Role of H_2S Catabolic Route

The multi-layered regulatory mechanisms described above, from metabolic feedback to redox-sensitive and enzyme-specific PTMs, allow cells to fine-tune H_2S catabolism in response to changing physiological and pathological conditions. Such dynamic control ensures that H_2S levels remain within a homeostatic window, preventing its cytotoxic accumulation while preserving its diverse signaling functions. Indeed, several pathological conditions have been associated with a dysregulation of the production or detoxification of H_2S. Chronic H_2S accumulation has been associated with (i) *neurotoxicity:* high levels of H_2S can inhibit CcOX, leading to reduced ATP production, causing neuronal damage, a condition that have been studied for Down Syndrome in which overdosage of CBS enzyme accounts for toxic overproduction of H_2S (Panagaki et al. 2022); (ii) *cardiovascular diseases:* dysregulation of H_2S has been linked to hypertension, atherosclerosis, and heart failure. It can disrupt NO signaling and alter vascular tone (Kolluru et al. 2023); (iii) *oncological diseases*: elevated levels of H_2S have been associated with tumor progression in different types of cancer (Szabo 2016). Accumulation of H_2S is commonly associated with increased levels of H_2S-producing enzymes such as

CBS, CSE, and mercaptopyruvate sulfurtransferase (3-MST) or with a dysregulation of the enzymes involved in H_2S degradation, namely SQR, ETEH1, TST, or SOx. Mutations in genes encoding for enzymes regulating sulfide catabolic route are a cause of enzymatic dysfunction, leading to a failure in H_2S detoxification and potentially inducing pathological accumulation. Several missense mutations have been described and associated with downregulation and/or with loss of function of SQR enzyme. For instance, the toxic accumulation of H_2S due to impairment of the sulfide oxidation linked to mutations in *SQRDL* gene has been associated with damages in tissues with high oxygen consumption, such as brain, muscles, and heart causing neurological issues (developmental delays, seizures), muscle weakness (myopathy), and lactic acidosis. Leigh disease is an example of SQR deficiency, representing a common neurological manifestation of mitochondrial bioenergetic defects. It is genetically heterogeneous, with over 75 causative genes identified, among which SQR mutations appear to play a significant role (Friederich et al. 2020). After sequencing analysis of families with cases of Leigh disease, p. Glu213Lys mutation has been described (Friederich et al. 2020). This mutation is a non-synonymous single nucleotide polymorphism (nsSNP) which involves the replacement of an acidic with a basic amino acid and disrupts hydrogen bonding interactions with neighboring residues, thus destabilizing the SQR protein. This protein instability resulted in a severe reduction of the protein expression levels and strongly decreased the enzymatic activity of SQR, thus limiting the ability to detoxify H_2S.

ETHE1 is another important enzyme in the H_2S detoxification pathway. Mutations in the *ETHE1* gene lead to the rare condition ethylmalonic encephalopathy (EE), a severe metabolic disorder. EE is an autosomal recessive disorder characterized by accumulation of toxic metabolites (such as H_2S, ethylmalonic acid, and lactic acid) which leads to progressive neurodegeneration, vascular lesions in the brain, muscle weakness, and severe metabolic acidosis. The inability to properly oxidize sulfur compounds in patients with EE leads to high levels of H_2S, which can inhibit key mitochondrial enzymes, further compromising energy metabolism. H_2S is also vasoactive and vasotoxic, thus explaining the symptoms of petechiae and orthostatic acrocyanosis in EE (Tiranti et al. 2009). Furthermore, vasculopathy is also associated with multiple necrotic brain lesions that result in hypotonia, psychomotor delay and regression, predominant pyramidal signs, seizures, and eventually global neurological failure (Walsh et al. 2010; Dionisi-Vici et al. 2016). EE always leads to death within the first decade of life in most of the children (Walsh et al. 2010). EE was first identified in 3 unrelated Italian children in 1991, 3 patients have been reported in China (Chen et al. 2020), but most of the patients have been diagnosed in Mediterranean and Arabic area (Tiranti et al. 2009). Only 37 disease-causing variants of ETHE1 are listed in Human Gene Mutation Database (HGMD) Professional, among them 19 are missense variants, 8 are insertion and deletion variants, 4 are exon deletion variants, and 5 are splicing variants, including 3 at canonical splicing sites. Most of them result in loss of protein function and some of them induce protein instability and low expression levels.

The *TST* gene encoding human rhodanese is located on chromosome 22 (Aita et al. 1997), and several nsSNPs have been reported (Billaut-Laden et al. 2006) with two of them, which lead to p. Glu102Asp and p. Pro285Ala substitutions, respectively, being the most prevalent (Billaut-Laden et al. 2006). The variant residues are located on loops in the N- and C-terminal domains, with Glu102, which resides at the lip of a cavity leading to the active site cysteine and could, in principle, influence substrate access and binding, and Pro285 that is solvent-exposed and distant from the active site (Libiad et al. 2015). Both resulted in increased protein stability and p. Glu102Asp was significantly more efficient than wild-type or p. Pro285Ala in converting cyanide to thiocyanate. Recently, it has been reported that TST significantly decreases in expression from normal tissue to tumor at every Dukes' stage A-D (Birkenkamp-Demtroder et al. 2002), and downregulation of *TST* gene expression has also been observed in diseases such as Friedreich's ataxia (Tan et al. 2003). Among the diseases associated with dysregulated levels of H_2S, cancer has an evident role. In cancer cells and tumor microenvironments, the enzymes deputed to the production of H_2S are often overexpressed or mutated in a way that the resulting enzyme is more prompt to produce H_2S. On the other hand, mutations also occur in the H_2S catabolizing enzymes thus leading to elevated H_2S levels. Increased H_2S in tumors enhances mitochondrial bioenergetics and supports the increased energy demands of rapidly growing cancer cells thus promoting cell proliferation. H_2S stimulates angiogenesis by promoting the formation of new blood vessels, which is critical for tumor growth and metastasis and, at the same time, protects cancer cells from apoptosis, contributing to tumor survival. Finally, H_2S promotes cell migration and invasion, which are essential steps for metastasis (Szabo et al. 2013; Zuhra et al. 2021).

4.2 Pharmacological Interventions to Regulate Hydrogen Sulfide

Pharmacological interventions targeting H_2S production or signaling have been shown to be valuable for managing pathological conditions, either in terms of normalizing aberrant elevated levels of H_2S using inhibitors of its generation or exploiting H_2S cytoprotectant effect using H_2S donors. Key targets include the H_2S-producing enzymes CBS, CSE, and 3-MST, whose inhibition can help manage diseases where elevated H_2S is harmful, such as cancer and cardiovascular disorders (Zuhra et al. 2020; Sun et al. 2009; Burnett et al. 1980; Asimakopoulou et al. 2013; Tokoro et al. 2003; Hanaoka et al. 2017). Although the physiological and biochemical mechanisms governing H_2S degradation are now well understood, pharmacological approaches to modulate H_2S catabolism remain in their infancy. In contrast to the numerous H_2S donors developed to increase endogenous levels for therapeutic benefit (Zhao et al. 2014), very few efforts have focused on modulating its enzymatic clearance. As the rate-limiting step in sulfide oxidation, SQR is considered a prime drug target for modulating this pathway. As a consequence, putative inhibitors of SQR may prove beneficial for the treatment of oncological diseases characterized by elevated H_2S levels, whereby blocking the H_2S degradation would lead to cell toxicity (Módis et al. 2013; Libiad et al. 2019) or cardiovascular diseases associated with reduced H_2S availability, where inhibiting its degradation could enhance cytoprotection (Calvert et al. 2010). In general, SQR inhibitors are expected to show maximal efficacy in tissues where the enzyme is highly expressed, such as the colon, liver, heart, and kidney (Lagoutte et al. 2010). SQR inhibition as a therapeutic strategy for targeting heart failure through modulation of H_2S signaling has been recently demonstrated. The authors of the research identified STI1 as a potent and selective inhibitor ($IC_{50} = 29$ nM) that competitively binds the Coenzyme Q (CoQ)-binding site of SQR. Structural insights from X-ray crystallography of human SQR (PDB: ID 6M06) revealed two distinct active regions: a hydrophilic site responsible for H_2S oxidation and a larger, hydrophobic CoQ-binding pocket. The latter was found to be more favorable for drug interaction. In a mouse model of pressure overload-induced heart failure (TAC model), STI1 treatment effectively prevented adverse cardiac remodeling, reduced pulmonary congestion, and preserved left ventricular function, thereby halting progression to heart failure with reduced ejection fraction (HFrEF) (Jackson et al. 2022). Within the cardiovascular system, the enzyme primarily responsible for H_2S production is CSE, which is expressed in cardiac muscle cells, vascular smooth muscle, and endothelial cells. A decline in H_2S levels has been linked to conditions like atherosclerosis and is associated with significant alterations in the persulfidation of various endothelial proteins, most notably those in the integrin family, observed in both native and cultured human endothelial cells (Bibli et al. 2021). In both human patients and mouse models of atherosclerosis, a reduction in the reactive thiol proteome has been documented, often accompanied by elevated levels of circulating cystathionine, indicating reduced CSE enzyme function. Experimental treatments using the H_2S donor compound SG1002 in cell cultures and mouse models, as well as administration of methylsulfonylmethane in clinical settings, have led to increased polysulfide concentrations in the bloodstream and markedly enhanced flow-mediated vasodilation (Bibli et al. 2021).

On the other side of the metabolic imbalance, elevated levels of H_2S have been shown to impair cellular bioenergetics due to its inhibitory effect on CcOX. A common therapeutic strategy to counteract this involves the use of inhibitors targeting key enzymes involved in H_2S production, CBS, CSE, and MST (Hou et al. 2025). However, in certain contexts, increased H_2S levels are not due to enhanced production but rather to impaired clearance, such as reduced SQR activity. This phenomenon is particularly relevant in primary and secondary CoQ deficiencies, which are clinically and genetically heterogeneous disorders affecting predominantly the muscle, kidney, and central nervous system. Given the essential role of CoQ as both an electron carrier in the mitochondrial respiratory chain and a lipid-soluble antioxidant, CoQ deficiencies lead to increased oxidative stress, impaired mitochondrial energy production, and defective pyrimidine biosynthesis (Quinzii et al. 2008, 2010). Notably, primary CoQ deficiency has been shown to reduce SQR activity in both murine models and human fibroblasts. As a consequence, diminished SQR activity disrupts the mitochondrial H_2S oxidation pathway, leading to elevated, cytotoxic, sulfide levels (Luna-Sánchez et al. 2017; Ziosi et al. 2017). Interestingly, supraphysiological levels of CoQ_{10} have been shown to upregulate SQR expression in skin fibroblasts from both healthy controls and patients. This upregulation of SQR was accompanied by a downregulation of CBS and CSE, enzymes of the transsulfuration pathway. Moreover, readaptations were observed in other sulfide-linked pathways, including the folate cycle, nucleotide metabolism, and the glutathione system. These findings provide new insights into how sulfide metabolism is integrated with one-carbon metabolism and may help explain some of the clinical benefits observed with CoQ_{10} supplementation in mitochondrial disorders (González-García et al. 2020). Although normal dietary intake and endogenous biosynthesis are generally sufficient to meet physiological demands for CoQ in healthy individuals, several cardiovascular disorders, including cardiomyopathies, have been associated with relatively low myocardial CoQ_{10} levels. Clinical studies, including randomized double-blind trials, have demonstrated the beneficial effects of oral CoQ_{10} supplementation, particularly in patients with chronic heart failure. These effects include improvements in cardiac function, symptoms, and overall hemodynamic performance (Overvad et al. 1999). In the Q-SYMBIO trial, for example, long-term CoQ_{10} supplementation was associated with a significant reduction in major adverse cardiovascular events and mortality (Mortensen et al. 2014). These findings support the rationale for considering CoQ_{10} as an adjunctive treatment in selected patients with cardiovascular disease.

Herein, we have described two possible strategies for modulating sulfide levels by either reducing or enhancing sulfide clearance. Reduction of sulfide clearance by SQR inhibitor proved to be effective in increasing H_2S levels in models characterized by reduced sulfides. It remains to be determined which of the two strategies, H_2S donation or inhibition of H_2S catabolism, is better tolerated for long-term treatment. Although the field of H_2S donation has advanced significantly in recent years, with successive generations of donors offering controlled and, in some cases, targeted H_2S release, this approach has thus far been the preferred method for elevating sulfide levels (Powell et al. 2018). In contrast, targeting enzymes involved in H_2S catabolism is a relatively new and emerging area of research with considerable promise. Compared to direct H_2S donors, inhibiting SQR may offer a more precise, mitochondria-focused strategy that minimizes the risk of uncontrolled H_2S release and the formation of potentially harmful by-products. Nevertheless, both approaches present important challenges. Broad enhancement of H_2S signaling or clearance carries the risk of disrupting its essential physiological roles, particularly in the cardiovascular and nervous systems. As such, any pharmacological intervention aimed at modulating H_2S metabolism must be finely tuned to maintain homeostatic levels while preventing pathological imbalances. Concerning the enhancement of H_2S clearance, the therapeutic benefits of CoQ_{10} are likely multifactorial, involving enhanced mitochondrial ATP production, reduced oxidative stress, and stabilization of cell membranes, thus not only related to the effect on SQR activity. Given its favorable safety profile and essential role in mitochondrial bioenergetics, CoQ_{10} continues to be explored as a therapeutic agent across a broad range of pathologies involving mitochondrial dysfunction and redox imbalance.

References

Abou-Hamdan A et al (2015) Oxidation of H2S in mammalian cells and mitochondria. Methods Enzymol 554:201–228
Aita N et al (1997) Cloning and expression of human liver rhodanese cDNA. Biochem Biophys Res Commun 231:56–60
Arndt S et al (2017) Assessment of H2S in vivo using the newly developed mitochondria-targeted mass spectrometry probe MitoA. J Biol Chem 292(19):7761–7773
Asimakopoulou A et al (2013) Selectivity of commonly used pharmacological inhibitors for cystathionine β synthase (CBS) and cystathionine γ lyase (CSE). Br J Pharmacol 169(4): 922–932
Balaban RS et al (1990) Regulation of oxidative phosphorylation in the mammalian cell. Am J Phys 258(3 Pt 1):C377–C389
Benhar M et al (2009) Protein denitrosylation: enzymatic mechanisms and cellular functions. Nat Rev Mol Cell Biol 10(10):721–732
Bibli SI et al (2021) Mapping the endothelial cell S-sulfhydrome highlights the crucial role of integrin sulfhydration in vascular function. Circulation 143(9):935–948
Billaut-Laden I et al (2006) Evidence for a functional genetic polymorphism of the human thiosulfate sulfurtransferase (Rhodanese), a cyanide and H2S detoxification enzyme. Toxicology 225(1):1–11
Birkenkamp-Demtroder K et al (2002) Geneexpression in colorectal cancer. Cancer Res 62:4352–4363
Brown GC (1992) Control of respiration and ATP synthesis in mammalian mitochondria and cells. Biochem J 284(Pt 1):1–13
Brown GC (2001) Regulation of mitochondrial respiration by nitric oxide inhibition of cytochrome c oxidase. Biochim Biophys Acta 1504(1):46–57
Burlina A et al (1991) New clinical phenotype of branched-chain acyl-CoA oxidation defect. Lancet 338(8781):1522–1523
Burnett G et al (1980) Mechanism-based inactivation of pig heart L-alanine transaminase by L-propargylglycine. Half-site reactivity. J Biol Chem 255(8):3487–3491
Calvert JW et al (2010) Novel insights into hydrogen sulfide--mediated cytoprotection. Antioxid Redox Signal 12(10):1203–1217
Cárdenas C et al (2010) Essential regulation of cell bioenergetics by constitutive InsP3 receptor Ca2+ transfer to mitochondria. Cell 142(2):270–283
Chen X et al (2017) Hydrogen Sulphide treatment increases insulin sensitivity and improves oxidant metabolism through the CaMKKbeta-AMPK pathway in PA-induced IR C2C12 cells. Sci Rep 7:13248
Chen X et al (2020) Novel compound heterozygous variants of ETHE1 causing ethylmalonic encephalopathy in a Chinese patient: a case report. Front Genet 11:341
Cirino G, Szabo C, Papapetropoulos A (2023) Physiological roles of hydrogen sulfide in mammalian cells, tissues, and organs. Physiol Rev 103(1):31–276
Dionisi-Vici C et al (2016) Liver transplant in ethylmalonic encephalopathy: a new treatment for an otherwise fatal disease. Brain 139(Pt 4):1045–1051
Fagerberg L et al (2014) Analysis of the human tissue-specific expression by genome-wide integration of transcriptomics and antibody-based proteomics. Mol Cell Proteomics 13(2): 397–406
Fernandes DGF et al (2021) Human cystathionine γ-lyase is inhibited by s-nitrosation: a new crosstalk mechanism between NO and H2S. Antioxidants 10(9):1391
Filipovic MR et al (2018) Chemical biology of H2S signaling through persulfidation. Chem Rev 118(3):1253–1337
Friederich MW et al (2020) Pathogenic variants in SQOR encoding sulfide:quinone oxidoreductase are a potentially treatable cause of Leigh disease. J Inherit Metab Dis 43(5):1024–1036

Furne J et al (2008) Whole tissue hydrogen sulfide concentrations are orders of magnitude lower than presently accepted values. Am J Physiol Regul Integr Comp Physiol 295(5):R1479–R1485

Gheibi S et al (2014) Hydrogen sulfide protects the brain against ischemic reperfusion injury in a transient model of focal cerebral ischemia. J Mol Neurosci 54(2):264–270

Gibbons C, Montgomery MG, Leslie AG, Walker JE (2000) The structure of the central stalk in bovine F(1)-ATPase at 2.4 A resolution. Nat Struct Biol 7(11):1055–1061

Giuffrè A et al (2018) Hydrogen sulfide biochemistry and interplay with other gaseous mediators in mammalian physiology. Oxid Med Cell Longev 2018:6290931

Giuffrè A et al (2020) Hydrogen sulfide metabolism and signaling in the tumor microenvironment. Adv Exp Med Biol 1219:335–353

González-García P et al (2020) Coenzyme Q10 modulates sulfide metabolism and links the mitochondrial respiratory chain to pathways associated to one carbon metabolism. Hum Mol Genet 29(19):3296–3311

Goubern M et al (2007) Sulfide, the first inorganic substrate for human cells. FASEB J 21(8): 1699–1706

Goudarzi S et al (2018) Spectroscopic and electronic structure study of ETHE1: elucidating the factors influencing sulfur oxidation and oxygenation in mononuclear nonheme iron enzymes. J Am Chem Soc 140(44):14887–14902

Hanaoka K et al (2017) Discovery and mechanistic characterization of selective inhibitors of H2S-producing enzyme: 3-Mercaptopyruvate Sulfurtransferase (3MST) targeting active-site cysteine persulfide. Sci Rep 7:40227

Hara MR et al (2005) S-nitrosylated GAPDH initiates apoptotic cell death by nuclear translocation following Siah1 binding. Nat Cell Biol 7(7):665–674

Henriques BJ et al (2014) Ethylmalonic encephalopathy ETHE1 R163W/R163Q mutations alter protein stability and redox properties of the iron centre. PLoS One 9(9):e107157

Hou Y et al (2025) Sulfide regulation and catabolism in health and disease. Signal Transduct Target Ther 10(1):174

Jackson MR et al (2012) Human sulfide:quinone oxidoreductase catalyzes the first step in hydrogen sulfide metabolism and produces a sulfane sulfur metabolite. Biochemistry 51(34):6804–6815

Jackson MR et al (2019) X-ray structure of human sulfide:quinone oxidoreductase: insights into the mechanism of mitochondrial hydrogen sulfide oxidation. Structure 27(5):794–805.e4

Jackson MR et al (2022) Discovery of a first-in-class inhibitor of sulfide:quinone oxidoreductase that protects against adverse cardiac remodelling and heart failure. Cardiovasc Res 118(7): 1771–1784

Johnson JL et al (2002) Isolated sulfite oxidase deficiency: identification of 12 novel SOx mutations in 10 patients. Hum Mutat 20(1):74

Johnson-Winters K et al (2010) Elucidating the catalytic mechanism of sulfite oxidizing enzymes using structural, spectroscopic, and kinetic analyses. Biochemistry 49(34):7242–7254

Kabil O, Banerjee R (2014) Enzymology of H2S biogenesis, decay and signaling. Antioxid Redox Signal 20(5):770–782

Katsouda A et al (2022) MPST sulfurtransferase maintains mitochondrial protein import and cellular bioenergetics to attenuate obesity. J Exp Med 219(7):e20211894

Keenan EK et al (2021) Discovering the landscape of protein modifications. Mol Cell 81(9): 1868–1878

Kolluru GK et al (2023) Sulfide regulation of cardiovascular function in health and disease. Nat Rev Cardiol 20(2):109–125

Krishnan N et al (2011) H2S-induced sulfhydration of the phosphatase PTP1B and its role in the endoplasmic reticulum stress response. Sci Signal 4(203):ra86

Kwiecień I et al (2003) Inhibition of the catalytic activity of rhodanese by S-nitrosylation using nitric oxide donors. Int J Biochem Cell Biol 35(12):1645–1657

Lagoutte E et al (2010) Oxidation of hydrogen sulfide remains a priority in mammalian cells and causes reverse electron transfer in colonocytes. Biochim Biophys Acta 1797(8):1500–1511

Landry AP et al (2017) H2S oxidation by nanodisc-embedded human sulfide quinone oxidoreductase. J Biol Chem 292(28):11641–11649
Landry AP et al (2018) Modulation of catalytic promiscuity during hydrogen sulfide oxidation. ACS Chem Biol 13(6):1651–1658
Landry AP et al (2019) Catalytic trisulfide in human sulfide quinone oxidoreductase catalyzes coenzyme a persulfide synthesis and inhibits butyrate oxidation. Cell Chem Biol 26(11): 1515–1525.e4
Leschelle X et al (2005) Adaptative metabolic response of human colonic epithelial cells to the adverse effects of the luminal compound sulfide. Biochim Biophys Acta 1725(2):201–212
Libiad M et al (2014) Organization of the human mitochondrial hydrogen sulfide oxidation pathway. J Biol Chem 289(45):30901–30910
Libiad M et al (2015) Polymorphic variants of human rhodanese exhibit differences in thermal stability and sulfur transfer kinetics. J Biol Chem 290(39):23579–23588
Libiad M et al (2019) Hydrogen sulfide perturbs mitochondrial bioenergetics and triggers metabolic reprogramming in colon cells. J Biol Chem 294(32):12077–12090
Linden DR et al (2012) Sulphide quinone reductase contributes to hydrogen sulphide metabolism in murine peripheral tissues but not in the CNS. Br J Pharmacol 165(7):2178–2190
Luna-Sánchez M et al (2017) CoQ deficiency causes disruption of mitochondrial sulfide oxidation, a new pathomechanism associated with this syndrome. EMBO Mol Med 9(1):78–95
Luo Y et al (2025) Thiosulphate sulfurtransferase: biological roles and therapeutic potential. Redox Biol 82:103595
Malagrinò F et al (2019) Hydrogen sulfide oxidation: adaptive changes in mitochondria of SW480 colorectal cancer cells upon exposure to hypoxia. Oxid Med Cell Longev 2019:8102936
Martínez-Ruiz A, Lamas S (2004) S-nitrosylation: a potential new paradigm in signal transduction. Cardiovasc Res 62(1):43–52
Marutani E et al (2021) Sulfide catabolism ameliorates hypoxic brain injury. Nat Commun 12(1): 3108
Módis K et al (2013) Intramitochondrial hydrogen sulfide production by 3-mercaptopyruvate sulfurtransferase maintains mitochondrial electron flow and supports cellular bioenergetics. FASEB J 27(2):601–611
Módis K et al (2014) Regulation of mitochondrial bioenergetic function by hydrogen sulfide. Part II. Pathophysiological and therapeutic aspects. Br J Pharmacol 171(8):2123–2146
Módis K et al (2016) S-Sulfhydration of ATP synthase by hydrogen sulfide stimulates mitochondrial bioenergetics. Pharmacol Res 113(Pt A):116–124
Mortensen SA et al (2014) The effect of coenzyme Q10 on morbidity and mortality in chronic heart failure: results from Q-SYMBIO: a randomized double-blind trial. JACC Heart Fail 2(6): 641–649
Murray CI, Kane LA, Uhrigshardt H, Wang SB, Van Eyk JE (2011) Site-mapping of in vitro S-nitrosation in cardiac mitochondria: implications for cardioprotection. Mol Cell Proteomics 10(3):M110.004721
Mustafa AK et al (2009) H2S signals through protein S-sulfhydration. Sci Signal 2(96):ra72
Olson KR (2015) Hydrogen sulfide as an oxygen sensor. Antioxid Redox Signal 22(5):377–397
Overvad K et al (1999) Coenzyme Q10 in health and disease. Eur J Clin Nutr 53(10):764–770
Panagaki T et al (2022) Overproduction of hydrogen sulfide, generated by cystathionine β-synthase, disrupts brain wave patterns and contributes to neurobehavioral dysfunction in a rat model of down syndrome. Redox Biol 51:102233
Panieri E et al (2020) Potential applications of NRF2 modulators in cancer therapy. Antioxidants 9(3):193
Paul BD, Snyder SH (2012) H_2S signalling through protein sulfhydration and beyond. Nat Rev Mol Cell Biol 13(8):499–507
Petersen LC (1997) The effect of inhibitors on the oxygen kinetics of cytochrome c oxidase. Biochim Biophys Acta 460(2):299–307

Pettinati I et al (2015) Crystal structure of human persulfide dioxygenase: structural basis of ethylmalonic encephalopathy. Hum Mol Genet 24(9):2458–2469
Popov D (2013) An outlook on vascular hydrogen sulphide effects, signalling, and therapeutic potential. Arch Physiol Biochem 119(5):189–194
Powell CR et al (2018) A review of hydrogen sulfide (H2S) donors: chemistry and potential therapeutic applications. Biochem Pharmacol 149:110–123
Quinzii CM et al (2008) Respiratory chain dysfunction and oxidative stress correlate with severity of primary CoQ10 deficiency. FASEB J 22(6):1874–1885
Quinzii CM et al (2010) Reactive oxygen species, oxidative stress, and cell death correlate with level of CoQ10 deficiency. FASEB J 24(10):3733–3743
Sen N et al (2008) Nitric oxide-induced nuclear GAPDH activates p300/CBP and mediates apoptosis. Nat Cell Biol 10(7):866–873
Stram AR, Payne RM (2016) Post-translational modifications in mitochondria: protein signaling in the powerhouse. Cell Mol Life Sci 73(21):4063–4073
Suman SG, Gretarsdottir JM (2019) Chemical and clinical aspects of metal-containing antidotes for poisoning by cyanide. Met Ions Life Sci 19
Sun J, Morgan M, Shen RF, Steenbergen C, Murphy E (2007) Preconditioning results in S-nitrosylation of proteins involved in regulation of mitochondrial energetics and calcium transport. Circ Res 101(11):1155–1163
Sun Q et al (2009) Structural basis for the inhibition mechanism of human cystathionine gamma-lyase, an enzyme responsible for the production of H(2)S. J Biol Chem 284(5):3076–3085
Szabo C (2016) Gasotransmitters in cancer: from pathophysiology to experimental therapy. Nat Rev Drug Discov 15(3):185–203
Szabo C et al (2013) Tumor-derived hydrogen sulfide, produced by cystathionine-β-synthase, stimulates bioenergetics, cell proliferation, and angiogenesis in colon cancer. Proc Natl Acad Sci USA 110(30):12474–12479
Szabo C et al (2014) Regulation of mitochondrial bioenergetic function by hydrogen sulfide. Part I. Biochemical and physiological mechanisms. Br J Pharmacol 171(8):2099–2122
Tan G et al (2003) Decreased expression of genes involved in sulfur amino acid metabolism in frataxin-deficient cells. Hum Mol Genet 12(14):1699–1711
Teng H et al (2013) Oxygen-sensitive mitochondrial accumulation of cystathionine β-synthase mediated by Lon protease. Proc Natl Acad Sci USA 110(31):12679–12684
Tiranti V et al (2009) Loss of ETHE1, a mitochondrial dioxygenase, causes fatal sulfide toxicity in ethylmalonic encephalopathy. Nat Med 15(2):200–205
Tokoro M et al (2003) Identification and characterization of two isoenzymes of methionine gamma-lyase from Entamoeba histolytica: a key enzyme of sulfur-amino acid degradation in an anaerobic parasitic protist that lacks forward and reverse trans-sulfuration pathways. J Biol Chem 278(43):42717–42727
Velayutham M et al (2016) Sulfite oxidase activity of cytochrome c: role of hydrogen peroxide. Biochem Biophys Rep 5:96–104
Vicente JB et al (2016) Bioenergetic relevance of hydrogen sulfide and the interplay between gasotransmitters at human cystathionine β-synthase. Biochim Biophys Acta 1857(8): 1127–1138
Vignane T, Filipovic MR (2023) Emerging chemical biology of protein persulfidation. Antioxid Redox Signal 39(1–3):19–39
Vo TTT et al (2022) The potential implications of hydrogen sulfide in aging and age-related diseases through the lens of Mitohormesis. Antioxidants 11(8):1619
Walsh DJ et al (2010) Novel ETHE1 mutation in a carrier couple having prior offspring affected with ethylmalonic encephalopathy: genetic analysis, clinical management and reproductive outcome. Mol Med Rep 3(2):223–226
Wang R (2012) Physiological implications of hydrogen sulfide: a whiff exploration that blossomed. Physiol Rev 92(2):791–896
Wang R (2014) Gasotransmitters: growing pains and joys. Trends Biochem Sci 39(5):227–232

Wang M et al (2014) Regulation of cystathionine γ-lyase in mammalian cells by hypoxia. Biochem Genet 52(1–2):29–37

Westley J (1973) Rhodanese. Adv Enzymol Relat Areas Mol Biol 39:327–368

Yang G et al (2013) Hydrogen sulfide protects against cellular senescence via S-sulfhydration of Keap1 and activation of Nrf2. Antioxid Redox Signal 18(15):1906–1919

Zhang H et al (2020) Hydrogen sulfide regulates insulin secretion and insulin resistance in diabetes mellitus, a new promising target for diabetes mellitus treatment? A review. J Adv Res 27:19–30

Zhang MY et al (2021) Hydrogen sulfide metabolite, sodium thiosulfate: clinical applications and underlying molecular mechanisms. Int J Mol Sci 22(12):6452

Zhao Y et al (2014) Hydrogen sulfide (H2S) releasing agents: chemistry and biological applications. Chem Commun 50(80):11788–11805

Ziosi M et al (2017) Coenzyme Q deficiency causes impairment of the sulfide oxidation pathway. EMBO Mol Med 9(1):96–111

Zivanovic J et al (2019) Selective persulfide detection reveals evolutionarily conserved antiaging effects of S-sulfhydration. Cell Metab 30(6):1152–1170.e13

Zuhra K, Szabo C (2022) The two faces of cyanide: an environmental toxin and a potential novel mammalian gasotransmitter. FEBS J 289(9):2481–2515

Zuhra K et al (2019) N-acetylcysteine serves as substrate of 3-mercaptopyruvate sulfurtransferase and stimulates sulfide metabolism in colon cancer cells. Cells 8(8):828

Zuhra K et al (2020) Cystathionine-β-synthase: molecular regulation and pharmacological inhibition. Biomolecules 10(5):697

Zuhra K et al (2021) the multifaceted roles of sulfane sulfur species in cancer-associated processes. Biochim Biophys Acta Bioenerg 1862(2):148338

Zuhra K et al (2025) Regulation of mammalian cellular metabolism by endogenous cyanide production. Nat Metab 7(3):531–555

The Role of Hydrogen Sulfide in Peripheral Nervous System Modulation

Saúl Huerta de la Cruz, Diana L. Silva-Velasco, Jesus H. Beltran-Ornelas, and David Centurión

Contents

Authors Saúl Huerta de la Cruz, Diana L. Silva-Velasco, and Jesus H. Beltran-Ornelas have equally contributed to this chapter.

S. Huerta de la Cruz
Department of Pharmacology, University of Vermont, Burlington, VT, USA

D. L. Silva-Velasco
Cardiovascular Translational Research Center, University of South Carolina, Columbia, SC, USA

J. H. Beltran-Ornelas
NextGen Precision Health, University of Missouri, Columbia, MO, USA

D. Centurión (✉)
Departamento de Farmacobiología, Cinvestav Sede Sur, Ciudad de México, Mexico
e-mail: dcenturi@cinvestav.mx

A. Papapetropoulos, C. Szabo (eds.), *Hydrogen Sulfide: A Toxicant with Healing Potential*, Handbook of Experimental Pharmacology 292,
https://doi.org/10.1007/164_2025_777

Abstract

Hydrogen sulfide (H_2S) is a gasotransmitter that contributes to the regulation of peripheral nervous system (PNS) function. H_2S is produced by several enzymes whose expression changes under different physiological and pathological conditions, influencing how peripheral neurons respond to environmental and internal signals. H_2S modulates neuronal excitability through its actions on ion channels and through interactions with other gasotransmitters, shaping sensory, autonomic, and pain-related pathways. In autonomic circuits, H_2S adjusts sympathetic and parasympathetic activity. Through these actions, it affects cardiovascular control, gastrointestinal motility, and respiratory rhythm. In pain pathways, H_2S can modulate nociception in either direction, with its effectsshaped by the physiological or pathological state. H_2S participates in multiple pain conditions and contributes to changes in peripheral and spinal processing that influence pain sensitivity. Overall, H_2S influences several components of peripheral neurobiology and represents a potential target for strategies aimed at treating autonomic dysfunction and chronic pain.

Keywords

Autonomic regulation · Hydrogen sulfide · Pain modulation · Peripheral nervous system

1 Introduction

Gasotransmitters have received increasing attention due to their vital roles in the body. Among these, hydrogen sulfide (H_2S) has emerged as a critical regulator of neuronal function. In the peripheral nervous system (PNS), H_2S influences sensory processing, autonomic regulation, and neuronal communication. Like nitric oxide (NO) and carbon monoxide (CO), H_2S is a small gaseous molecule that acts as a signaling messenger in many biological systems.

The PNS is fundamental to sensory, motor, and autonomic functions, serving as the communication link between the central nervous system and the rest of the body. Through this connection, the PNS enables the perception of external stimuli, coordination of movement, and regulation of vital functions. When disrupted, PNS dysfunction can lead to pathological conditions, including chronic pain and autonomic disorders. Within the PNS, H_2S modulates sensory and autonomic pathways. It influences key physiological processes such as pain signaling and cardiovascular regulation. Thus, understanding interactions between H_2S and PNS components may lead to new therapeutic strategies.

This chapter explores how H_2S affects PNS function, with a focus on neuronal excitability, pain modulation, and autonomic control. It begins with an overview of H_2S metabolism and signaling mechanisms. Then, it examines the effects of H_2S on autonomic regulation, followed by its role in sensory neurons and pain pathways. Together, these sections provide a comprehensive view of H_2S in peripheral neurobiology.

2 Mechanisms of Hydrogen Sulfide Signaling in Peripheral Nerves

H_2S is synthesized from L-cysteine by several enzymes: (a) cystathionine-γ-lyase (CSE), (b) cystathionine-β-synthase (CBS), and (c) cysteine aminotransferase (CAT) coupled with 3-mercaptopyruvate sulfurtransferase (3MST). CBS and CSE are predominantly localized in the cytosol, whereas 3MST and CAT are found in both mitochondria and the cytosol (Cirino et al. 2023; Rangel-Galvan et al. 2023).

CBS is the most studied H_2S-producing enzyme in the nervous system. It was initially thought to exist only in the central nervous system, but later studies show that it is also expressed throughout the PNS. CBS protein has been found in various parts of the PNS, including colon dorsal root ganglia (DRG) neurons (Wang et al. 2018; Yan et al. 2016; Yuan et al. 2015; Zhu et al. 2015), enteric nerve cells (Liao et al. 2020), the carotid sinus (Yu et al. 2017), duodenal myenteric neurons (Lu et al. 2014), and trigeminal ganglion neurons (Feng et al. 2013). However, CBS mRNA is absent from Schwann cells in the sciatic nerve (Park et al. 2015), indicating cell-type-specific expression patterns within the PNS.

CSE is also expressed in peripheral nerves, with protein presence confirmed in adrenal chromaffin cells (Gridina et al. 2021; Peng et al. 2010), the carotid sinus (Barrera et al. 2021; Peng et al. 2010; Yu et al. 2017), the sciatic nerve (Park et al. 2015), neurons and glial cells of the superior mesenteric ganglion (Sha et al. 2013), as well as in duodenal myenteric neurons (Lu et al. 2014). CSE plays a key role in carotid body function, particularly in modulating responses to hypoxia but not anoxia (Peng et al. 2019). Hypoxia enhances H_2S production in glomus cells, the primary O_2-sensing sites in the carotid body, and promotes catecholamine secretion – an effect attenuated by CSE inhibitors and absent in CSE knockout mice (Peng et al. 2010). Interestingly, H_2S perfusion in the carotid sinus reduces sympathetic outflow (Guo et al. 2016), suggesting a complex interplay between endogenous and exogenous H_2S in hypoxic responses.

3MST has been less extensively studied in the PNS. However, mRNA expression of 3MST has been detected in Schwann cells of the sciatic nerve (Park et al. 2015) as well as in DRG and spinal cord tissue (Syhr et al. 2015). Protein expression of 3MST has been identified in the adrenal glands (Tomita et al. 2016) and DRG (Miyamoto et al. 2014), implicating its potential role in peripheral neuronal function.

2.1 Regulation of H_2S Production in Peripheral Neurons

The expression of H_2S-producing enzymes within the PNS is dynamically regulated in pathological conditions. CBS expression increases in several pathological conditions. For example, it is upregulated in enteric nerve cells during acute pancreatitis (Liao et al. 2020), in colon DRG neurons following nerve constriction injury (Zhu et al. 2015), and in thoracolumbar DRG neurons under prenatal maternal stress (Wang et al. 2018). CBS is closely linked to pain processing, with its inhibition modulating nociceptive pathways (Yan et al. 2016). Repeated NaHS (sodium hydrosulfide) administration enhances CBS expression in DRG neurons of diabetic animals, whereas hydroxylamine (a CBS inhibitor) produces the opposite effect (Roa-Coria et al. 2019).

CSE expression is also altered by pathophysiological states. CSE mRNA levels increase in Schwann cells of the sciatic nerve following injury (Park et al. 2015), and in DRG neurons after zymosan injection or spared nerve injury (Syhr et al. 2015). Additionally, CSE protein expression rises in the carotid body during intermittent hypoxia (Barrera et al. 2021) and in DRG neurons under inflammatory and neuropathic pain conditions (Syhr et al. 2015). Unlike CBS, CSE may not be critical for pain signaling. Mice lacking CSE still show normal pain responses in various models (Syhr et al. 2015), suggesting that other H_2S-producing enzymes might compensate.

3MST is generally overlooked in studies of PNS-related pathologies. However, its expression in DRG and spinal cord tissue remains unchanged in CSE-deficient mice (Syhr et al. 2015). This enzyme, along with cysteine aminotransferase, helps produce H_2S synthesis in neurons. Its activity appears to be influenced by mitochondrial metabolism (Miyamoto et al. 2014).

In the PNS, H_2S exerts its effects through multiple mechanisms, including direct ion channel modulation and interactions with other gasotransmitters (Szabo, 2017).

2.2 Ion Channel Modulation

H_2S plays a crucial role in modulating ion channels within the PNS, influencing both physiological and pathological processes (Kimura et al. 2005). This gasotransmitter can modulate neuronal activity by interacting with several ion channels, including transient receptor potential (TRP), voltage-gated sodium (NaV) and calcium (CaV), as well as potassium (K^+) channels. H_2S plays a well-established role in pain signaling. It increases pain sensitivity by activating TRP and NaV channels, contributing to hyperalgesia in disease conditions (Terada et al. 2015). Additionally, H_2S affects vascular and autonomic functions by modulating CaV and K^+ channels, further influencing sensory and sympathetic responses. The following sections will delve into the most common effects of H_2S on these ion channels, highlighting the mechanisms by which it regulates neuronal excitability and pain perception within the PNS.

2.2.1 TRP Channels

TRP channels constitute a superfamily of non-selective cation channels predominantly expressed in unmyelinated C-fiber nociceptors, where they play a key role in nociceptive signal transduction. Among them, transient receptor potential ankyrin 1 channel (TRPA1) is activated by temperatures below 17 °C (Sekiguchi et al. 2024), whereas other TRP channels respond to distinct stimuli.

H_2S interacts with TRP channels within the PNS. In diabetic animals, CBS colocalizes with the transient receptor potential vanilloid 1 channel (TRPV1), TRPA1, and transient receptor potential cation channel 6 (TRPC6) in DRG neurons (Roa-Coria et al. 2019). Pharmacological inhibition of CBS with aminooxyacetic acid (AOAA) reduces TRPV1 expression in colonic DRG neurons following chronic nerve injury, while upregulation of CBS enhances TRPV1 activation, contributing to chronic visceral hypersensitivity (Zhu et al. 2015).

H_2S also sensitizes capsaicin-sensitive lung vagal fibers through TRPA1 activation at afferent nerve endings (Hsu et al. 2013). Furthermore, this mechanism mediates skin vasodilation via releasing vasoactive sensory neuropeptides, calcitonin gene-related peptide (CGRP), and substance P, which subsequently activate their receptors and ATP-sensitive potassium channels (K_{ATP}) (Hajna et al. 2016). Mechanistically, H_2S enhances TRPV1 activity through S-persulfidation and increased TRPV1 expression. These effects are reversed by hydroxylamine, a CBS inhibitor (Yu et al. 2017). Collectively, these findings suggest that CBS-derived H_2S, acting via TRPA1, TRPV1, and TRPC channels, may represent a therapeutic target for pathologies that affect the PNS, including peripheral neuropathy (Roa-Coria et al. 2019).

2.2.2 Sodium Channels

NaV channels are critical for action potential generation and propagation in peripheral neurons and serve as key targets of H_2S. H_2S increases sodium current density in DRG neurons. This is supported by findings that AOAA, a CBS inhibitor, reduces NaV1.7 and NaV1.8 expression, which in turn lowers sodium current density (Wang et al. 2018; Yan et al. 2016). However, in vitro assessments suggest that AOAA's effect on sodium currents does not fully translate to in vivo models.

CBS colocalizes with NaV1.7 and NaV1.8 in DRG neurons, and H_2S upregulates NaV1.7 expression via the MEK/ERK signaling pathway (Tian et al. 2020). Because NaV1.7 is strongly linked to neuropathic pain (Vasylyev et al. 2024), these findings suggest that CBS-derived H_2S contributes to pain by increasing NaV1.7 activity in DRG neurons.

2.2.3 Calcium Channels

H_2S modulates CaV channels, particularly CaV3.2 (T-type) channels, without direct post-translational modification. Instead, H_2S enhances CaV3.2 activity by scavenging Zn^{2+} bound to His191 in the extracellular domain of the channel (Sekiguchi et al. 2024). This mechanism was demonstrated using β-cyano-l-alanine (a CSE inhibitor), which prevented the upregulation of CaV3.2 (Ozaki et al. 2018).

Physiologically, H_2S acts as a pronociceptive molecule within the PNS by modulating CaV3.2. This role is particularly evident in pancreatic nociception, where exogenous H_2S sensitizes neurons via CaV3.2 activation (Terada et al. 2015).

2.2.4 Potassium Channels

K^+ channels are classic targets of H_2S in various physiological systems, including the PNS. Typically, H_2S activates K^+ channels, which causes neurons to hyperpolarize and become less excitable. This is often linked to reduced activity of peripheral sympathetic fibers. Pharmacological inhibition of CBS with AOAA has been shown to reduce H_2S production while increasing potassium currents in trigeminal ganglion neurons, supporting the role of H_2S in modulating neuronal excitability via K^+ channels (Miao et al. 2014).

In anesthetized rats, perfusion of NaHS into isolated carotid sinus resulted in a dose-dependent reduction of renal sympathetic nerve activity via K_{ATP} activation, while inhibition of CSE with DL-propargylglycine (PAG) led to the opposite effect, supporting the role of endogenous H_2S in sympathetic inhibition (Guo et al. 2016). However, conflicting findings suggest a facilitatory role for H_2S in sympathetic signaling. In this sense, in the mouse superior mesenteric ganglion, CSE inhibition led to a reduction in ganglionic transmission (Sha et al. 2013). To clarify this discrepancy, Dominguez-Rodríguez et al. (2017) examined the effects of H_2S on cultured sympathetic neurons. They found that H_2S had a dual effect, where H_2S enhanced the electrically evoked release of [^{3}H]-noradrenaline from postganglionic terminal, but also caused hyperpolarization and reduced firing in ganglionic neurons (Dominguez-Rodriguez et al. 2017). These findings suggest that the effects of H_2S on sympathetic transmission are context-dependent and may vary depending on the specific neuronal population and signaling pathways involved.

Additionally, baroreflex sensitivity is enhanced by NaHS, an effect attributed to the upregulation and activation of K_{ATP} channels, specifically through the Kir6.2 pore-forming subunit (Li et al. 2021). Further supporting the role of K^+ channels, we have demonstrated that H_2S-induced sympathoinhibition is mediated by the activation of multiple K^+ channels, including voltage-gated (Kv), inward-rectifying (K_{IR}), K_{ATP}, and large-conductance Ca^{2+}-activated (BK_{Ca}) potassium channels (Medina-Terol et al. 2022). Interestingly, in obese rats, H_2S-induced sympathoinhibition appears to rely exclusively on K_{IR} channel activation, suggesting that metabolic conditions may influence the specific K^+ channel subtypes involved in H_2S signaling (Gomez et al. 2024).

Collectively, these findings underscore the complex and context-dependent nature of H_2S-mediated K^+ channel modulation within the PNS. By influencing sympathetic and sensory neuronal excitability, H_2S contributes to autonomic regulation, positioning K^+ channels as key players in its neuromodulatory effect.

2.3 Crosstalk with Other Gasotransmitters

H_2S, NO, and CO are recognized as gasotransmitters involved in neuromodulation. These molecules mutually influence their expression and activity, a phenomenon termed "crosstalk" (Rangel-Galvan et al. 2023).

For instance, H_2S modulates catecholamine secretion by inhibiting CO-mediated suppression of hypoxia-induced responses in adrenal chromaffin cells (Gridina et al. 2021). In carotid body chemosensory reflexes, which regulate respiration and blood pressure, reduced CO levels lead to increased H_2S production (Peng et al. 2014; Prabhakar et al. 2018). The interaction between H_2S and NO is particularly significant in pain modulation. H_2S facilitates the formation of S-nitrosothiols, which act on extracellular cysteines in CaV3.2 channels to regulate neuronal excitability. Additionally, H_2S influences the NO/cGMP/PKG signaling pathway, exerting antinociceptive effects (Rangel-Galvan et al. 2023).

Evidence suggests that H_2S acts as a neuromodulator of blood vessels through activation of the nitroxyl (HNO)-TRPA1-CGRP pathway. HNO is produced by interaction of H_2S and NO, and it activates TRPA1 channels in perivascular sensory nerves, releasing CGRP, a potent vasodilator (Eberhardt et al. 2014). This signaling cascade is a key neuronal mechanism that helps regulate vascular tone. It also shows how H_2S and NO depend on each other in controlling blood vessel function.

3 Hydrogen Sulfide and Autonomic Regulation

3.1 Effects of H_2S on the Sympathetic System

H_2S has emerged as a significant gaseous signaling molecule with diverse physiological roles, including its impact on the autonomic nervous system. The autonomic nervous system, composed of sympathetic and parasympathetic branches, regulates vital functions, including cardiovascular activity, digestion, and respiration. Recent research has highlighted the dual role of H_2S in modulating these autonomic functions, particularly in vascular tone regulation and autonomic dysfunction-related diseases.

3.1.1 Heart Rate

H_2S plays a pivotal role in cardiovascular regulation by modulating sympathetic nervous system activity. Its effects on heart rate depend on the specific brain region and differ between normal and disease conditions. Notably, H_2S can either suppress or enhance sympathetic activity, leading to a decrease or increase in heart rate.

One key inhibitory mechanism of H_2S occurs in the rostral ventrolateral medulla. Here, H_2S activates K_{ATP} channels, which reduce neuronal excitability and lower both heart rate and blood pressure (Duan et al. 2015). Similarly, in the nucleus of the solitary tract (NTS), H_2S attenuates angiotensin II-induced sympathetic excitation by regulating angiotensin II type 1 receptor (AT1R) and AT1R-associated protein, as well as by reducing oxidative stress (Feng et al. 2020). Moreover, subchronic H_2S

treatment mitigates hypertension following traumatic brain injury (TBI) by decreasing sympathetic hyperactivity. Giving NaHS for 7 days after injury prevents the TBI-related increase in heart rate and sympathetic overactivation. This likely occurs by inhibiting vasopressor responses to noradrenaline and α_2-adrenoceptor agonists (Huerta de la Cruz et al. 2022).

Peripheral mechanisms also contribute to H_2S-mediated inhibition of sympathetic nerve activity. Studies indicate that carotid sinus perfusion with H_2S donors produces a dose-dependent reduction in renal sympathetic nerve activity (RSNA). This effect is mediated by the activation of KATP channels and inhibition of L-type calcium channels, which together suppress sympathetic outflow and promote vasodilation (Guo et al. 2016). Studies also show that H_2S inhibits both cardioaccelerator and vasopressor sympathetic outflow in pithed rats. This effect occurs regardless of whether the animals have normal weight or diet-induced obesity (Centurion et al. 2018; Centurión et al. 2016; Gomez et al. 2019). These effects are primarily mediated by K^+ channel activation (Medina-Terol et al. 2022).

The bradycardic effects of H_2S also arise from its central actions. Intracerebroventricular injection of NaHS lowers heart rate and blood pressure by activating the K_{ATP} channels. This activation hyperpolarizes neurons in cardiovascular control centers, thereby reducing sympathetic nerve activity. The reversal of these effects by glibenclamide further confirms the involvement of these channels (Liu et al. 2011). Interestingly, this mechanism does not involve calcium channels, HCN channels, or nitric oxide signaling. This suggests that other, unknown pathways may be responsible (Swan et al. 2017).

Conversely, in certain pathological conditions, H_2S enhances sympathetic activity. In the paraventricular nucleus of the hypothalamus (PVN), exogenous H_2S (GYY4137, $\geq$1 nmol) increases heart rate and RSNA, amplifying cardiac sympathetic reflexes. Notably, while 1–2 nmol significantly enhances activity, higher doses (4 nmol) do not induce further increases, suggesting a saturation effect (Gan et al. 2012). Additionally, endogenous H_2S levels in the PVN appear to decline in chronic heart failure, indicating a complex and yet-to-be fully elucidated regulatory mechanism.

Preclinical studies further suggest the therapeutic potential of H_2S modulation in heart failure with preserved ejection fraction. Treatment with SG1002, a slow-releasing H_2S donor, particularly in combination with sodium-glucose cotransporter 2 inhibitors, enhances diastolic function and exercise capacity while mitigating multiorgan pathology (Doiron et al. 2024). However, additional research is required to elucidate the precise role of the sympathetic system in these effects.

3.1.2 Digestion

H_2S plays a crucial role in gastrointestinal physiology by modulating motility and neural signaling within the digestive tract. Notably, gut microbiota-derived H_2S serves as a key regulator of intestinal homeostasis and communication between the digestive system and the central nervous system (Donertas Ayaz and Zubcevic 2020).

In the esophagus, H_2S inhibits contractions triggered by the vagus nerve but does not affect direct muscle contractions. This suggests a neuronal mechanism via inhibition of vagal motor activity (Shiina et al. 2016). Additionally, H_2S in the NTS suppresses gastric motility via transient TRPV1 activation, which triggers nuclear factor-kappa B and neurokinin-1 receptor signaling pathways (Sun et al. 2020, 2023).

Within the colon, H_2S acts as a prosecretory neuromodulator. Functional assays indicate that NaHS induces chloride secretion via TRPV1 activation on extrinsic afferent terminals, which subsequently excites enteric neurons (Schicho et al. 2006). H_2S also affects preganglionic nicotinic synapses in the superior mesenteric ganglion. This suggests another way it may regulate intestinal motility and gut-brain communication. The enzyme CSE, responsible for H_2S production, is expressed in neurons and glial cells, further indicating an endogenous role for H_2S in sympathetic and digestive function. Notably, H_2S production inhibition using PAG significantly reduced the amplitude of excitatory postsynaptic potentials, suggesting that endogenous H_2S production is essential for synaptic transmission in this ganglion (Sha et al. 2013).

3.1.3 Respiratory Function

Recent studies show that H_2S affects respiratory and sympathetic function in complex, region-specific ways. In heart failure models, inhibition of H_2S synthesis using PAG reduces excessive sympathetic activation, improving autonomic balance by enhancing heart rate variability and decreasing systolic blood pressure variability. Additionally, PAG administration influences the carotid chemoreceptor reflex, mitigating apnea episodes and respiratory interval variability (Del Rio et al. 2013).

H_2S also modulates respiratory rhythm by region-specific effects within the brainstem. NaHS administration in neonatal rats increases respiratory impulse frequency in the pre-Bötzinger complex, whereas a biphasic response – initial suppression followed by increased frequency – is observed in the parafacial respiratory group, indicating site-dependent modulation (Chen et al. 2013). Furthermore, H_2S demonstrates neuroprotective properties against hypoxia, mitigating respiratory suppression, reducing oxidative stress, and inhibiting neuronal apoptosis through preservation of Bcl-2 mRNA expression (Li et al. 2016). Collectively, these findings highlight the complex and region-specific role of H_2S in modulating respiratory and sympathetic function.

3.2 Effects of H_2S on the Parasympathetic System

There is substantial evidence indicating that H_2S modulates the parasympathetic system. In this regard, Donertas Ayaz et al. (2021) studied how H_2S affects autonomic function and inflammation. They administered NaHS into the brains of animals with angiotensin II-induced hypertension. The findings demonstrated that NaHS reduced blood pressure and attenuated microglial activation within the PVN. The authors proposed that these effects were mediated via modulation of the autonomic nervous system, primarily through the attenuation of sympathetic overactivation and neuroinflammation, thereby contributing to improved autonomic balance (Donertas Ayaz et al. 2021). Similarly, Xia et al. (2023) showed that overexpressing CBS, a major H_2S-producing enzyme, lowered blood pressure and reduced neuronal excitability in the PVN. This effect was attributed to activation of the nuclear factor erythroid 2-related factor (Nrf2) pathway, which mitigates oxidative stress and inflammation, ultimately enhancing parasympathetic dominance (Xia et al. 2023).

However, the precise relationship between H_2S and the parasympathetic system remains a subject of debate. For instance, in anesthetized rats, the administration of sodium sulfide (Na_2S) and NaHS resulted in a dose-dependent reduction in blood pressure and heart rate. Notably, this bradycardic effect was not abolished by atropine, suggesting that it was not mediated by parasympathetic activation but rather by a direct effect on the sinoatrial node (Yoo et al. 2015). Although H_2S modulates cardiovascular functions typically associated with vagal tone, its mechanisms of action may be independent of direct parasympathetic involvement.

3.3 H_2S in Vascular Tone Modulation

H_2S helps regulate vascular tone through several mechanisms. It acts on smooth muscle cells, endothelial cells, and perivascular nerves. These mechanisms interact with sympathetic autonomic signaling at both central and peripheral levels and may also stimulate perivascular sensory nerves (Huerta de la Cruz et al. 2023).

One primary mechanism by which H_2S induces vasodilation involves activation of K_{ATP} channels in vascular smooth muscle cells (Liang et al. 2011). Furthermore, H_2S mediates the vasodilatory effects of endothelin-1 within the cerebral circulation. In neonatal pig arterioles, low levels of endothelin-1 cause vasodilation by triggering H_2S production. This effect involves activating KATP and BKCa channels in vascular smooth muscle cells (Patel et al. 2018).

In coronary circulation, H_2S modulates blood flow through the activation of K_{ATP} channels in smooth muscle, independent of NO synthesis (Casalini et al. 2014). H_2S also affects TRP channels, especially TRPV4 channels in endothelial cells. This promotes calcium influx and the production of vasodilators like nitric oxide and prostacyclin, helping to regulate vascular tone (Munteanu et al. 2024). Notably, TRPV4 activation in endothelial cells has been linked to increased calcium influx and K^+ channel activation, promoting vascular relaxation (Naik et al. 2016).

H_2S is widely recognized for its vasodilatory effects, primarily mediated by the activation of K_{ATP} channels. However, some studies have also reported vasoconstrictive effects under certain conditions. These contrasting findings suggest that the vascular actions of H_2S are highly context-dependent and vary according to the vascular bed and physiological state (O'Sullivan, 2006). This dual behavior underscores the complexity of H_2S signaling in the cardiovascular system. Notably, H_2S also interacts with the NO/cGMP pathway in several ways. It stabilizes eNOS dimers, promotes eNOS phosphorylation through PI3K/Akt signaling, and inhibits

PDE5. These actions increase cGMP levels and enhance vasodilation. Additionally, H_2S inhibits cytochrome C oxidase (Gheibi et al. 2018), which reduces mitochondrial ATP production (Dongo et al. 2018). In parallel, it promotes vasodilation by activating sensory nerves and triggering CGRP release (Eberhardt et al. 2014; Hajna et al. 2016; White et al. 2013).

Moreover, we have demonstrated that NaHS administration mitigates vascular dysfunction by reducing oxidative stress in models of insulin resistance (Silva-Velasco et al. 2024), chronic stress (Beltran-Ornelas et al. 2024), hyperglycemia (Silva-Velasco et al. 2023), and severe traumatic brain injury (Lopez-Preza et al. 2023). These findings further highlight the therapeutic potential of H_2S in vascular pathophysiology and underscore its role in maintaining vascular integrity under pathological conditions.

3.4 H_2S in Autonomic Dysfunction and Potential Therapeutic Targets

Autonomic dysfunction is a critical factor in various cardiovascular, metabolic, and neurodegenerative disorders. In this context, H_2S has emerged as a signaling molecule with modulatory properties on the autonomic nervous system (Table 1). One well-studied effect of H_2S is its ability to lower blood pressure. In hypertensive rats, injecting NaHS into the brain reduced both blood pressure and sympathetic activity. This occurred by inhibiting microglial activation in the PVN (Donertas Ayaz et al. 2021).

Furthermore, in models of hypertension induced by a high-salt diet, GYY4137 administration (H_2S donor) in the PVN reduced sympathetic activity and the hypertensive response through the modulation of oxidative stress and inflammation (Liang et al. 2017). Similarly, in spontaneously hypertensive rats, overexpression of the CBS enzyme increased H_2S levels within the PVN, thereby activating the Nrf2 pathway (Xia et al. 2023). These findings suggest that modulation of H_2S signaling could be a potential therapeutic strategy for hypertension.

H_2S is also important in aging and cardiovascular disease. One major mechanism is its regulation of the Klotho protein, which protects blood vessels. H_2S increases Klotho levels, which helps reduce oxidative stress and inflammation in the cardiovascular system. This may prevent age-related autonomic dysfunction. By preserving endothelial integrity and modulating sympathetic nervous system activity, the interaction between H_2S and Klotho represents a promising therapeutic target for age-related cardiovascular diseases (Yu et al. 2023).

Additionally, in metabolic disorders such as obesity and insulin resistance, H_2S also shows significant therapeutic potential. Overexpressing CBS in the PVN affects energy metabolism. It alters neuroendocrine hormone release and activates the mTOR pathway. This activation contributes to metabolic improvements, including enhanced glucose regulation and lipid profile optimization (Zheng et al. 2018). Furthermore, continuous NaHS infusion in obese rats inhibited vasopressor sympathetic outflow via activation of K_{IR} channels, contributing to the modulation of

Table 1 Effects of H_2S donors on autonomic impairments in different diseases

Pathology	H_2S dosage and administration scheme	Effect	Mechanism	Reference
Hypertension induced by a high-salt diet	GYY4137 (2 nmol/h) was injected into the bilateral PVN for 6 weeks in male Dahl rats	Endogenous or exogenous H_2S in the PVN reduces sympathetic activity and the hypertensive response	H_2S reduced inflammation and oxidative stress in the PVN	Liang et al. (2017)
Heart failure	Male Sprague-Dawley rats, Sham and with CHF, were treated with microinjection of GYY4137 (0.01, 0.1, 1, 2, and 4 nmol) into the PVN	Treatment with GYY4137 increased renal sympathetic nerve activity, raised MAP, and elevated HR	Exogenous administration of H_2S in the PVN increases sympathetic activity and enhances the cardiac sympathetic afferent reflex. Endogenous H_2S production in the PVN does not appear to be involved in the excessive sympathetic activation observed in rats with CHF	Gan et al. (2012)
Aging-related cardiovascular dysfunction	Eight-week-old male Sprague-Dawley rats treated with NaHS (56 μmol/kg per day i.p.) for 12 weeks	NaHS treatment lowered blood pressure, reduced renal sympathetic nerve activity, and decreased norepinephrine levels	H_2S attenuates sympathetic activity by increasing Klotho protein levels and reducing inflammation in the PVN of the hypothalamus	Yu et al. (2023)
Obesity and insulin resistance	Overexpression of CBS through bilateral microinjection of a lentiviral vector into the PVN of Sprague-Dawley rats	Overexpression of the CBS enzyme in the hypothalamic PVN reduced food intake, body weight, and fat mass. Additionally, it lowered blood glucose levels, improved lipid	H_2S regulates the release of neuroendocrine hormones and activates the mTOR pathway to improve energy metabolism	Zheng et al. (2018)

(continued)

Table 1 (continued)

Pathology	H_2S dosage and administration scheme	Effect	Mechanism	Reference
		profiles, and reduced insulin resistance in obese rats		
Hypertension	NaHS (30 and 60 nmol/h) was administered ICV in male Sprague-Dawley rats for 4 weeks	Central administration of NaHS reduced blood pressure and sympathetic activity in rats with angiotensin II-induced hypertension	Central H_2S attenuates hypertension by reducing microglial activation in the hypothalamic PVN, which in turn decreases sympathetic nervous system activity	Donertas Ayaz et al. (2021)
Obesity	Male Wistar rats fed with a high-fat diet. A continuous infusion of NaHS at 310 μg/kg·min was used	NaHS induces inhibition of vasopressor sympathetic outflow in obese rats	NaHS-induced sympathetic inhibition appears to be mediated by inward K_{IR}	Gomez et al. (2024)
Cardiovascular regulation	Anesthetized Wistar Kyoto rats bilaterally microinjected with NaHS (0.2–2000 pmol/side) into the RVLM	Microinjection of NaHS into the RVLM and PVN did not produce significant changes in blood pressure, HR, or LSNA compared to the vehicle	H_2S in those regions is not playing a critical role in the regulation of blood pressure, HR, and LSNA	Streeter et al. (2011)

PVN paraventricular nucleus of the hypothalamus, *CHF* chronic heart failure, *MAP* mean arterial pressure, *HR* heart rate, *NaHS* sodium hydrosulfide, *CBS* cystathionine beta-synthase, *ICV* intracerebroventricularly, *SHR* spontaneously hypertensive rats, *WKY* Wistar Kyoto, *Nrf2* nuclear factor erythroid 2-related factor, K_{IR} rectifier potassium channels, *LSNA* lumbar sympathetic nerve activity, *RVLM* rostral ventrolateral medulla

autonomic tone (Gomez et al. 2024). These findings underscore the importance of H_2S in the regulation of energy metabolism and autonomic responses in obesity.

Beyond its well-characterized cardiovascular effects, H_2S influences various organ systems through its interaction with both the sympathetic and parasympathetic nervous systems (Fig. 1). Collectively, these studies highlight the principal molecular targets of H_2S, including microglia modulation in the PVN, activation of the Nrf2 pathway, regulation of Klotho protein, stimulation of the mTOR pathway, and interaction with K_{IR} channels. Targeting the H_2S pathway through exogenous donors, inhibitors, or genetic interventions may represent a promising avenue for

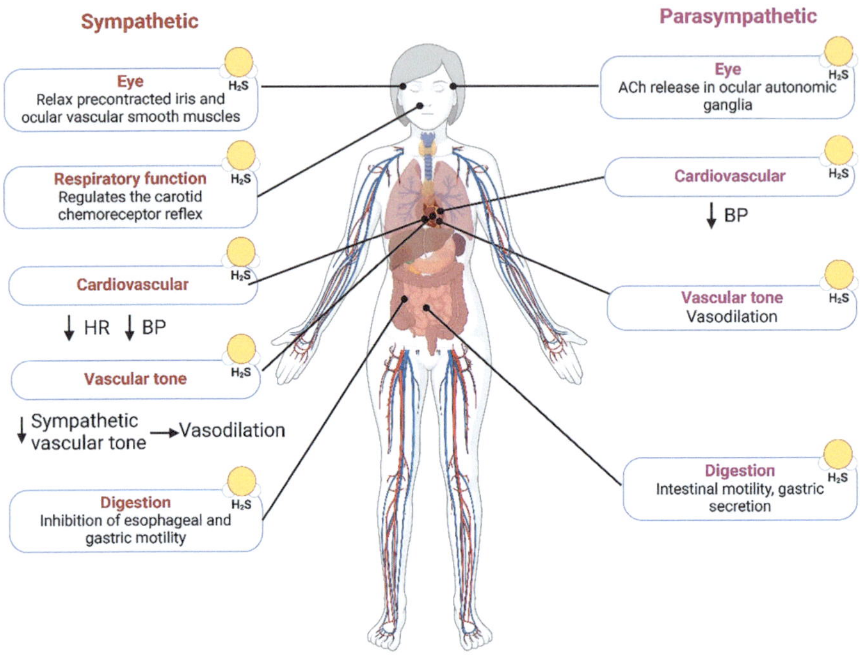

Fig. 1 Effects of H_2S on sympathetic and parasympathetic systems. *HR* Heart rate, *BP* blood pressure, *ACh* acetylcholine

drug development aimed to mitigate autonomic dysfunction and its associated pathologies.

4 Hydrogen Sulfide in Pain Modulation

4.1 H_2S Modulation of Nociceptors and Pain Perception

Nociception, the sensory process by which pain is perceived, begins when peripheral primary afferent neurons detect noxious stimuli and relay signals to second-order spinal neurons. These primary sensory neurons, known as nociceptors, detect mechanical, thermal, and chemical stimuli and convert them into electrical signals via specific membrane receptors and ion channels (Dubin and Patapoutian 2010). Their cell bodies are located in the dorsal root ganglia (DRG) or the trigeminal ganglia. The peripheral fibers project to target tissues, while their central fibers synapse with spinal neurons, transmitting pain information to higher brain centers.

H_2S has recently been recognized as an important modulator of nociceptive pathways, particularly through its effects on sensory neurons involved in pain transmission. Through the enzymatic production by CSE and CBS, H_2S levels are regulated in peripheral neurons, particularly nociceptors (Eberhardt et al. 2014; Syhr et al. 2015; Takahashi et al. 2010; Xu et al. 2009). Notably, CBS is expressed in sensory neurons that also express TRPV1 and TRPA1 channels—both key mediators of pain signaling (Eberhardt et al. 2014; Xu et al. 2009). Certain pain-inducing conditions, such as inflammation or peripheral nerve injury, increase CSE expression in DRG neurons. However, mice lacking CSE—either globally or selectively in sensory neurons—still display normal pain behaviors in inflammatory and neuropathic pain models. This suggests that CSE may not be essential for nociceptive behavior in chronic pain, and that other sources of H_2S might mediate pain responses in these conditions (Syhr et al. 2015).

Studies using H_2S donors have highlighted the capacity of exogenous H_2S to modulate pain thresholds and induce hyperalgesia. For example, intraplanar administration of NaHS, an H_2S donor, significantly lowered the mechanical pain threshold, which was associated with an increased number of c-Fos-positive cells in the superficial layers of the L5 spinal dorsal horn, an indicator of neuronal activation in response to pain stimuli. Moreover, L-cysteine administration induced hyperalgesia in rats, an effect that was blocked by β-cyano-L-alanine or PAG, both inhibitors of H_2S biosynthesis (Kawabata et al. 2007). H_2S also affects central pain processing. Intrathecal (spinal) administration of NaHS reproduces the hyperalgesic effects seen with peripheral injections (Maeda et al. 2009). Moreover, intraplantar NaHS causes pain not only in the injected paw but also in the opposite paw, indicating that H_2S can affect pain processing across both sides of the spinal cord (Zhao et al. 2016).

Notably, the hyperalgesic effects of H_2S are due to sensitization of DRG and spinal neurons by acting on various ion channels, including T-type calcium channels and TRP channels, such as TRPA1. Indeed, previous studies showed that NaHS sensitizes Cav3.2 channels in both peripheral and spinal pain pathways, contributing to amplified pain perception, while other channel subtypes like Cav3.1 and Cav3.3 remain unaffected, suggesting that Cav3.2 channels play a specific role in H_2S-mediated hyperalgesia (Donatti et al. 2014; Fukami et al. 2017; Kawabata et al. 2007; Sekiguchi et al. 2014). Moreover, the activation of TRPA1 channels by H_2S results in a significant influx of calcium ions (Ca^{2+}) into sensory neurons, leading to increased excitability and sensitization, correlating with increased pain sensitivity (Ogawa et al. 2012; Okubo et al. 2012). Additionally, the contralateral hyperalgesia observed after intraplanar NaHS injection is linked to the upregulation of the NMDA receptor subunit NR2B in the spinal cord, further illustrating the complex central mechanisms through which H_2S can influence pain perception (Zhao et al. 2016).

4.2 H_2S in Chronic Pain Conditions and Potential Therapeutic Targets

H_2S has a complex role in chronic pain, influencing both neuropathic and inflammatory conditions. It can promote or reduce pain depending on the context, acting through signaling pathways, ion channels, and enzymes.

In inflammatory pain models, H_2S often increases pain sensitivity by upregulating the CBS/H_2S pathway. In a model of temporomandibular joint inflammation induced by complete Freund's adjuvant (CFA), inhibiting CBS with AOAA reduced both hyperalgesia and depolarizing current density in trigeminal ganglion neurons. This highlights the role of CBS/H_2S in inflammatory sensitization (Miao et al. 2014). Similarly, in CFA-induced peripheral inflammation, CBS upregulation in DRG neurons contributed to hyperalgesia, which was mitigated by inhibitors of CBS like AOAA and hydroxylamine. In this context, the pronociceptive role of H_2S may be linked to sodium channel hyperexcitability mediated by PKA activation (Qi et al. 2013). In addition, H_2S shows a dependency on the intensity of nociceptive stimuli, where PAG reduced formalin-induced inflammation and c-Fos expression in the spinal cord only at higher formalin doses (Lee et al. 2008). Inhibiting H_2S production may also reduce inflammatory cell leakage and disrupt substance P signaling, which plays a role in neurogenic inflammation. The findings suggest that H_2S contributes to the persistence of pain by enhancing vascular permeability and promoting the recruitment of inflammatory cells to the site of injury (Garattini et al. 2019; Santos et al. 2018). In another study focusing on inflammatory pain induced by lipopolysaccharide and prostaglandin E2, NaHS, a donor of H_2S, reduced mechanical hyperalgesia but did not alter thermal nociception. The effect of NaHS was blocked by glibenclamide, an inhibitor of K_{ATP} channels, suggesting that H_2S acts through the modulation of these channels. In contrast, the inhibition of CSE using PAG did not affect the production of proinflammatory cytokines like TNF-α and IL-1β, indicating that the antinociceptive effects of H_2S are independent of its effects on cytokine production but rather through the direct modulation of nociceptors via K_{ATP} channels (Cunha et al. 2008).

Conversely, H_2S donors can reduce pain in neuropathic models. In a chronic constriction injury (CCI) model, systemic NaHS administration reduced neuropathic pain. This effect was likely due to inhibition of pCREB expression in the spinal cord (Lin et al. 2014). Similarly, the inhalation of NaHS prevented the development of pain behavior in the CCI model by suppressing the upregulation of IL-6 and CCL2, potentially through the inhibition of microglia activation (Kida et al. 2015). Conversely, the overexpression of CBS has been reported in the CCI model. Indeed, systemic treatment with AOAA, a CBS inhibitor, reversed both the pain observed in this model and the activation of NF-kβ (p65), ERK1/2, and CREB, suggesting that CBS-generated H_2S contributes to the activation of these pain-regulating pathways (Gui et al. 2016). In line with this, the inhibition of CSE using β-cyanolanine or PAG reduced hyperalgesia in a neuropathic pain model induced by the L5 spinal nerve cutting (Takahashi et al. 2010).

In chemotherapy-induced neuropathy models, systemic or central administration of isothiocyanate-based H_2S donors alleviated hypersensitivity caused by paclitaxel and oxaliplatin, effects reversed by Kv7 channel blockers, indicating that the activation of Kv7 channels is a key mechanism (Di Cesare Mannelli et al. 2017). Other studies showed that Cav3.2 T-type calcium channels are involved in paclitaxel-induced pain. Inhibiting CSE with β-cyanolanine or PAG reduced hyperalgesia by limiting both Cav3.2 activation and H_2S overproduction (Okubo et al. 2011).

The role of H_2S in diabetic neuropathy has also been investigated. In diabetic rats, NaHS and L-cysteine increased nociceptive behavior, while inhibitors of CBS and CSE reduced hyperalgesia, suggesting that H_2S exacerbates nociceptive responses in this model. Interestingly, L-cysteine-induced hyperalgesia was mitigated by CBS and CSE inhibitors, indicating a mechanism involving H_2S overproduction (Velasco-Xolalpa et al. 2013).

Genetic and pharmacological studies further support the complex role of H_2S. For example, mice lacking CSE showed normal pain behavior in both inflammatory and neuropathic models, suggesting that CSE may not be essential for pain processing (Syhr et al. 2015). On the other hand, treatment with L-cysteine in a model of insulin resistance effectively inhibited tactile and cold allodynia, linked to reduced B1 bradykinin receptor expression, suggesting a beneficial role of H_2S modulation (Ismael et al. 2008).

In conclusion, H_2S influences chronic pain through diverse mechanisms, including modulation of ion channels and proinflammatory signaling pathways. Because of its dual effects, H_2S presents both challenges and opportunities for therapy. Thus, targeting H_2S pathways may lead to more precise pain treatments.

5 Conclusion

H_2S has emerged as a key modulator of the PNS function, affecting both sensory and autonomic neurons through its interactions with ion channels, receptors, and intracellular signaling pathways. Increasing evidence supports its dual role as a neuroprotective or pronociceptive agent, depending on the physiological or pathological context. This complexity highlights H_2S as a central factor in pain signaling, inflammation, and autonomic regulation.

Despite growing knowledge of H_2S biology, many of its molecular mechanisms remain poorly understood. Future research should focus on clarifying how H_2S interacts with its molecular targets, particularly through post-translational modifications of ion channels and downstream signaling pathways. Studies are also needed to explore the long-term effects of H_2S modulation in chronic pain and autonomic dysfunction to facilitate clinical translation.

Targeting H_2S signaling holds considerable therapeutic promise for treating chronic pain, neuropathies, and autonomic disorders. The main challenge is achieving selective and controlled regulation of H_2S levels to maximize benefits while minimizing side effects. Developing precise pharmacological strategies to modulate H_2S synthesis and signaling could open new avenues for managing PNS-related diseases. Moving forward, interdisciplinary approaches—integrating electrophysiology, imaging, molecular biology, and translational research—will be essential to fully realize the therapeutic potential of H_2S.

Acknowledgments This study was supported by the National Council of Humanities, Sciences, and Technologies, Conahcyt, 252702 to DC, and the American Heart Association Postdoctoral Fellowship 24POST1188081 and the Cardiovascular Research Institute of Vermont ECAC Award to SHC.

References

Barrera A, Morales-Loredo H, Garcia JM, Fregoso G, Pace CE, Mendiola PJ, Naik JS, Gonzalez Bosc LV, Kanagy NL (2021) Simulated sleep apnea alters hydrogen sulfide regulation of blood flow and pressure. Am J Physiol Heart Circ Physiol 320(2):H511–H519. https://doi.org/10.1152/ajpheart.00672.2019

Beltran-Ornelas JH, Silva-Velasco DL, Tapia-Martinez JA, Sanchez-Lopez A, Cano-Europa E, Huerta de la Cruz S, Centurion D (2024) Sodium hydrosulfide reverts chronic stress-induced cardiovascular alterations by reducing oxidative stress. J Cardiovasc Pharmacol 83(4):317–329. https://doi.org/10.1097/FJC.0000000000001538

Casalini ED, Goodwill AG, Owen MK, Moberly SP, Berwick ZC, Tune JD (2014) Contribution of hydrogen sulfide to the control of coronary blood flow. Microcirculation 21(2):104–111. https://doi.org/10.1111/micc.12083

Centurion D, de la Cruz SH, Castillo-Santiago SV, Becerril-Chacon ME, Torres-Perez JA, Sanchez-Lopez A (2018) NaHS prejunctionally inhibits the cardioaccelerator sympathetic outflow in pithed rats. Eur J Pharmacol 823:35–40. https://doi.org/10.1016/j.ejphar.2018.01.030

Centurión D, De la Cruz SH, Gutiérrez-Lara EJ, Beltran-Ornelas JH, Sánchez-López A (2016) Pharmacological evidence that NaHS inhibits the vasopressor responses induced by stimulation of the preganglionic sympathetic outflow in pithed rats. Eur J Pharmacol 770:40–45. https://doi.org/10.1016/j.ejphar.2015.11.057

Chen L, Zhang J, Ding Y, Li H, Nie L, Zhou H, Tang Y, Zheng Y (2013) Site-specific hydrogen sulfide-mediated central regulation of respiratory rhythm in medullary slices of neonatal rats. Neuroscience 233:118–126. https://doi.org/10.1016/j.neuroscience.2012.12.047

Cirino G, Szabo C, Papapetropoulos A (2023) Physiological roles of hydrogen sulfide in mammalian cells, tissues, and organs. Physiol Rev 103(1):31–276. https://doi.org/10.1152/physrev.00028.2021

Cunha TM, Dal-Secco D, Verri WA Jr, Guerrero AT, Souza GR, Vieira SM, Lotufo CM, Neto AF, Ferreira SH, Cunha FQ (2008) Dual role of hydrogen sulfide in mechanical inflammatory hypernociception. Eur J Pharmacol 590(1–3):127–135. https://doi.org/10.1016/j.ejphar.2008.05.048

Del Rio R, Marcus NJ, Schultz HD (2013) Inhibition of hydrogen sulfide restores normal breathing stability and improves autonomic control during experimental heart failure. J Appl Physiol (1985) 114(9):1141–1150. https://doi.org/10.1152/japplphysiol.01503.2012

Di Cesare Mannelli L, Lucarini E, Micheli L, Mosca I, Ambrosino P, Soldovieri MV, Martelli A, Testai L, Taglialatela M, Calderone V, Ghelardini C (2017) Effects of natural and synthetic isothiocyanate-based H(2)S-releasers against chemotherapy-induced neuropathic pain: role of Kv7 potassium channels. Neuropharmacology 121:49–59. https://doi.org/10.1016/j.neuropharm.2017.04.029

Doiron JE, Xia H, Yu X, Nevins AR, LaPenna KB, Sharp TE 3rd, Goodchild TT, Allerton TD, Elgazzaz M, Lazartigues E, Shah SJ, Li Z, Lefer DJ (2024) Adjunctive therapy with an oral H(2)S donor provides additional therapeutic benefit beyond SGLT2 inhibition in cardiometabolic heart failure with preserved ejection fraction. Br J Pharmacol 181(21):4294–4310. https://doi.org/10.1111/bph.16493

Dominguez-Rodriguez M, Drobny H, Boehm S, Salzer I (2017) Electrophysiological investigation of the subcellular fine tuning of sympathetic neurons by hydrogen sulfide. Front Pharmacol 8:522. https://doi.org/10.3389/fphar.2017.00522

Donatti AF, Araujo RM, Soriano RN, Azevedo LU, Leite-Panissi CA, Branco LG (2014) Role of hydrogen sulfide in the formalin-induced orofacial pain in rats. Eur J Pharmacol 738:49–56. https://doi.org/10.1016/j.ejphar.2014.05.023

Donertas Ayaz B, Oliveira AC, Malphurs WL, Redler T, de Araujo AM, Sharma RK, Sirmagul B, Zubcevic J (2021) Central administration of hydrogen sulfide donor NaHS reduces Iba1-positive cells in the PVN and attenuates rodent angiotensin II hypertension. Front Neurosci 15:690919. https://doi.org/10.3389/fnins.2021.690919

Donertas Ayaz B, Zubcevic J (2020) Gut microbiota and neuroinflammation in pathogenesis of hypertension: a potential role for hydrogen sulfide. Pharmacol Res 153:104677. https://doi.org/10.1016/j.phrs.2020.104677

Dongo E, Beliczai-Marosi G, Dybvig AS, Kiss L (2018) The mechanism of action and role of hydrogen sulfide in the control of vascular tone. Nitric Oxide 81:75–87. https://doi.org/10.1016/j.niox.2017.10.010

Duan XC, Guo R, Liu SY, Xiao L, Xue HM, Guo Q, Jin S, Wu YM (2015) Gene transfer of cystathionine beta-synthase into RVLM increases hydrogen sulfide-mediated suppression of sympathetic outflow via KATP channel in normotensive rats. Am J Physiol Heart Circ Physiol 308(6):H603–H611. https://doi.org/10.1152/ajpheart.00693.2014

Dubin AE, Patapoutian A (2010) Nociceptors: the sensors of the pain pathway. J Clin Invest 120(11):3760–3772. https://doi.org/10.1172/JCI42843

Eberhardt M, Dux M, Namer B, Miljkovic J, Cordasic N, Will C, Kichko TI, de la Roche J, Fischer M, Suarez SA, Bikiel D, Dorsch K, Leffler A, Babes A, Lampert A, Lennerz JK, Jacobi J, Marti MA, Doctorovich F et al (2014) H2S and NO cooperatively regulate vascular tone by activating a neuroendocrine HNO-TRPA1-CGRP signalling pathway. Nat Commun 5: 4381. https://doi.org/10.1038/ncomms5381

Feng X, Guo Q, Xue H, Duan X, Jin S, Wu Y (2020) Hydrogen sulfide attenuated angiotensin II-induced sympathetic excitation in offspring of renovascular hypertensive rats. Front Pharmacol 11:565726. https://doi.org/10.3389/fphar.2020.565726

Feng X, Zhou YL, Meng X, Qi FH, Chen W, Jiang X, Xu GY (2013) Hydrogen sulfide increases excitability through suppression of sustained potassium channel currents of rat trigeminal ganglion neurons. Mol Pain 9:4. https://doi.org/10.1186/1744-8069-9-4

Fukami K, Fukami K, Sekiguchi F, Sekiguchi F, Kawabata A, Kawabata A (2017) Hydrogen sulfide and T-type Ca2+ channels in pain processing, neuronal differentiation and neuroendocrine secretion. Pharmacology 99(3–4):196–203. https://doi.org/10.1159/000449449

Gan XB, Liu TY, Xiong XQ, Chen WW, Zhou YB, Zhu GQ (2012) Hydrogen sulfide in paraventricular nucleus enhances sympathetic activity and cardiac sympathetic afferent reflex in chronic heart failure rats. PLoS One 7(11):e50102. https://doi.org/10.1371/journal.pone.0050102

Garattini EG, Santos BM, Ferrari DP, Capel CP, Francescato HDC, Coimbra TM, Leite-Panissi CRA, Branco LGS, Nascimento GC (2019) Propargylglycine decreases neuro-immune interaction inducing pain response in temporomandibular joint inflammation model. Nitric Oxide 93: 90–101. https://doi.org/10.1016/j.niox.2019.10.001

Gheibi S, Jeddi S, Kashfi K, Ghasemi A (2018) Regulation of vascular tone homeostasis by NO and H2S: implications in hypertension. Biochem Pharmacol 149:42–59. https://doi.org/10.1016/j.bcp.2018.01.017

Gomez CB, de la Cruz SH, Medina-Terol GJ, Beltran-Ornelas JH, Sánchez-López A, Silva-Velasco DL, Centurión D (2019) Chronic administration of NaHS and L-cysteine restores cardiovascular changes induced by high-fat diet in rats. Eur J Pharmacol 863:172707. https://doi.org/10.1016/j.ejphar.2019.172707

Gomez CB, Sanchez-Lopez A, Carvajal K, Centurion D (2024) Effect of hydrogen sulfide on sympathoinhibition in obese pithed rats and participation of K(+) channel. Int J Hypertens 2024: 5848352. https://doi.org/10.1155/2024/5848352

Gridina A, Su X, Khan SA, Peng YJ, Wang B, Nanduri J, Fox AP, Prabhakar NR (2021) Gaseous transmitter regulation of hypoxia-evoked catecholamine secretion from murine adrenal chromaffin cells. J Neurophysiol 125(5):1533–1542. https://doi.org/10.1152/jn.00669.2020

Gui Y, Li A, Qiu B, Chen F, Chen L, Liu D, Chen S, Zhou W, Zhou H (2016) Endogenous CBS-H2S pathway contributes to the development of CCI-induced neuropathic pain. Neurochem Res 41(6):1381–1389. https://doi.org/10.1007/s11064-016-1842-z

Guo Q, Wu Y, Xue H, Xiao L, Jin S, Wang R (2016) Perfusion of isolated carotid sinus with hydrogen sulfide attenuated the renal sympathetic nerve activity in anesthetized male rats. Physiol Res 65(3):413–423. https://doi.org/10.33549/physiolres.933050

Hajna Z, Saghy E, Payrits M, Aubdool AA, Szoke E, Pozsgai G, Batai IZ, Nagy L, Filotas D, Helyes Z, Brain SD, Pinter E (2016) Capsaicin-sensitive sensory nerves mediate the cellular and microvascular effects of H2S via TRPA1 receptor activation and neuropeptide release. J Mol Neurosci 60(2):157–170. https://doi.org/10.1007/s12031-016-0802-z

Hsu CC, Lin RL, Lee LY, Lin YS (2013) Hydrogen sulfide induces hypersensitivity of rat capsaicin-sensitive lung vagal neurons: role of TRPA1 receptors. Am J Physiol Regul Integr Comp Physiol 305(7):R769–R779. https://doi.org/10.1152/ajpregu.00202.2013

Huerta de la Cruz S, Medina-Terol GJ, Tapia-Martinez JA, Silva-Velasco DL, Beltran-Ornelas JH, Sanchez-Lopez A, Sancho M, Centurion D (2023) Hydrogen sulfide as a neuromodulator of the vascular tone. Eur J Pharmacol 940:175455. https://doi.org/10.1016/j.ejphar.2022.175455

Huerta de la Cruz S, Rocha L, Santiago-Castaneda C, Sanchez-Lopez A, Pinedo-Rodriguez AD, Medina-Terol GJ, Centurion D (2022) Hydrogen sulfide subchronic treatment improves hypertension induced by traumatic brain injury in rats through vasopressor sympathetic outflow inhibition. J Neurotrauma 39(1–2):181–195. https://doi.org/10.1089/neu.2020.7552

Ismael MA, Talbot S, Carbonneau CL, Beausejour CM, Couture R (2008) Blockade of sensory abnormalities and kinin B(1) receptor expression by N-acetyl-L-cysteine and ramipril in a rat model of insulin resistance. Eur J Pharmacol 589(1–3):66–72. https://doi.org/10.1016/j.ejphar.2008.05.006

Kawabata A, Ishiki T, Nagasawa K, Yoshida S, Maeda Y, Takahashi T, Sekiguchi F, Wada T, Ichida S, Nishikawa H (2007) Hydrogen sulfide as a novel nociceptive messenger. Pain 132(1–2):74–81. https://doi.org/10.1016/j.pain.2007.01.026

Kida K, Marutani E, Nguyen RK, Ichinose F (2015) Inhaled hydrogen sulfide prevents neuropathic pain after peripheral nerve injury in mice. Nitric Oxide 46:87–92. https://doi.org/10.1016/j.niox.2014.11.014

Kimura H, Nagai Y, Umemura K, Kimura Y (2005) Physiological roles of hydrogen sulfide: synaptic modulation, neuroprotection, and smooth muscle relaxation. Antioxid Redox Signal 7(5–6):795–803. https://doi.org/10.1089/ars.2005.7.795

Lee AT, Shah JJ, Li L, Cheng Y, Moore PK, Khanna S (2008) A nociceptive-intensity-dependent role for hydrogen sulphide in the formalin model of persistent inflammatory pain. Neuroscience 152(1):89–96. https://doi.org/10.1016/j.neuroscience.2007.11.052

Li H, Chen L, Hou X, Zhou H, Zheng Y (2016) Hydrogen sulfide attenuates hypoxia-induced respiratory suppression in anesthetized adult rats. Respir Physiol Neurobiol 220:1–9. https://doi.org/10.1016/j.resp.2015.09.001

Li Y, Feng Y, Liu L, Li X, Li XY, Sun X, Li KX, Zha RR, Wang HD, Zhang MD, Fan XX, Wu D, Fan Y, Zhang HC, Qiao GF, Li BY (2021) The baroreflex afferent pathway plays a critical role in H(2)S-mediated autonomic control of blood pressure regulation under physiological and hypertensive conditions. Acta Pharmacol Sin 42(6):898–908. https://doi.org/10.1038/s41401-020-00549-5

Liang GH, Adebiyi A, Leo MD, McNally EM, Leffler CW, Jaggar JH (2011) Hydrogen sulfide dilates cerebral arterioles by activating smooth muscle cell plasma membrane KATP channels. Am J Physiol Heart Circ Physiol 300(6):H2088–H2095. https://doi.org/10.1152/ajpheart.01290.2010

Liang YF, Zhang DD, Yu XJ, Gao HL, Liu KL, Qi J, Li HB, Yi QY, Chen WS, Cui W, Zhu GQ, Kang YM (2017) Hydrogen sulfide in paraventricular nucleus attenuates blood pressure by regulating oxidative stress and inflammatory cytokines in high salt-induced hypertension. Toxicol Lett 270:62–71. https://doi.org/10.1016/j.toxlet.2017.02.004

Liao R, Xue L, Qiang Z, Zhang C, Liu Y (2020) Release of endogenous hydrogen sulfide in enteric nerve cells suppresses intestinal motility during severe acute pancreatitis. Acta Biochim Biophys Sin Shanghai 52(1):64–71. https://doi.org/10.1093/abbs/gmz139

Lin JQ, Luo HQ, Lin CZ, Chen JZ, Lin XZ (2014) Sodium hydrosulfide relieves neuropathic pain in chronic constriction injured rats. Evid Based Complement Alternat Med 2014:514898. https://doi.org/10.1155/2014/514898

Liu WQ, Chai C, Li XY, Yuan WJ, Wang WZ, Lu Y (2011) The cardiovascular effects of central hydrogen sulfide are related to K(ATP) channels activation. Physiol Res 60(5):729–738. https://doi.org/10.33549/physiolres.932092

Lopez-Preza FI, Huerta de la Cruz S, Santiago-Castaneda C, Silva-Velasco DL, Beltran-Ornelas JH, Tapia-Martinez J, Sanchez-Lopez A, Rocha L, Centurion D (2023) Hydrogen sulfide prevents the vascular dysfunction induced by severe traumatic brain injury in rats by reducing reactive oxygen species and modulating eNOS and H(2)S-synthesizing enzyme expression. Life Sci 312:121218. https://doi.org/10.1016/j.lfs.2022.121218

Lu W, Li J, Gong L, Xu X, Han T, Ye Y, Che T, Luo Y, Li J, Zhan R, Yao W, Liu K, Cui S, Liu C (2014) H2 S modulates duodenal motility in male rats via activating TRPV1 and K(ATP) channels. Br J Pharmacol 171(6):1534–1550. https://doi.org/10.1111/bph.12562

Maeda Y, Aoki Y, Sekiguchi F, Matsunami M, Takahashi T, Nishikawa H, Kawabata A (2009) Hyperalgesia induced by spinal and peripheral hydrogen sulfide: evidence for involvement of Cav3.2 T-type calcium channels. Pain 142(1–2):127–132. https://doi.org/10.1016/j.pain.2008.12.021

Medina-Terol GJ, Huerta de la Cruz S, Beltran-Ornelas JH, Sanchez-Lopez A, Centurion D (2022) Pharmacological evidence that potassium channels mediate hydrogen sulfide-induced inhibition of the vasopressor sympathetic outflow in pithed rats. Eur J Pharmacol 931:175160. https://doi.org/10.1016/j.ejphar.2022.175160

Miao X, Meng X, Wu G, Ju Z, Zhang HH, Hu S, Xu GY (2014) Upregulation of cystathionine-beta-synthetase expression contributes to inflammatory pain in rat temporomandibular joint. Mol Pain 10:9. https://doi.org/10.1186/1744-8069-10-9

Miyamoto R, Otsuguro K, Yamaguchi S, Ito S (2014) Contribution of cysteine aminotransferase and mercaptopyruvate sulfurtransferase to hydrogen sulfide production in peripheral neurons. J Neurochem 130(1):29–40. https://doi.org/10.1111/jnc.12698

Munteanu C, Popescu C, Vladulescu-Trandafir AI, Onose G (2024) Signaling paradigms of H(2)S-induced vasodilation: a comprehensive review. Antioxidants (Basel) 13(10). https://doi.org/10.3390/antiox13101158

Naik JS, Osmond JM, Walker BR, Kanagy NL (2016) Hydrogen sulfide-induced vasodilation mediated by endothelial TRPV4 channels. Am J Physiol Heart Circ Physiol 311(6):H1437–H1444. https://doi.org/10.1152/ajpheart.00465.2016

O'Sullivan SE (2006) What is the significance of vascular hydrogen sulphide (H2S)? Br J Pharmacol 149(6):609–610. https://doi.org/10.1038/sj.bjp.0706907

Ogawa H, Takahashi K, Miura S, Imagawa T, Saito S, Tominaga M, Ohta T (2012) H(2)S functions as a nociceptive messenger through transient receptor potential ankyrin 1 (TRPA1) activation. Neuroscience 218:335–343. https://doi.org/10.1016/j.neuroscience.2012.05.044

Okubo K, Matsumura M, Kawaishi Y, Aoki Y, Matsunami M, Okawa Y, Sekiguchi F, Kawabata A (2012) Hydrogen sulfide-induced mechanical hyperalgesia and allodynia require activation of both Cav3.2 and TRPA1 channels in mice. Br J Pharmacol 166(5):1738–1743. https://doi.org/10.1111/j.1476-5381.2012.01886.x

Okubo K, Takahashi T, Sekiguchi F, Kanaoka D, Matsunami M, Ohkubo T, Yamazaki J, Fukushima N, Yoshida S, Kawabata A (2011) Inhibition of T-type calcium channels and hydrogen sulfide-forming enzyme reverses paclitaxel-evoked neuropathic hyperalgesia in rats. Neuroscience 188:148–156. https://doi.org/10.1016/j.neuroscience.2011.05.004

Ozaki T, Matsuoka J, Tsubota M, Tomita S, Sekiguchi F, Minami T, Kawabata A (2018) Zinc deficiency promotes cystitis-related bladder pain by enhancing function and expression of Ca(v) 3.2 in mice. Toxicology 393:102–112. https://doi.org/10.1016/j.tox.2017.11.012
Park BS, Kim HW, Rhyu IJ, Park C, Yeo SG, Huh Y, Jeong NY, Jung J (2015) Hydrogen sulfide is essential for Schwann cell responses to peripheral nerve injury. J Neurochem 132(2):230–242. https://doi.org/10.1111/jnc.12932
Patel S, Fedinec AL, Liu J, Weiss MA, Pourcyrous M, Harsono M, Parfenova H, Leffler CW (2018) H(2)S mediates the vasodilator effect of endothelin-1 in the cerebral circulation. Am J Physiol Heart Circ Physiol 315(6):H1759–H1764. https://doi.org/10.1152/ajpheart.00451.2018
Peng YJ, Makarenko VV, Gridina A, Chupikova I, Zhang X, Kumar GK, Fox AP, Prabhakar NR (2019) H(2)S mediates carotid body response to hypoxia but not anoxia. Respir Physiol Neurobiol 259:75–85. https://doi.org/10.1016/j.resp.2018.08.001
Peng YJ, Makarenko VV, Nanduri J, Vasavda C, Raghuraman G, Yuan G, Gadalla MM, Kumar GK, Snyder SH, Prabhakar NR (2014) Inherent variations in CO-H2S-mediated carotid body O2 sensing mediate hypertension and pulmonary edema. Proc Natl Acad Sci USA 111(3): 1174–1179. https://doi.org/10.1073/pnas.1322172111
Peng YJ, Nanduri J, Raghuraman G, Souvannakitti D, Gadalla MM, Kumar GK, Snyder SH, Prabhakar NR (2010) H2S mediates O2 sensing in the carotid body. Proc Natl Acad Sci USA 107(23):10719–10724. https://doi.org/10.1073/pnas.1005866107
Prabhakar NR, Peng YJ, Nanduri J (2018) Recent advances in understanding the physiology of hypoxic sensing by the carotid body. F1000Res 7:10.12688/f1000research.16247.1
Qi F, Zhou Y, Xiao Y, Tao J, Gu J, Jiang X, Xu GY (2013) Promoter demethylation of cystathionine-beta-synthetase gene contributes to inflammatory pain in rats. Pain 154(1): 34–45. https://doi.org/10.1016/j.pain.2012.07.031
Rangel-Galvan M, Rangel-Galvan V, Rangel-Huerta A (2023) T-type calcium channel modulation by hydrogen sulfide in neuropathic pain conditions. Front Pharmacol 14:1212800. https://doi.org/10.3389/fphar.2023.1212800
Roa-Coria JE, Pineda-Farias JB, Barragan-Iglesias P, Quinonez-Bastidas GN, Zuniga-Romero A, Huerta-Cruz JC, Reyes-Garcia JG, Flores-Murrieta FJ, Granados-Soto V, Rocha-Gonzalez HI (2019) Possible involvement of peripheral TRP channels in the hydrogen sulfide-induced hyperalgesia in diabetic rats. BMC Neurosci 20(1):1. https://doi.org/10.1186/s12868-018-0483-3
Santos BM, Garattini EG, Branco LGS, Leite-Panissi CRA, Nascimento GC (2018) The therapeutic potential of cystathionine gamma-lyase in temporomandibular inflammation-induced orofacial hypernociception. Physiol Behav 188:128–133. https://doi.org/10.1016/j.physbeh.2018.02.007
Schicho R, Krueger D, Zeller F, Von Weyhern CW, Frieling T, Kimura H, Ishii I, De Giorgio R, Campi B, Schemann M (2006) Hydrogen sulfide is a novel prosecretory neuromodulator in the Guinea-pig and human colon. Gastroenterology 131(5):1542–1552. https://doi.org/10.1053/j.gastro.2006.08.035
Sekiguchi F, Miyamoto Y, Kanaoka D, Ide H, Yoshida S, Ohkubo T, Kawabata A (2014) Endogenous and exogenous hydrogen sulfide facilitates T-type calcium channel currents in Cav3.2-expressing HEK293 cells. Biochem Biophys Res Commun 445(1):225–229. https://doi.org/10.1016/j.bbrc.2014.01.185
Sekiguchi F, Tsubota M, Kawabata A (2024) Sulfide and polysulfide as pronociceptive mediators: focus on Ca(v)3.2 function enhancement and TRPA1 activation. J Pharmacol Sci 155(3): 113–120. https://doi.org/10.1016/j.jphs.2024.04.007
Sha L, Linden DR, Farrugia G, Szurszewski JH (2013) Hydrogen sulfide selectively potentiates central preganglionic fast nicotinic synaptic input in mouse superior mesenteric ganglion. J Neurosci 33(31):12638–12646. https://doi.org/10.1523/JNEUROSCI.4429-12.2013
Shiina T, Shima T, Horii K, Naitou K, Nakamori H, Sano Y, Shimizu Y (2016) Inhibitory action of hydrogen sulfide on esophageal striated muscle motility in rats. Eur J Pharmacol 771:123–129. https://doi.org/10.1016/j.ejphar.2015.12.018

Silva-Velasco DL, Beltran-Ornelas JH, Tapia-Martinez J, Sanchez-Lopez A, de la Cruz SH, Cervantes-Perez LG, Del Valle-Mondragon L, Sanchez-Mendoza A, Centurion D (2023) NaHS restores the vascular alterations in the renin-angiotensin system induced by hyperglycemia in rats. Peptides 164:171001. https://doi.org/10.1016/j.peptides.2023.171001

Silva-Velasco DL, Hong E, Beltran-Ornelas JH, Sanchez-Lopez A, Huerta de la Cruz S, Tapia-Martinez JA, Gomez CB, Centurion D (2024) Hydrogen sulfide ameliorates hypertension and vascular dysfunction induced by insulin resistance in rats by reducing oxidative stress and activating eNOS. Eur J Pharmacol 963:176266. https://doi.org/10.1016/j.ejphar.2023.176266

Streeter E, Al-Magableh M, Hart JL, Badoer E (2011) Hydrogen sulfide in the RVLM and PVN has no effect on cardiovascular regulation. Front Physiol 2:55. https://doi.org/10.3389/fphys.2011.00055

Sun H, Yang J, Shi Y, Wang Y, Li C, Zhu M (2020) Hydrogen sulfide in the nucleus tractus solitarii regulates gastric acid secretion in rats. J Physiol Pharmacol 71(4). https://doi.org/10.26402/jpp.2020.4.05

Sun HZ, Li CY, Shi Y, Li JJ, Wang YY, Han LN, Zhu LJ, Zhang YF (2023) Effect of exogenous hydrogen sulfide in the nucleus tractus solitarius on gastric motility in rats. World J Gastroenterol 29(29):4557–4570. https://doi.org/10.3748/wjg.v29.i29.4557

Swan KW, Song BM, Chen AL, Chen TJ, Chan RA, Guidry BT, Katakam PVG, Kerut EK, Giles TD, Kadowitz PJ (2017) Analysis of decreases in systemic arterial pressure and heart rate in response to the hydrogen sulfide donor sodium sulfide. Am J Physiol Heart Circ Physiol 313(4): H732–H743. https://doi.org/10.1152/ajpheart.00729.2016

Syhr KMJ, Boosen M, Hohmann SW, Longen S, Kohler Y, Pfeilschifter J, Beck KF, Geisslinger G, Schmidtko A, Kallenborn-Gerhardt W (2015) The H2S-producing enzyme CSE is dispensable for the processing of inflammatory and neuropathic pain. Brain Res 1624:380–389. https://doi.org/10.1016/j.brainres.2015.07.058

Szabo C (2017) Hydrogen sulfide, an enhancer of vascular nitric oxide signaling: mechanisms and implications. Am J Physiol Cell Physiol 312(1):C3–C15. https://doi.org/10.1152/ajpcell.00282.2016

Takahashi T, Aoki Y, Okubo K, Maeda Y, Sekiguchi F, Mitani K, Nishikawa H, Kawabata A (2010) Upregulation of Ca(v)3.2 T-type calcium channels targeted by endogenous hydrogen sulfide contributes to maintenance of neuropathic pain. Pain 150(1):183–191. https://doi.org/10.1016/j.pain.2010.04.022

Terada Y, Fujimura M, Nishimura S, Tsubota M, Sekiguchi F, Kawabata A (2015) Roles of Cav3.2 and TRPA1 channels targeted by hydrogen sulfide in pancreatic nociceptive processing in mice with or without acute pancreatitis. J Neurosci Res 93(2):361–369. https://doi.org/10.1002/jnr.23490

Tian JJ, Tan CY, Chen QY, Zhou Y, Qu ZW, Zhang M, Ma KT, Shi WY, Li L, Si JQ (2020) Upregulation of Nav1.7 by endogenous hydrogen sulfide contributes to maintenance of neuropathic pain. Int J Mol Med 46(2):782–794. https://doi.org/10.3892/ijmm.2020.4611

Tomita M, Nagahara N, Ito T (2016) Expression of 3-Mercaptopyruvate sulfurtransferase in the mouse. Molecules 21(12). https://doi.org/10.3390/molecules21121707

Vasylyev DV, Zhao P, Schulman BR, Waxman SG (2024) Interplay of Nav1.8 and Nav1.7 channels drives neuronal hyperexcitability in neuropathic pain. J Gen Physiol 156(11). https://doi.org/10.1085/jgp.202413596

Velasco-Xolalpa ME, Barragan-Iglesias P, Roa-Coria JE, Godinez-Chaparro B, Flores-Murrieta FJ, Torres-Lopez JE, Araiza-Saldana CI, Navarrete A, Rocha-Gonzalez HI (2013) Role of hydrogen sulfide in the pain processing of non-diabetic and diabetic rats. Neuroscience 250:786–797. https://doi.org/10.1016/j.neuroscience.2013.06.053

Wang HJ, Xu X, Xie RH, Rui YY, Zhang PA, Zhu XJ, Xu GY (2018) Prenatal maternal stress induces visceral hypersensitivity of adult rat offspring through activation of cystathionine-beta-synthase signaling in primary sensory neurons. Mol Pain 14:1744806918777406. https://doi.org/10.1177/1744806918777406

White BJ, Smith PA, Dunn WR (2013) Hydrogen sulphide-mediated vasodilatation involves the release of neurotransmitters from sensory nerves in pressurized mesenteric small arteries isolated from rats. Br J Pharmacol 168(4):785–793. https://doi.org/10.1111/j.1476-5381.2012.02187.x

Xia WJ, Liu KL, Wang XM, Yang Y, Meng T, Qiao JA, Zhang N, Sun YJ, Kang YM, Yu XJ (2023) Hypothalamic paraventricular nucleus hydrogen sulfide exerts antihypertensive effects in spontaneously hypertensive rats via the Nrf2 pathway. Am J Hypertens 36(6):306–315. https://doi.org/10.1093/ajh/hpad012

Xu GY, Winston JH, Shenoy M, Zhou S, Chen JD, Pasricha PJ (2009) The endogenous hydrogen sulfide producing enzyme cystathionine-beta synthase contributes to visceral hypersensitivity in a rat model of irritable bowel syndrome. Mol Pain 5:44. https://doi.org/10.1186/1744-8069-5-44

Yan J, Hu S, Zou K, Xu M, Wang Q, Miao X, Yu SP, Xu GY (2016) Inhibition of cystathionine beta-synthetase suppresses sodium channel activities of dorsal root ganglion neurons of rats with lumbar disc herniation. Sci Rep 6:38188. https://doi.org/10.1038/srep38188

Yoo D, Jupiter RC, Pankey EA, Reddy VG, Edward JA, Swan KW, Peak TC, Mostany R, Kadowitz PJ (2015) Analysis of cardiovascular responses to the H2S donors Na2S and NaHS in the rat. Am J Physiol Heart Circ Physiol 309(4):H605–H614. https://doi.org/10.1152/ajpheart.00171.2015

Yu H, Yu Q, Mi Y, Wang P, Jin S, Xiao L, Guo Q, Wu Y (2023) Hydrogen sulfide inhibited sympathetic activation in D-galactose-induced aging rats by upregulating klotho and inhibiting inflammation in the paraventricular nucleus. Biomedicine 11(2). https://doi.org/10.3390/biomedicines11020566

Yu W, Liao Y, Huang Y, Chen SY, Sun Y, Sun C, Wu Y, Tang C, Du J, Jin H (2017) Endogenous hydrogen sulfide enhances carotid sinus baroreceptor sensitivity by activating the transient receptor potential cation channel subfamily V member 1 (TRPV1) Channel. J Am Heart Assoc 6(5). https://doi.org/10.1161/JAHA.116.004971

Yuan B, Tang WH, Lu LJ, Zhou Y, Zhu HY, Zhou YL, Zhang HH, Hu CY, Xu GY (2015) TLR4 upregulates CBS expression through NF-kappaB activation in a rat model of irritable bowel syndrome with chronic visceral hypersensitivity. World J Gastroenterol 21(28):8615–8628. https://doi.org/10.3748/wjg.v21.i28.8615

Zhao S, Liu FF, Wu YM, Jiang YQ, Guo YX, Wang XL (2016) Upregulation of spinal NMDA receptors mediates hydrogen sulfide-induced hyperalgesia. J Neurol Sci 363:176–181. https://doi.org/10.1016/j.jns.2016.02.058

Zheng F, Han J, Lu H, Cui C, Yang J, Cui Q, Cai J, Zhou Y, Tang C, Xu G, Geng B (2018) Cystathionine beta synthase-hydrogen sulfide system in paraventricular nucleus reduced high fatty diet induced obesity and insulin resistance by brain-adipose axis. Biochim Biophys Acta Mol basis Dis 1864(10):3281–3291. https://doi.org/10.1016/j.bbadis.2018.07.014

Zhu L, Zhao L, Qu R, Zhu HY, Wang Y, Jiang X, Xu GY (2015) Adrenergic stimulation sensitizes TRPV1 through upregulation of cystathionine beta-synthetase in a rat model of visceral hypersensitivity. Sci Rep 5:16109. https://doi.org/10.1038/srep16109

Hydrogen Sulfide Signaling in Neurodegenerative Movement Disorders

Andrew A. Pieper and Bindu D. Paul

Contents

A. A. Pieper (✉)
Department Department of Psychiatry, Case Western Reserve University School of Medicine, Cleveland, OH, USA

Brain Health Medicines Center, Harrington Discovery Institute, University Hospitals Cleveland Medical Center, Cleveland, OH, USA

Geriatric Psychiatry, GRECC, Louis Stokes VA Medical Center, Cleveland, USA

Institute for Transformative Molecular Medicine, School of Medicine, Case Western Reserve University School of Medicine, Cleveland, OH, USA

Department of Neurosciences, Case Western Reserve University School of Medicine, Cleveland, OH, USA

Department of Pathology, Case Western Reserve University School of Medicine, Cleveland, OH, USA
e-mail: bpaul8@jhmi.edu

B. D. Paul (✉)
Department of Physiology, Pharmacology and Therapeutics, Johns Hopkins University, Baltimore, MD, USA

Department of Psychiatry and Behavioral Sciences, Johns Hopkins University, Baltimore, MD, USA

The Solomon H. Snyder Department of Neuroscience, Johns Hopkins University, Baltimore, MD, USA

The Lieber Institute for Brain Development, Baltimore, MD, USA

A. Papapetropoulos, C. Szabo (eds.), *Hydrogen Sulfide: A Toxicant with Healing Potential*, Handbook of Experimental Pharmacology 292,
https://doi.org/10.1007/164_2025_757

Abstract

Hydrogen sulfide (H_2S) is a gaseous signaling molecule, also known as a gasotransmitter, present in nearly all mammalian organs. It plays crucial roles in regulating various physiological processes in both the brain and peripheral systems. The body maintains tight control over H_2S levels, as both excessive and deficient levels can disrupt normal physiological functions and lead to disease. H_2S has a significant impact on cognitive and motor functions, which are often compromised in neurodegenerative disorders. It modulates signaling and metabolism primarily by post-translationally modifying reactive cysteine residues on proteins through sulfhydration, also known as persulfidation. This chapter reviews the signaling mechanisms regulated by H_2S in neurodegenerative diseases that significantly affect motor function, specifically focusing on Parkinson's disease (PD), Huntington's disease (HD), amyotrophic lateral sclerosis (ALS), spinocerebellar ataxia (SCA), and Leigh syndrome (LS), as well as other mitochondrial disorders. While PD, HD, and SCA are linked to decreased levels of H_2S, elevated levels of H_2S are associated with ALS, DS, and LS. We also explore potential therapeutic applications of modulating H_2S levels in the brain.

Keywords

Amyotrophic lateral sclerosis · Cysteine · Huntington's disease · Hydrogen sulfide · Parkinson's disease · Spinocerebellar ataxia · Sulfhydration/persulfidation

1 Introduction

Hydrogen sulfide (H_2S) is commonly recognized as a foul-smelling gas found in the environment, yet it is also produced endogenously at low levels in mammals. Here, it plays a crucial role in signaling events that regulate various physiologic processes. The earliest recorded studies of this gas, from the 1700s, focused on its environmental toxicity, as its biological significance was not known at the time.

One of the earliest contributors to H_2S research was the Italian physician Bernardino Ramazzini. In his book *De Morbis Artificum Diatriba*, he attributed eye irritation, inflammation, and bacterial infections in sewer workers to what he termed "sewer gas" (Ramazzini 1700). Ramazzini noted the gas's pungent odor from cesspits and hypothesized that it also led to the tarnishing of copper coins carried by these workers. Subsequent studies confirmed its harmful effects on the ocular system (Lambert et al. 2006). The toxicity of this "sewer gas" was further depicted by Victor Hugo in his classic novel *Les Miserables*, where he described the sewers of Paris as "the intestine of the Leviathan." In 1806, the gas in question was identified as H_2S (Dupuyturen 1806).

Before this recognition, Carl Wilhelm Scheele had synthesized H_2S in the laboratory in 1750 by demonstrating that it was released when ferrous sulfide reacted with mineral acids. This was followed by determination of its chemical composition in 1776 by the French chemist Claude Louis Berthollet, who also noted its acidic nature (Mitchell and Davenport 1924). Later research revealed that H_2S is produced by microbes and mammalian tissues, although it was initially considered merely a metabolic waste product (Rizzo 1967; Gayon 1877; Clarke 1953; Myers 1920; Shatalin et al. 2011; Mironov et al. 2017).

Much later pioneering work of Rui Wang and associates, in the early 2000s, however, suggested important roles in the vascular system (Zhao et al. 2001; Cheng et al. 2004). Specifically, H_2S was recognized as an endothelial derived relaxation factor (EDRF) with important roles in the circulatory system (Zhao et al. 2001; Yang et al. 2008). Similar to nitric oxide (NO), generation of H_2S from vascular endothelial cells was demonstrated to be triggered by muscarinic cholinergic activation, with depletion of the endothelium or depletion of CSE shown to abrogate this effect (Yang et al. 2008; Wang 2009; Tang et al. 2013; Tang et al. 2005; Mustafa et al. 2011). The endogenous production of H_2S in the brain was first observed in 1989 by Warenycia, while its function in neurotransmission was shown later by Kimura and colleagues to play a role in N-methyl D aspartate (NMDA) receptor–mediated long-term potentiation (LTP) in the brain (Abe and Kimura 1996; Warenycia et al. 1989).

It is now well understood that H_2S mediates its effects largely through sulfhydration or persulfidation, in which –SH groups on reactive cysteine residues of proteins are converted to persulfide (-SSH) groups (Filipovic 2015; Mustafa et al. 2009; Sen et al. 2012). Notably, altered H_2S metabolism and sulfhydration have been observed in several neurodegenerative conditions. This chapter focuses on the current understanding of H_2S signaling in neurodegenerative diseases that are highly associated with motor dysfunction: (1) Parkinson's disease (PD), (2) Huntington's disease (HD), (3) amyotrophic lateral sclerosis (ALS), (4) ataxias, and (5) Leigh syndrome (LS), as well as other mitochondrial disorders.

1.1 H_2S Biosynthesis

Due to the importance of maintaining proper homeostasis of H_2S, its production in the body is tightly regulated at multiple levels. H_2S is generated either directly or indirectly from the amino acid cysteine through the reverse trans-sulfuration pathway, with additional contributions from various non-enzymatic systems (Paul and Snyder 2012; Paul and Snyder 2015). Cysteine is a semi-essential amino acid produced within cells that can also be obtained from dietary sources via various cellular transporters (Paul et al. 2018).

The primary enzymes responsible for H_2S production in the body are cystathionine γ-lyase (CSE), cystathionine β-synthase (CBS), and 3-mercaptopyruvate sulfurtransferase (3-MST) (Paul and Snyder 2012; Stipanuk 2020; Stipanuk and Beck 1982; Kuo et al. 1983). CSE, also known as cystathionase and encoded by the *CTH* gene on chromosome 1, utilizes cysteine or homocysteine as substrates to generate H_2S (Paul and Snyder 2012; Filipovic et al. 2018). It is the only mammalian biosynthetic enzyme for cysteine, and its depletion can be lethal in mice fed a cysteine-deficient diet (Mani et al. 2011; Ishii et al. 2010). CBS, encoded by the *CBS* gene on human chromosome 21, catalyzes the condensation of serine and homocysteine to form cystathionine, which CSE then converts to cysteine (Paul and Snyder 2012; Paul and Snyder 2015). CBS preferentially generates H_2S through a combination of homocysteine and cysteine, or homocysteine alone (Kruger 2017; Chen et al. 2004). Notably, excessively elevated CBS levels have been linked to Down syndrome due to an additional copy of the gene (Szabo 2020; Kamoun 2001). 3-MST, also known as tRNA thiouridine modification protein, human liver rhodanese 2, and thiosulfate sulfurtransferase 2, is encoded on human chromosome 22 (Dunham et al. 1999). In the brain, 3-MST is present in both neurons and astrocytes (Ishigami et al. 2009; Shibuya et al. 2009), while CSE is predominantly localized to neurons and CBS to glial cells (Morikawa et al. 2012; Linden et al. 2008; Enokido et al. 2005). Additional sources of H_2S synthesis include the acid-labile sulfide pool, sulfane sulfur, and microbial flora, as well as non-enzymatic generation of H_2S via a reaction involving iron and vitamin B6, which also operate in the brain (Ishigami et al. 2009; Shen et al. 2012, 2013; Singh and Lin 2015; Yang et al. 2019).

1.2 H_2S Catabolism

The pharmacological profile of H_2S follows a bell-shaped curve, with lower concentrations being protective and higher concentrations being toxic. Low levels of H_2S stimulate mitochondrial bioenergetics and protect neurons from oxidative stress, while higher concentrations of H_2S are toxic as they inhibit complex IV (cytochrome c oxidase) of the electron transport chain (Paul et al. 2021; Szabo et al.

2014; Modis et al. 2014; Modis et al. 2016; Kimura et al. 2010; Sbodio et al. 2019; Tripathi et al. 2023). To prevent toxicity, the body balances H_2S synthesis and catabolism to maintain homeostatic levels (Paul et al. 2021). The two primary pathways for H_2S degradation are methylation and oxidation, with methylation being the minor pathway (Filipovic et al. 2018).

H_2S oxidation primarily occurs through the enzyme sulfide:quinone oxidoreductase (SQOR), which is part of the sulfide oxidizing unit (SOU) in the mitochondria that also includes persulfide dioxygenase (ETHE), sulfite oxidase (SO), and thiosulfate sulfurtransferase/rhodanese (TST). This unit connects sulfide catabolism to oxidative phosphorylation (Paul et al. 2021; Hildebrandt and Grieshaber 2008). SQOR, located on the inner mitochondrial membrane, catalyzes the first step in H_2S oxidation, generating a persulfide intermediate while releasing electrons transferred to coenzyme Q (CoQ) via flavin adenine dinucleotide (FAD) (Jackson et al. 2012; Landry et al. 2021). CoQ relays these electrons to the electron transport chain, while the persulfide formed on SQOR can be transferred to acceptors like the antioxidant glutathione (GSH) to form GSSH, or sulfite (SO_3^{2}) to form thiosulfate ($S_2O_3^{2-}$). GSSH can be subsequently converted into GSH and sulfite by persulfide dioxygenase (ETHE)/persulfide dioxygenase (Paul et al. 2021). The resulting sulfite can be acted upon by either thiosulfate sulfurtransferase (TST) to produce $S_2O_3^{2-}$ or by sulfite oxidase (SO) to produce sulfate (SO_4^{2-}).

2 Parkinson's Disease

2.1 Background of PD

PD is the second most common neurodegenerative disorder, following Alzheimer's disease. It is characterized by classic symptoms of resting tremor, bradykinesia, rigidity, sleep disturbance, anxiety, depression, and cognitive decline (Poewe et al. 2017; Aarsland et al. 2021). While rarely occurring before the age of 50, the incidence of PD increases with age, peaking between 85 and 89 years. PD primarily affects the substantia nigra pars compacta and striatum regions of the brain, leading to the selective degeneration of dopaminergic neurons. This degeneration results in decreased dopamine levels and pathological accumulation of α-synuclein within intraneuronal inclusions, known as Lewy bodies (Bloem et al. 2021).

PD can occur both genetically and sporadically, with most cases thought to arise from a combination of genetic susceptibility and environmental factors such as exposure to pesticides, herbicides, solvents, or heavy metals (Collaborators GBDPsD 2018). Although the exact molecular mechanisms underlying PD remain unclear, several processes influenced by H_2S are implicated in disease progression. These include elevated neuroinflammation, mitochondrial dysfunction, redox imbalance, decreased sulfhydration, degeneration of dopaminergic neurons, deposition of α-synuclein, and impaired autophagy (Absalyamova et al. 2024; Hou et al. 2020; Liu et al. 2022) (Fig. 1). Furthermore, research has shown that cysteine-dopamine conjugates and related metabolites are elevated in the postmortem substantia nigra

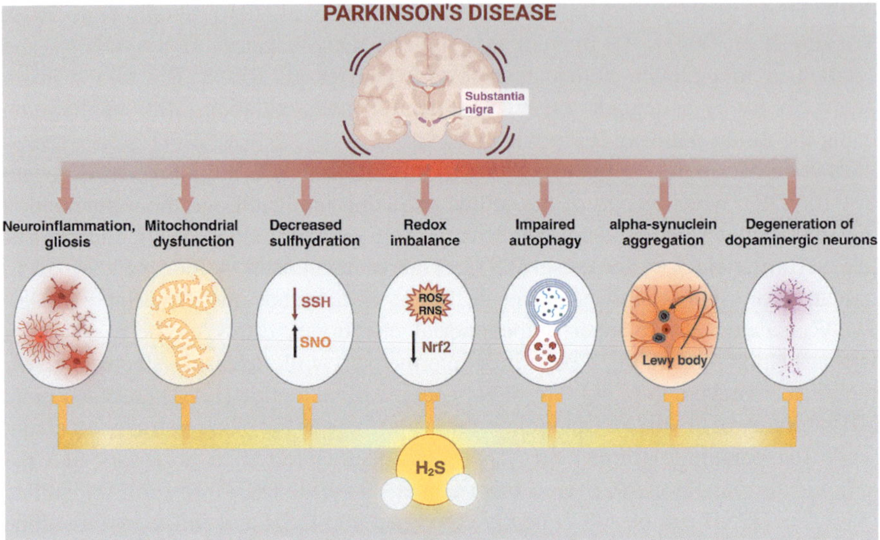

Fig. 1 Neuroprotective effects of H_2S homeostasis in Parkins on's disease. Parkinson's disease profoundly affects the substantia nigra of the brain and is disrupts several brain functions. The brains of PD patients exhibit heightened neuroinflammation, mitochondrial dysfunction, redox imbalance with impaired antioxidant stress responses involving the nuclear factor erythroid 2-related factor 2 (Nrf2) signaling cascade, leading to elevated reactive oxygen species (ROS) and reactive nitrogen species (RNS) accumulation. Abnormal gasotransmitter signaling involving H_2S and nitric oxide (NO) is also observed accompanied by reduced sulfhydration and increased nitrosylation of cysteine residues. Degeneration of dopaminergic neurons follows leading to abnormal dopaminergic neurotransmission. All these abnormalities are prevented by administration of H 2S donors

of PD patients, suggesting a decreased abundance of cysteine and glutathione that would be expected to contribute to oxidative damage (Spencer et al. 1998).

2.2 H_2S Signaling in PD and Therapeutic Interventions

Numerous preclinical studies using chemical and genetic PD models have demonstrated dysregulation of H_2S homeostasis. For example, reduced striatal levels of endogenous H_2S have been observed in the 6-hydroxydopamine (6-OHDA) rat model of PD (Sun et al. 2023). In this model, administration of sodium hydrosulfide (NaHS), an H_2S donor, has been shown to reduce oxidative stress and prevent motor dysfunction (Sun et al. 2023). Additionally, NaHS alleviates pathology in the rotenone model of PD, where it inhibits microglial activation and neuroinflammation in the substantia nigra that occurs in response to rotenone-mediated inhibition of mitochondrial complex I (Hu et al. 2010).

Research also indicates that overexpression of CBS using lentiviral constructs provides protection against 6-OHDA-induced PD (Yin et al. 2017). In addition, we have previously reported decreased sulfhydration in the brains of mice exposed to MPTP, as well in postmortem brain samples from PD patients, that is accompanied by excessive nitrosylation, a cysteine modification alternatively derived from NO that has detrimental effects on neuronal health (Vandiver et al. 2013). We also found decreased sulfhydration and increased nitrosylation of the E3-ubiquitin ligase parkin, a protein that clears misfolded proteins, in PD patients (Vandiver et al. 2013). Autosomal-recessive mutations in the *PARK2* gene, which encodes parkin, are the most common cause of inherited PD, while heterozygous mutations are implicated in the more prevalent sporadic form (Funayama et al. 2023; Dawson and Dawson 2010; Klein and Lohmann 2009; Sun et al. 2006).

In addition to its role in clearing damaged and misfolded protein aggregates, parkin is also essential for mitochondrial health (Clark et al. 2006). Modifications to critical cysteine residues in parkin influence its activity and, consequently, mitochondrial homeostasis. Our findings suggest that sulfhydration enhances parkin's catalytic activity, while nitrosylation reduces it, potentially contributing to the pathological accumulation of misfolded proteins in PD, including α-synuclein (Vandiver et al. 2013; Chung et al. 2004). Furthermore, sulfhydration supports protective mitophagy, a specialized form of autophagy that eliminates damaged or unnecessary mitochondria (Sun et al. 2020; Wang et al. 2008).

Inhaling H_2S has also shown neuroprotective effects in models of PD, such as the 1-methyl-4-phenyl-1,2,3,6-tetrahydropyridine (MPTP) model. Specifically, inhaling H_2S at a concentration of 40 ppm for 8 h a day over 1 week effectively prevents MPTP-induced neurodegeneration and motor deficits (Kida et al. 2011). The neuroprotective effects of H_2S inhalation are linked to reduced neuroinflammation and increased nuclear factor erythroid 2-related factor 2 (Nrf2)-mediated signaling in the substantia nigra and striatum (Chen et al. 2004). Interestingly, higher ambient levels of H_2S in Rotorua, New Zealand, due to its geothermal vents, correlate with faster motor response times in the finger tapping component of the motor function rating scale used to monitor PD progression in patients (Cakmak 2017; Reed et al. 2014). The reported neuroprotective effects to date of H_2S-mediated therapy in preclinical models of PD are summarized in Table 1, suggesting that administering H_2S or boosting its production by stimulating the trans-sulfuration pathway might be beneficial to PD patients.

3 Huntington's Disease

3.1 Background of HD

Huntington's disease (HD) is a devastating neurodegenerative condition that is inherited in an autosomal dominant manner. It primarily affects the corpus striatum of the brain and is characterized by involuntary movements (chorea), motor impairments, and a range of neuropsychiatric and cognitive impairments (Bates et al. 2015). HD is classified as a triplet repeat disorder and is caused by variable expansion of codons that encode a polyglutamine tract in the *Huntingtin (Htt)* gene

Table 1 Therapeutic effects of H_2S donors on Parkinson's disease

PD model	Intervention	Physiological efficacy	References
6-Hydroxydopamine (6-OHDA), rats	NaHS in conjunction with the HDAC inhibitor TSA	NaHS decreased neuronal cell death, increased dopamine (DA), tyrosine hydroxylase (TH), and DOPAC levels, decreased glial fibrillary acidic protein (GFAP), neuroinflammatory cytokines, and HDAC activity	Sun et al. (2023)
6-OHDA, rats	NaHS (30 or 100 μmol/kg/d, i.p).	Decreased activity of caspase-3 and expression of Bax, prevented the decrease of Bcl-2 in substantia nigra, promoted autophagy, and prevented the downregulation of leptin	Jiang et al. (2022)
6-OHDA, rats	NaHS (1.68 or 5.6 mg kg^{-1}) for 3 weeks	Prevented dopaminergic neuronal loss in the substantia nigra and striatum, decreased malondialdehyde (MDA) and NADPH oxidase levels and ATP depletion in striatum.	Hu et al. (2010)
6-OHDA, rats	Overexpression of CBS	Increased H_2S production in the substantia nigra, decreased loss of the nigral DA neurons, inhibited 6-OHDA-induced oxidative stress injury, and downregulated the expression of α-synuclein	Yin et al. (2017)
Rotenone, rats	Entinostat (20 mg/kg), i.p.	Administration of entinostat (20 mg/kg) intraperitoneally restored H_2S levels, DA, DOPAC, and TH levels	Li et al. (2024)
Rotenone, rats	NaHS (1.68 mg kg^{-1}) for 3 day before and 4 weeks after rotenone	Ameliorated motor deficits, prevented microglial activation in the SN, and accumulation of pro-inflammatory factors (TNF-α and nitric oxide) in the striatum via NF-κB pathway	Hu et al. (2010)
Rotenone, rats	NaHS (30–100 μmol/kg/day for 7d pre- and post-rotenone injection)	Ameliorated depression, enhanced expression of synaptic plasticity proteins in the hippocampus, and preserved the ultrastructure of synapses	Liu et al. (2024)
1-methyl-4-phenyl-1,2,3,6-tetrahydropyridine (MPTP)	Inhalation of H_2S at 40 ppm for 8 h/day for a week	Inhibition of MPTP-induced loss of dopaminergic neurons, neuroinflammation in substantia nigra and striatum, and upregulation of Nrf2 signaling	Kida et al. (2011)
MPTP, mice	Air mixed with 80 ppm H_2S for 4 h/day for 5 day	Prevented motor deficits, anxiety, and degeneration of TH neurons in the nigrostriatal region	Nagashima et al. (2023)

(The Huntington's Disease Collaborative Research Group 1993). The length of this polyglutamine tract critically influences the age of onset and severity of HD, although other genetic modifiers also play a role (Keum et al. 2016; Andrew et al. 1993; Pengo and Squitieri 2024; Gusella et al. 2014; Genetic Modifiers of Huntington's Disease C 2015; Wright et al. 2019; Gu et al. 2022; Sawant et al. 2021). The expansion of polyglutamine repeats leads to the misfolding and aggregation of mutant Htt (mHtt) protein, resulting in neurotoxicity through various mechanisms, including dysregulation of transcription and translation, defects in nucleocytoplasmic transport, mitochondrial dysfunction, neuroinflammation, impaired stress responses, and insufficient autophagy, among other aberrations (Gu et al. 2022; Davies et al. 1997; DiFiglia et al. 1997; Ament et al. 2018; Banez-Coronel et al. 2015; Grima et al. 2017; Ament et al. 2017; Paul and Snyder 2019; Paul 2022; Sbodio et al. 2016; Oh et al. 2022; Ravikumar et al. 2002).

3.2 Disrupted Cysteine and H_2S Signaling in HD and Related Therapeutic Interventions

Research by our team and others has shown that cysteine homeostasis is disrupted in HD. This includes impaired trafficking of the neuronal cysteine transporter (excitatory amino acid transporter 3; EAAT3/EAAC1) and decreased activity of the system Xc^- transporter, which imports cystine, the oxidized form of cysteine (Paul 2022; Li et al. 2010; Frederick et al. 2014). Additionally, the CSE enzyme, responsible for producing cysteine and H_2S, is transcriptionally depleted in HD due to sequestration of specificity protein 1 (SP1) by mHtt (Paul et al. 2014; Paul and Snyder 2014). Consequently, there is an overall cysteine deficit in HD striatal cells, which rely on cysteine supplementation in culture media for survival (Paul et al. 2014). In healthy wild-type cells, expression of the *cth* gene (encoding CSE) increases during cysteine deprivation, but this does not occur in HD cells due to their inability to induce activating transcription factor 4 (ATF4), thereby further exacerbating oxidative stress (Sbodio et al. 2016). Interestingly, reduced CSE expression appears to correlate to mHtt levels (Li et al. 2016). Beyond these regulatory nodes, dysregulation of cysteine metabolism occurs post-transcriptionally in HD cells, with mHtt binding both CSE and CBS and inhibiting their catalytic activity (Paul et al. 2022). Thus, cysteine homeostasis in HD is affected at multiple levels, leading to redox imbalance (Fig. 2).

Sulfhydration is likewise reduced in HD cells, and supplementation with cysteine or its prodrugs has been shown to enhance cell survival (Paul et al. 2014; Zivanovic et al. 2019). For example, administration of N-acetylcysteine (NAC) and cysteine prevents striatal degeneration and mitigates motor deficits and oxidative damage in the R6/2 mouse model of HD (Paul et al. 2014). Chronic administration of NAC also delays the onset and progression of motor deficits in the R6/1 mouse model of HD and unexpectedly provides an antidepressant effect in both R6/1 and wild-type mice as well (Wright et al. 2015). Notably, NAC improves the decreased mitochondrial respiratory capacity of striatal cells in R6/1 mice and reduces markers of oxidative

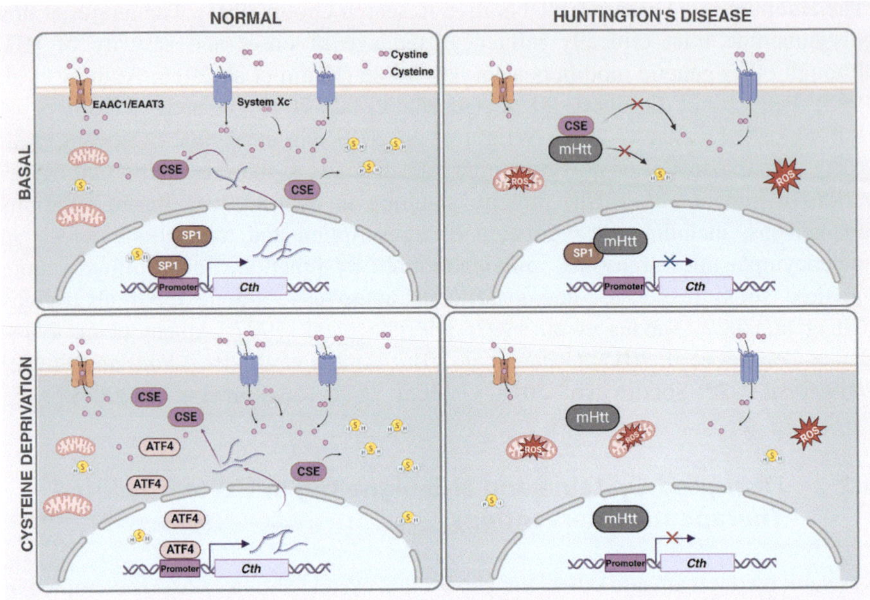

Fig. 2 Disrupted cysteine and H_2S homeostasis in Huntington's disease. In normal cells under basal conditions, transcription of CSE is regulated by specificity protein 1 (SP1). CSE produces both cysteine and hydrogen sulfide (H_2S) which modulate redox balance and multiple signaling pathways. Cysteine is also imported through the excitatory amino acid transporter 3 (EAAT3/EAAC1). The oxidized form of cysteine, cystine, is taken up by the system Xc^- transporter. The cystine imported is reduced to cysteine in the reducing atmosphere of the cell. In HD cells, SP1 is sequestered by mutant huntingtin (mHtt), which prevents transcription and expression of CSE. The activities of the EAAC1 and system Xc^- transporters are also decreased in HD, leading to reduced uptake of this amino acid. Furthermore, mHtt binds to CSE and inhibits its catalytic activity, leading to an overall cysteine deficit and attendant oxidative stress though increased accumulation of reactive oxygen species (ROS) in the mitochondria and cytosol. During conditions of cysteine scarcity, in normal cells, activating transcription factor 4 (ATF4) is induced, which regulates expression of CSE to augment cysteine and H_2S production. These processes are compromised in HD, further elevating oxidative stress

damage in their mitochondria. Additional studies have suggested that baseline reductions in cysteine contribute to glutamatergic dysfunction and depressive-like behavior in HD, which can be alleviated with NAC treatment (Wright et al. 2016). Given that most cysteine and cystine transporters operate as antiporters for cysteine/cystine and the excitatory amino acid glutamate, it is not surprising that NAC administration can influence glutamatergic transmission. The beneficial effects of NAC on behavior have also been observed in some human cases of HD as well (Berk 2015; Berk et al. 2013). Additionally, we have shown that small molecule stimulators of the trans-sulfuration pathway in cell culture models of HD are also protective, identifying another neuroprotective strategy (Sbodio et al. 2018).

4 Amyotrophic Lateral Sclerosis

4.1 Background of ALS

ALS, commonly known as Lou Gehrig's disease, is characterized by degeneration of upper motor neurons, which connect the cortex to the brainstem and spinal cord, as well as lower motor neurons, which extend from the brainstem and spinal cord to the muscles. This leads to progressively worsening symptoms of muscle weakness, cramps, dysarthria, dysphagia, and ultimately paralysis (Hardiman et al. 2017). Approximately 50% of ALS patients also experience cognitive deficits (Phukan et al. 2012; Elamin et al. 2013).

The disease presents in two major forms: spinal-onset ALS, which typically begins with early limb weakness, and bulbar-onset ALS, which is marked by dysarthria and dysphagia. About 90% of ALS cases are sporadic, while 10% are currently understood to be inherited (Rowland and Shneider 2001). The most frequently implicated genes in familial ALS include mutations in chromosome 9 open reading frame 72 (C9orf72), Cu/Zn superoxide dismutase (SOD1), transactive response DNA binding protein 43 kDa (TDP-43), fused in sarcoma (FUS), and TANK-binding kinase 1 (TBK1) (Neumann et al. 2006; Marangi and Traynor 2015; Gurney et al. 1994; DeJesus-Hernandez et al. 2011; Brown and Al-Chalabi 2017; Rosen et al. 1993).

Multiple physiologic pathways affected by H_2S show dysfunction in ALS, and H_2S levels are most prominently elevated in patients with spinal cord onset as opposed to bulbar onset (Davoli et al. 2015). As detailed below, the role of H_2S has been most extensively characterized in models of SOD1 and C9orf72 ALS. Mutated SOD1 protein accumulates in pathological aggregates that are directly toxic to motor neurons (Gurney et al. 1994; Rosen et al. 1993), while the *C9ORF72* mutation involves expansion of the hexanucleotide repeat G4C2 and is responsible for 30–50% of familial ALS in Caucasians, termed C9-ALS patients (Cruts et al. 2013). In normal individuals, the wild-type *C9ORF72* gene typically contains fewer than 20–25 copies of this repeat, while in C9-ALS patients, this number can reach into the thousands (DeJesus-Hernandez et al. 2011). These repeated sequences can undergo repeat-associated non-ATG (RAN) translation, producing dipeptide repeat proteins (DPRs) such as poly Gly-Ala (GA), Gly-Pro (GP), Gly-Arg (GR), Pro-Ala (PA), and Pro-Arg (PR). Among these, poly-GA is expressed at the highest levels in the brains of C9-ALS patients and is associated with toxicity resulting from disruptions in several cellular processes, including nucleocytoplasmic transport, ER-Golgi vesicular trafficking, synaptic function, and proteostasis (Sultana et al. 2024; Jafarinia et al. 2024; Jensen et al. 2020; Beckers et al. 2021; Zhang et al. 2015).

4.2 H_2S Signaling in ALS

In contrast to the decreased H_2S observed in PD and HD, ALS is associated with elevated H_2S levels, first identified in the cerebrospinal fluid (CSF) of ALS patients and the tissues of G93A SOD1 mice (Davoli et al. 2015). The G93A SOD1 mice exhibit increased CBS levels in the mitochondria of astrocytes, as well as decreased electron transport chain complex IV activity in the brain (Davoli et al. 2015; Kirkinezos et al. 2005). This is consistent with previous research linking H_2S to the inhibition of electron transport chain complex IV. Moreover, G93A SOD1 mice showed remarkably low levels of plasma cysteine, which could indicate increased cysteine conversion to H_2S (Bame et al. 2014).

These investigators also demonstrated that motor neurons are especially susceptible to the toxic effects of H_2S (Davoli et al. 2015). A proteomic analysis of primary spinal cord cultures treated with NaHS further revealed an upregulation of oxidative stress and necroptosis pathways (Greco et al. 2018). Additionally, H_2S exposure induced morphological changes in astrocytes, decreased ATP production, reduced mitochondrial respiration rates, and caused mitochondrial depolarization (De Stefano et al. 2024). Correspondingly, when G93ASOD1 mice were treated with the CBS inhibitor amino-oxyacetic acid, H_2S levels in muscle and brain decreased, leading to improved survival in female, but not male, mice (Spalloni et al. 2019).

While H_2S levels are elevated in G93A SOD1 mice, the relationship between *C9ORF72* expansion and H_2S signaling remains less clear. Recent studies have indicated that the glycine-alanine dipeptide repeat protein (GA50), elevated in these mice, can bind SQOR, a crucial component of the sulfide oxidation pathway that detoxifies sulfide in mitochondria, which can accumulate due to H_2S production (Fu et al. 2023). The brain and spinal cord have lower levels of SQOR, which limits their ability to metabolize sulfide, making them more vulnerable to sulfide toxicity. The reaction catalyzed by SQOR is a pivotal first step in sulfide oxidation, leading to the formation of persulfide species that are further oxidized to sulfite, thiosulfate, and sulfate through subsequent enzymatic reactions involving ETHE1 or SDO, TST, and SO, as described in the section on H_2S catabolism (Paul et al. 2021).

In human microglial cells, both the expression of GA50 and the depletion of SQOR increased the accumulation of reactive oxygen species (ROS), as well as cytoplasmic escape of mitochondrial DNA that served to activate the NLRP3 inflammasome (Fig. 3). Notably, the compound irisflorentin (IFT) inhibited the interaction between GA50 and SQOR and reduced activation of the NLRP3 inflammasome (Fu et al. 2023). IFT, a traditional Chinese medicine that has been used for centuries, is a botanically derived compound isolated from the roots of *Belamcanda chinensis* (L.) DC that possesses anti-inflammatory and anti-tumor properties (Zhang et al. 2016).

It is anticipated that SQOR depletion may contribute to higher H_2S levels (Landry et al. 2021; Kanemaru and Ichinose 2024), which aligns with the observation of elevated H_2S in ALS patients. Thus, treatments aimed at normalizing H_2S levels may prove beneficial for managing ALS. Future research examining the sulfhydration

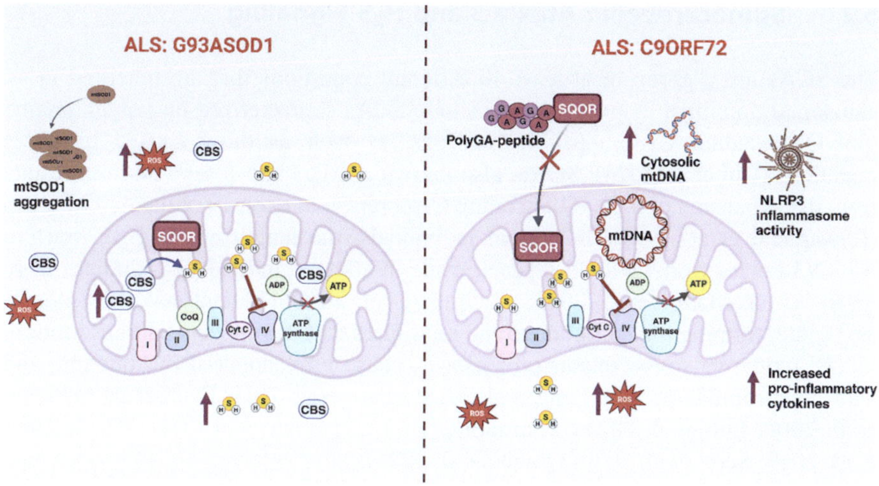

Fig. 3 H_2S homeostasis in amyotrophic lateral sclerosis. In ALS caused by the G93ASOD1 mutation, mtSOD1 aggregates and causes elevated oxidative stress due to increased CBS activity that elevates H_2S, in turn dangerously inhibiting complex IV of the mitochondrial electron transport chain. In ALS caused by an expansion of hexanucleotide repeats in the *C9ORF72* locus, production of dipeptide repeats such as polyGA peptide increases levels of reactive oxygen species (ROS), activates the NLRP3 inflammasome, and increases levels of proinflammatory cytokines. Additionally, the polyGA aggregates also bind sulfide-quinone oxidoreductase (SQOR) and inhibit its activity, leading to further increase in H_2S levels

state and the expression of the other H_2S biosynthetic enzymes (CSE and 3-MST) could provide further insights into the mechanisms involved, as well as the role of H_2S in other genetic forms of ALS.

5 Spinocerebellar Ataxia

5.1 Background of SCA

The word "ataxia' is derived from the Greek words "a," meaning without, and "taxis," meaning order. Ataxia can present as a primary disorder or as a symptom of various conditions. It typically manifests through impaired coordination, balance difficulties, slurred speech, abnormally wide gait, difficulty with fine motor tasks, and slowed or involuntary eye movements, among other symptoms. The primary cause of ataxia is neurodegeneration in the cerebellum, and it can occur as either an inherited or sporadic condition (Klockgether and Paulson 2011; Ashizawa et al. 2018). Hereditary ataxias are classified as autosomal dominant, autosomal recessive, X-linked, and those involving mitochondrial inheritance (Cui et al. 2024). In this section, we discuss the role of H_2S signaling in the two forms of ataxia in which its dysregulation has been characterized: spinocerebellar ataxia 3 (SCA3) and ataxia telangiectasia, representing autosomal dominant and autosomal recessive forms of ataxia, respectively.

5.2 Spinocerebellar Ataxia 3 and H_2S Signaling

The SCAs are a group of at least 40 different conditions that are inherited in an autosomal dominant manner. These include SCAs characterized by polyglutamine (polyQ)-encoding CAG repeat expansions, as well as those caused by point mutations (Cui et al. 2024). SCA3, also known as Machado-Joseph disease, results from the expansion of a polyQ-encoding CAG repeat in exon 10 of the *ATXN3* gene (Kawaguchi et al. 1994). In normal individuals, the length of the CAG tract in ATXN3 ranges from 13 to 44 repeats, while in affected individuals, it ranges from 55 to 90 (de Mattos et al. 2019; Akcimen et al. 2020). Mutant ataxin-3 tends to aggregate, forming toxic oligomers and multimers that disrupt genetic transcription, induce oxidative stress, impair proteostasis, cause mitochondrial dysfunction, and lead to neuroinflammation (Zeng et al. 2018; Matos et al. 2019; Evert et al. 2006; Yu et al. 2009; Luo et al. 2023a; Harmuth et al. 2022; Wiatr et al. 2021; Klockgether et al. 2019; Scott et al. 2020; Quinet et al. 2024).

Notably, the trans-sulfuration pathway is disrupted in SCA3 (Fig. 4). In these patients, CSE levels and sulfhydration are reduced in the pontine tissue, an area where ataxin-3 aggregates are particularly concentrated (Snijder et al. 2016).

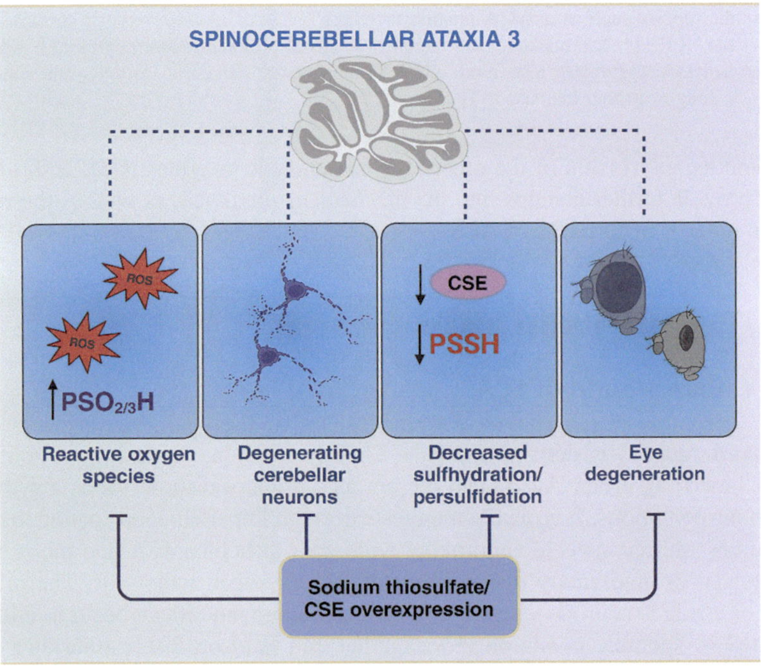

Fig. 4 Dysregulated H_2S signaling in spinocerebellar ataxia 3. Depletion of CSE was observed in both human SCA3 and a *Drosophila* model expressing mutant ataxin 3, leading to elevated ROS levels and protein oxidation accompanied by decreased sulfhydration and degeneration of cerebellar neurons. In *Drosophila*, degeneration of the eye was observed. All these abnormalities were ameliorated either by overexpression of CSE or by administering the slow release H_2S donor, sodium thiosulfate (STS)

Notably, overexpressing CSE restored normal sulfhydration levels, decreased oxidative stress, attenuated neuroinflammation, and provided neuroprotection in a *Drosophila* model of SCA3 (Snijder et al. 2016). Additionally, administering sodium thiosulfate (STS), an H_2S donor, also brought sulfhydration levels back to normal and protected against neurodegeneration (Snijder et al. 2016). Therefore, both genetic (increasing CSE expression) and pharmacological (administering STS) restoration of normal sulfhydration effectively prevented neurodegeneration in the SCA3 flies.

5.3 Ataxia Telangiectasia and H_2S Signaling

Ataxia telangiectasia (A-T) is an autosomal recessive disease caused by mutations in the *ATM* gene that cause cerebellar neurodegeneration. The name is derived from ataxia (unsteady gait) and dilation of blood vessels (telangiectases) that are visible in the conjunctiva of the eyes and facial skin of A-T patients (Savitsky et al. 1995). Other features of A-T include immunodeficiency, recurrent sinopulmonary infections, radiation sensitivity, premature senescence and aging, chromosomal instability, predisposition to cancer development (particularly of lymphoid origin), and immunodeficiency (Rothblum-Oviatt et al. 2016; Paull and Woolley 2024). Symptoms arise at an early age (1–4 years), and patients usually live only until early adulthood.

The *ATM* gene encodes a 350 kDa protein belonging to the phosphatidylinositol 3-kinase-related protein kinase family (PIKK) (Savitsky et al. 1995). The ATM protein is activated by oxidative stress and double-stranded DNA breaks (DSBs) and plays important roles in maintaining redox balance and DNA repair (Ditch and Paull 2012; Lee and Paull 2007; Cirotti et al. 2021). In response to DNA damage, the Mre11-Rad50-Nbs1 (MRN) complex is recruited to the site of DSBs, followed by autophosphorylation and activation of ATM, leading to conversion of inactive ATM dimers to the active monomeric forms to stimulate DNA repair (Bakkenist and Kastan 2003; Guo et al. 2010). Cells and animal models lacking ATM exhibit elevated oxidative stress, impaired antioxidant stress response, and higher accumulation of DNA damage (Guo et al. 2010; Barzilai et al. 2002; Meredith and Dodson 1987; Kamsler et al. 2001; Quick and Dugan 2001; Fang et al. 2016; McDonald et al. 2011).

Notably, aberrant H_2S signaling has been linked to this condition, as A-T patients exhibit impaired cysteine and GSH metabolism and also display lower cystine/glutamate exchanger subunit xCT, glutathione-S-transferase (GST), glutathione reductase (GR) levels, glutathione peroxidase (GPx) activity, and GSH levels in their cerebellar astrocytes (Meredith and Dodson 1987; Campbell et al. 2016). Interestingly, levels of both CSE and CBS are higher in cells lacking ATM, at both transcriptional and translational levels, and replenishing ATM suppresses the expression of both enzymes while also increasing the labile sulfide pool (Islam et al. 2022). Additionally, CSE activity has been reported to be higher in $ATM^{-/-}$ cells compared to $ATM^{+/+}$ cells, indicating that ATM is a negative regulator of CSE enzymatic activity. The effect of ATM on CSE and CBS appears to be at the post-translational level and further studies will clarify the exact relationship.

6 Leigh Syndrome

6.1 Background of LS

LS was first described by Denis Leigh in 1951 in a 7-month-old infant and is the most commonly inherited mitochondrial disease affecting children, often leading to mortality within the early years of life (Yagensky et al. 2019; Leigh 1951). More than 75 mutations have been identified as causes of LS, predominantly affecting genes involved in mitochondrial bioenergetics (Lake et al. 2016; Chang et al. 2020). A key contributor to LS is impaired function of complex IV, a critical component of the mitochondrial electron transport chain (DiMauro et al. 1987).

Clinically, LS presents with both neurological and non-neurological symptoms, including developmental delays, epilepsy, weakness, hypotonia, feeding difficulties, and ataxia (Rahman et al. 1996). Neuropathological features of LS include focal and bilateral lesions in the brainstem, basal ganglia, cerebellum, and other regions of the brain. Pathologically, these lesions exhibit demyelination, spongiosis, gliosis, necrosis, and capillary proliferation.

6.2 H_2S Homeostasis in Leigh Syndrome

Similar to ALS, mutations in SQOR have been linked to LS, causing elevated H_2S levels and subsequent inhibition of Complex IV in the mitochondrial electron transport chain (Friederich et al. 2020). The role of H_2S in LS was confirmed in a study using a mouse LS model, which involved deficiency of mitochondrial SQOR. This model was developed by replacing the normal SQOR gene with a version lacking the sequence encoding the N-terminal mitochondrial localization sequence (SQORΔN) (Marutani et al. 2021).

By virtue of lacking mitochondrial SQOR, SQOR$^{\Delta N/\Delta N}$ mice recapitulate the clinical characteristics of LS, including progressive neurological deficits, brain lesions, and a reduced lifespan. These mice also display elevated blood lactate levels, significantly decreased complex IV activity, and increased H_2S levels in the plasma, brain, liver, and muscle (Fig. 5). Since SQOR expression is typically lower in the brain than in peripheral tissues, reduced enzyme activity in the brain can have particularly harmful effects (Linden et al. 2012).

SQOR$^{\Delta N/\Delta N}$ mice also exhibit an upregulation of the integrated stress response (ISR), evidenced by increased levels of ATF4 and DNA damage–inducible transcript 3 protein (DDIT3), both key proteins in the ISR pathway, along with elevated NADH-reductive stress. Treatment with the broad-spectrum antibiotic metronidazole reduced microbiota-produced H_2S levels, leading to increased survival of the mice. Additionally, maintaining these mice on a sulfur-restricted diet decreased

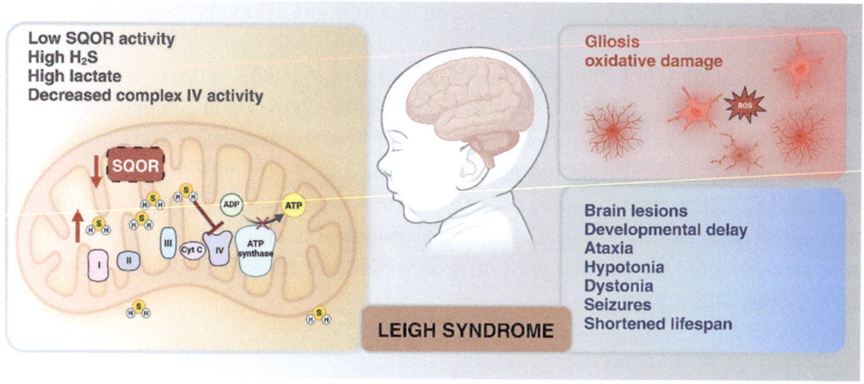

Fig. 5 Elevated H_2S production in Leigh syndrome. Leigh's disease is an infantile mitochondrial disease caused by mutations in genes involved in mitochondrial function. The disease is characterized by brain lesions, developmental delay, ataxia, hypotonia, dystonia, seizures, and shortened lifespan. Gliosis, oxidative damage, lactate accumulation, and neuroinflammation are pathological hallmarks. Cases with loss-of-function mutations in sulfide quinone oxidoreductase (SQOR) are associated with the complications of decreased catabolism, accumulation of H_2S, decreased complex IV activity, and reduced ATP production

systemic H_2S levels, alleviated disease symptoms, and improved lifespan (Kanemaru et al. 2024).

7 Other Mitochondrial Diseases

Mutations in any of the proteins of the sulfide oxidation unit can elevate H_2S. For instance, mutations in *ETHE1,* which encodes ETHE1, causes ethylmalonic encephalopathy (EE). EE is an infantile autosomal recessive disorder affecting the brain, gastrointestinal system, and blood vessels (Tiranti et al. 2004; Zafeiriou et al. 2007). In addition to increased H_2S, accumulation of lactate and ethylmalonic acid occurs as well. In a mouse model of ETHE deficiency, increase in thiosulfate levels was found, which reflects increased H_2S production (Tiranti et al. 2004). Similarly, TST depletion results in high circulating sulfide levels alongside increased gluconeogenesis, hypertriglyceridemia, and fatty liver (Carter et al. 2021). However, in TST-depleted mice, there were no changes in hepatic sulfide levels and these mice were viable, unlike the *Ethe*$^{-/-}$ mice (Morton et al. 2016). TST also generates H_2S from thiosulfate by utilizing dihydrolipoic acid (DHLA) and thus is involved in both production and detoxification of H_2S, although its role inH_2S catabolism seems to be its predominant function (Villarejo and Westley 1963; Libiad et al. 2015). The decreased persulfidation reported in these mice may stem from the elevated oxidative stress and suboptimal Nrf2 signaling that oxidizes cysteine residues irreversibly, a feature seen in Down syndrome (Luo et al. 2023b; Panagaki et al. 2024). Similarly, SO deficiency caused by pathogenic loss-of-function variants or loss of its molybdenum cofactor (Moco) is usually lethal (Schwarz 2016; Kaczmarek et al. 2022).

Loss of SO will likely lead to decreased persulfidation levels and aberrant redox homeostasis.

Deficiency of CoQ, the level at which H_2S enters the mitochondrial electron transport chain, can also cause perturb sulfide catabolism (Luna-Sanchez et al. 2017; Ziosi et al. 2017), while supplementation of CoQ10 has been shown to protect against nephrotic syndrome by normalizing the sulfide oxidation pathway (Kleiner et al. 2018). Thus, the beneficial effects of CoQ10, which is commonly administered to patients with primary or secondary mitochondrial dysfunction, may stem from its role as a cofactor for sulfide:quinone oxidoreductase (SQOR) (Luna-Sanchez et al. 2017; Ziosi et al. 2017).

8 Concluding Remarks

Since the discovery and chemical synthesis of H_2S three centuries ago, research into the endogenous functions of this gasotransmitter has revealed its vital roles in both the peripheral and central nervous systems. It is now evident that H_2S is involved in nearly every biological process. However, both chronic decreases and increases in H_2S levels can be detrimental to cellular health, highlighting the importance of maintaining a balance between its biosynthesis and catabolism.

H_2S does not operate in isolation and may interact with other members of the gasotransmitter family, such as NO or other endogenous metabolites, and studying these interactions may deepen our understanding of signaling processes modulated by gasotransmitters (Coletta et al. 2012; Scrivner et al. 2021). Accordingly, an emerging focus within the field is the development of innovative therapeutics that utilize hybrid molecules (Xu et al. 2023; Dos Reis et al. 2024; Glowacka et al. 2023; Costa et al. 2020). For instance, H_2S-releasing moieties, such as 5-(4-hydroxyphenyl)-3H-1,2-dithiole-3-thione (ADT-OH) and allyldisulfide that are conjugated to L-DOPA methyl ester, have been shown to confer protective effects (Lee et al. 2010). These compounds released H_2S, reduced tumor necrosis factor-α (TNF-α), interleukin-6 (IL-6), and nitrite from stimulated human microglia, astrocytes, THP-1 and U373 cells. Among these hybrids, ACS84 (4-(3-thioxo-3H-1,2-dithiol-4-yl)-benzoic acid) dissociated into L-DOPA and ADT-OH within 1 h after injection into rats, with both metabolites being detected in the brain. ACS84 ameliorated motor deficits prevented neuronal loss in the substantia nigra and increased dopamine levels in the striatum in a rat model of PD (Xie et al. 2013).

Another hybrid molecule developed to counter neurodegeneration, S-memantine, is a hybrid of ADT and memantine, a clinically approved NMDA receptor antagonist used in the treatment of Alzheimer's disease (Marutani et al. 2012). Another strategy involved conjugation of memantine with isothiocyantes, which release H_2S, to generate a compound termed memit (Sestito et al. 2019; Martelli et al. 2014). Memit prevented Aβ (Abe and Kimura 1996; Chen et al. 2004; Cheng et al. 2004; Clarke 1953; Dunham et al. 1999; Dupuyturen 1806; Enokido et al. 2005; Filipovic 2015; Filipovic et al. 2018; Gayon 1877; Ishigami et al. 2009; Ishii et al. 2010; Kamoun 2001; Kruger 2017; Kuo et al. 1983; Lambert et al. 2006; Linden et al.

2008; Mani et al. 2011; Mironov et al. 2017; Mitchell and Davenport 1924; Morikawa et al. 2012; Mustafa et al. 2009, 2011; Myers 1920; Paul et al. 2018; Paul and Snyder 2012, 2015; Ramazzini 1700; Rizzo 1967; Sen et al. 2012; Shatalin et al. 2011; Shen et al. 2012; Shibuya et al. 2009; Stipanuk 2020; Stipanuk and Beck 1982; Szabo 2020; Tang et al. 2005, 2013; Wang 2009; Warenycia et al. 1989; Yang et al. 2008; Zhao et al. 2001) self-induced aggregation and Aβ oligomer-induced damage in human neurons and rat microglia cells (Sestito et al. 2019). Chimeric molecules that combine aspirin with H_2S and NO-releasing moieties have also been considered for treatment of neurodegenerative diseases (Sestito et al. 2019).

In conclusion, the pursuit of compounds or treatments that stimulate H_2S production is gaining traction. It is crucial to consider the brain penetrance of these molecules when developing drugs for neurologic diseases. Furthermore, studying H_2S homeostasis in relation to aging is of significant importance (Zivanovic et al. 2019), as aging is a substantial risk factor for several late-onset neurodegenerative diseases, including PD and HD (Hou et al. 2019). Understanding the role of H_2S in aging may pave the way for new insights and therapeutic strategies to combat these debilitating conditions.

Acknowledgments BDP and AAP were supported by the American Heart Association and Paul Allen Foundation Initiative in Brain Health and Cognitive Impairment (19PABH134580006) and by NIH/NIA 1R01AG071512. BDP acknowledges support from NIH/NIA 1R21AG073684, NIH NIDA, grant P50 DA044123, funding from the Solve-ME foundation and the Catalyst Award from Johns Hopkins University. AAP was supported by the Valour Foundation, the Rebecca E. Barchas, MD, Professor in Translational Psychiatry of Case Western Reserve University and the Morley-Mather Chair in Neuropsychiatry of University Hospitals of Cleveland Medical Center, and by the Wick Foundation, Department of Veterans Affairs Merit Award I01BX005976, NIH/NIA RO1AGs066707, NIH/NIA 1 U01 AG073323, the Louis Stokes VA Medical Center resources and facilities, the Lincoln Neurotherapeutics Research Fund, the Leonard Krieger Fund of the Cleveland Foundation, the Meisel & Pesses Family Foundation, and an anonymous donor.

References

Aarsland D, Batzu L, Halliday GM, Geurtsen GJ, Ballard C, Ray Chaudhuri K et al (2021) Parkinson disease-associated cognitive impairment. Nat Rev Dis Primers 7(1):47

Abe K, Kimura H (1996) The possible role of hydrogen sulfide as an endogenous neuromodulator. J Neurosci 16(3):1066–1071

Absalyamova M, Traktirov D, Burdinskaya V, Artemova V, Muruzheva Z, Karpenko M (2024) Molecular basis of the development of Parkinson's disease. Neuroscience 565:292–300

Akcimen F, Martins S, Liao C, Bourassa CV, Catoire H, Nicholson GA et al (2020) Genome-wide association study identifies genetic factors that modify age at onset in Machado-Joseph disease. Aging (Albany NY) 12(6):4742–4756

Ament SA, Pearl JR, Grindeland A, St Claire J, Earls JC, Kovalenko M et al (2017) High resolution time-course mapping of early transcriptomic, molecular and cellular phenotypes in Huntington's disease CAG knock-in mice across multiple genetic backgrounds. Hum Mol Genet 26(5): 913–922

Ament SA, Pearl JR, Cantle JP, Bragg RM, Skene PJ, Coffey SR et al (2018) Transcriptional regulatory networks underlying gene expression changes in Huntington's disease. Mol Syst Biol 14(3):e7435

Andrew SE, Goldberg YP, Kremer B, Telenius H, Theilmann J, Adam S et al (1993) The relationship between trinucleotide (CAG) repeat length and clinical features of Huntington's disease. Nat Genet 4(4):398–403
Ashizawa T, Oz G, Paulson HL (2018) Spinocerebellar ataxias: prospects and challenges for therapy development. Nat Rev Neurol 14(10):590–605
Bakkenist CJ, Kastan MB (2003) DNA damage activates ATM through intermolecular autophosphorylation and dimer dissociation. Nature 421(6922):499–506
Bame M, Grier RE, Needleman R, Brusilow WS (2014) Amino acids as biomarkers in the SOD1 (G93A) mouse model of ALS. Biochim Biophys Acta 1842(1):79–87
Banez-Coronel M, Ayhan F, Tarabochia AD, Zu T, Perez BA, Tusi SK et al (2015) RAN translation in Huntington disease. Neuron 88(4):667–677
Barzilai A, Rotman G, Shiloh Y (2002) ATM deficiency and oxidative stress: a new dimension of defective response to DNA damage. DNA Repair (Amst) 1(1):3–25
Bates GP, Dorsey R, Gusella JF, Hayden MR, Kay C, Leavitt BR et al (2015) Huntington disease. Nat Rev Dis Primers 1:15005
Beckers J, Tharkeshwar AK, Van Damme P (2021) C9orf72 ALS-FTD: recent evidence for dysregulation of the autophagy-lysosome pathway at multiple levels. Autophagy 17(11): 3306–3322
Berk M (2015) N-Acetylcysteine for Huntington's? Aust N Z J Psychiatry 49(11):1071–1072
Berk M, Malhi GS, Gray LJ, Dean OM (2013) The promise of N-acetylcysteine in neuropsychiatry. Trends Pharmacol Sci 34(3):167–177
Bloem BR, Okun MS, Klein C (2021) Parkinson's disease. Lancet 397(10291):2284–2303
Brown RH, Al-Chalabi A (2017) Amyotrophic lateral sclerosis. N Engl J Med 377(2):162–172
Cakmak YO (2017) Rotorua, hydrogen sulphide and Parkinson's disease-A possible beneficial link? N Z Med J 130(1455):123–125
Campbell A, Bushman J, Munger J, Noble M, Proschel C, Mayer-Proschel M (2016) Mutation of ataxia-telangiectasia mutated is associated with dysfunctional glutathione homeostasis in cerebellar astroglia. Glia 64(2):227–239
Carter RN, Gibbins MTG, Barrios-Llerena ME, Wilkie SE, Freddolino PL, Libiad M et al (2021) The hepatic compensatory response to elevated systemic sulfide promotes diabetes. Cell Rep 37(6):109958
Chang X, Wu Y, Zhou J, Meng H, Zhang W, Guo J (2020) A meta-analysis and systematic review of Leigh syndrome: clinical manifestations, respiratory chain enzyme complex deficiency, and gene mutations. Medicine (Baltimore) 99(5):e18634
Chen X, Jhee KH, Kruger WD (2004) Production of the neuromodulator H2S by cystathionine beta-synthase via the condensation of cysteine and homocysteine. J Biol Chem 279(50): 52082–52086
Cheng Y, Ndisang JF, Tang G, Cao K, Wang R (2004) Hydrogen sulfide-induced relaxation of resistance mesenteric artery beds of rats. Am J Physiol Heart Circ Physiol 287(5):H2316–H2323
Chung KK, Thomas B, Li X, Pletnikova O, Troncoso JC, Marsh L et al (2004) S-nitrosylation of parkin regulates ubiquitination and compromises parkin's protective function. Science 304(5675):1328–1331
Cirotti C, Rizza S, Giglio P, Poerio N, Allega MF, Claps G et al (2021) Redox activation of ATM enhances GSNOR translation to sustain mitophagy and tolerance to oxidative stress. EMBO Rep 22(1):e50500
Clark IE, Dodson MW, Jiang C, Cao JH, Huh JR, Seol JH et al (2006) Drosophila pink1 is required for mitochondrial function and interacts genetically with parkin. Nature 441(7097):1162–1166
Clarke PH (1953) Hydrogen sulphide production by bacteria. J Gen Microbiol 8(3):397–407
Coletta C, Papapetropoulos A, Erdelyi K, Olah G, Modis K, Panopoulos P et al (2012) Hydrogen sulfide and nitric oxide are mutually dependent in the regulation of angiogenesis and endothelium-dependent vasorelaxation. Proc Natl Acad Sci USA 109(23):9161–9166

Collaborators GBDPsD (2018) Global, regional, and national burden of Parkinson's disease, 1990-2016: a systematic analysis for the Global Burden of Disease Study 2016. Lancet Neurol 17(11): 939–953

Costa S, Muscara MN, Allain T, Dallazen J, Gonzaga L, Buret AG et al (2020) Enhanced analgesic effects and gastrointestinal safety of a novel, hydrogen sulfide-releasing anti-inflammatory drug (ATB-352): a role for endogenous cannabinoids. Antioxid Redox Signal 33(14):1003–1009

Cruts M, Gijselinck I, Van Langenhove T, van der Zee J, Van Broeckhoven C (2013) Current insights into the C9orf72 repeat expansion diseases of the FTLD/ALS spectrum. Trends Neurosci 36(8):450–459

Cui ZT, Mao ZT, Yang R, Li JJ, Jia SS, Zhao JL et al (2024) Spinocerebellar ataxias: from pathogenesis to recent therapeutic advances. Front Neurosci 18:1422442

Davies SW, Turmaine M, Cozens BA, DiFiglia M, Sharp AH, Ross CA et al (1997) Formation of neuronal intranuclear inclusions underlies the neurological dysfunction in mice transgenic for the HD mutation. Cell 90(3):537–548

Davoli A, Greco V, Spalloni A, Guatteo E, Neri C, Rizzo GR et al (2015) Evidence of hydrogen sulfide involvement in amyotrophic lateral sclerosis. Ann Neurol 77(4):697–709

Dawson TM, Dawson VL (2010) The role of parkin in familial and sporadic Parkinson's disease. Mov Disord 25(Suppl 1):S32–S39

de Mattos EP, Kolbe Musskopf M, Bielefeldt Leotti V, Saraiva-Pereira ML, Jardim LB (2019) Genetic risk factors for modulation of age at onset in Machado-Joseph disease/spinocerebellar ataxia type 3: a systematic review and meta-analysis. J Neurol Neurosurg Psychiatry 90(2): 203–210

De Stefano S, Tiberi M, Salvatori I, De Bardi M, Gimenez J, Pirshayan M et al (2024) Hydrogen sulfide modulates astrocytic toxicity in mouse spinal cord cultures: implications for amyotrophic lateral sclerosis. Antioxidants (Basel) 13(10)

DeJesus-Hernandez M, Mackenzie IR, Boeve BF, Boxer AL, Baker M, Rutherford NJ et al (2011) Expanded GGGGCC hexanucleotide repeat in noncoding region of C9ORF72 causes chromosome 9p-linked FTD and ALS. Neuron 72(2):245–256

DiFiglia M, Sapp E, Chase KO, Davies SW, Bates GP, Vonsattel JP et al (1997) Aggregation of huntingtin in neuronal intranuclear inclusions and dystrophic neurites in brain. Science 277(5334):1990–1993

DiMauro S, Servidei S, Zeviani M, DiRocco M, DeVivo DC, DiDonato S et al (1987) Cytochrome c oxidase deficiency in Leigh syndrome. Ann Neurol 22(4):498–506

Ditch S, Paull TT (2012) The ATM protein kinase and cellular redox signaling: beyond the DNA damage response. Trends Biochem Sci 37(1):15–22

Dos Reis RA, Sarkar I, Rodrigues MG, Matson JB, Seabra AB, Kashfi K (2024) NO- and H(2)S-releasing nanomaterials: a crosstalk signaling pathway in cancer. Nitric Oxide 151:17–30

Dunham I, Shimizu N, Roe BA, Chissoe S, Hunt AR, Collins JE et al (1999) The DNA sequence of human chromosome 22. Nature 402(6761):489–495

Dupuyturen M (1806) Rapport sur une espéce de mephitisme des fosses d'aisance, produite par le gas azote. J Med

Elamin M, Bede P, Byrne S, Jordan N, Gallagher L, Wynne B et al (2013) Cognitive changes predict functional decline in ALS: a population-based longitudinal study. Neurology 80(17): 1590–1597

Enokido Y, Suzuki E, Iwasawa K, Namekata K, Okazawa H, Kimura H (2005) Cystathionine beta-synthase, a key enzyme for homocysteine metabolism, is preferentially expressed in the radial glia/astrocyte lineage of developing mouse CNS. FASEB J 19(13):1854–1856

Evert BO, Schelhaas J, Fleischer H, de Vos RA, Brunt ER, Stenzel W et al (2006) Neuronal intranuclear inclusions, dysregulation of cytokine expression and cell death in spinocerebellar ataxia type 3. Clin Neuropathol 25(6):272–281

Fang EF, Kassahun H, Croteau DL, Scheibye-Knudsen M, Marosi K, Lu H et al (2016) NAD(+) replenishment improves lifespan and Healthspan in ataxia telangiectasia models via Mitophagy and DNA repair. Cell Metab 24(4):566–581

Filipovic MR (2015) Persulfidation (S-sulfhydration) and H2S. Handb Exp Pharmacol 230:29–59
Filipovic MR, Zivanovic J, Alvarez B, Banerjee R (2018) Chemical biology of H(2)S signaling through Persulfidation. Chem Rev 118(3):1253–1337
Frederick NM, Bertho J, Patel KK, Petr GT, Bakradze E, Smith SB et al (2014) Dysregulation of system xc(−) expression induced by mutant huntingtin in a striatal neuronal cell line and in R6/2 mice. Neurochem Int 76:59–69
Friederich MW, Elias AF, Kuster A, Laugwitz L, Larson AA, Landry AP et al (2020) Pathogenic variants in SQOR encoding sulfide:quinone oxidoreductase are a potentially treatable cause of Leigh disease. J Inherit Metab Dis 43(5):1024–1036
Fu RH, Chen HJ, Hong SY (2023) Glycine-alanine dipeptide repeat protein from C9-ALS interacts with sulfide Quinone oxidoreductase (SQOR) to induce the activity of the NLRP3 Inflammasome in HMC3 microglia: Irisflorentin Reverses This Interaction. Antioxidants (Basel) 12(10)
Funayama M, Nishioka K, Li Y, Hattori N (2023) Molecular genetics of Parkinson's disease: contributions and global trends. J Hum Genet 68(3):125–130
Gayon MU Note presentee par M. Pasteur. Sur les alterations des oeufs, a l'occasion d'une Note de MM. A. Bechamp et G. Eustache. Ann Ecole Normale Super 1877;Ser 2
Genetic Modifiers of Huntington's Disease C (2015) Identification of genetic factors that modify clinical onset of Huntington's disease. Cell 162(3):516–526
Glowacka U, Magierowski M, Sliwowski Z, Cieszkowski J, Szetela M, Wojcik-Grzybek D et al (2023) Hydrogen sulfide-releasing indomethacin-derivative (ATB-344) prevents the development of oxidative gastric mucosal injuries. Antioxidants (Basel) 12(8):1545
Greco V, Spalloni A, Corasolla Carregari V, Pieroni L, Persichilli S, Mercuri NB et al (2018) Proteomics and toxicity analysis of spinal-cord primary cultures upon hydrogen sulfide treatment. Antioxidants (Basel) 7(7)
Grima JC, Daigle JG, Arbez N, Cunningham KC, Zhang K, Ochaba J et al (2017) Mutant huntingtin disrupts the nuclear pore complex. Neuron 94(1):93–107 e6
Gu X, Richman J, Langfelder P, Wang N, Zhang S, Banez-Coronel M et al (2022) Uninterrupted CAG repeat drives striatum-selective transcriptionopathy and nuclear pathogenesis in human Huntingtin BAC mice. Neuron 110(7):1173–92 e7
Guo Z, Kozlov S, Lavin MF, Person MD, Paull TT (2010) ATM activation by oxidative stress. Science 330(6003):517–521
Gurney ME, Pu H, Chiu AY, Dal Canto MC, Polchow CY, Alexander DD et al (1994) Motor neuron degeneration in mice that express a human Cu,Zn superoxide dismutase mutation. Science 264(5166):1772–1775
Gusella JF, MacDonald ME, Lee JM (2014) Genetic modifiers of Huntington's disease. Mov Disord 29(11):1359–1365
Hardiman O, Al-Chalabi A, Chio A, Corr EM, Logroscino G, Robberecht W et al (2017) Amyotrophic lateral sclerosis. Nat Rev Dis Primers 3:17071
Harmuth T, Weber JJ, Zimmer AJ, Sowa AS, Schmidt J, Fitzgerald JC et al (2022) Mitochondrial dysfunction in spinocerebellar ataxia type 3 is linked to VDAC1 Deubiquitination. Int J Mol Sci 23(11)
Hildebrandt TM, Grieshaber MK (2008) Three enzymatic activities catalyze the oxidation of sulfide to thiosulfate in mammalian and invertebrate mitochondria. FEBS J 275(13):3352–3361
Hou Y, Dan X, Babbar M, Wei Y, Hasselbalch SG, Croteau DL et al (2019) Ageing as a risk factor for neurodegenerative disease. Nat Rev Neurol 15(10):565–581
Hou X, Watzlawik JO, Fiesel FC, Springer W (2020) Autophagy in Parkinson's disease. J Mol Biol 432(8):2651–2672
Hu LF, Lu M, Tiong CX, Dawe GS, Hu G, Bian JS (2010) Neuroprotective effects of hydrogen sulfide on Parkinson's disease rat models. Aging Cell 9(2):135–146
Ishigami M, Hiraki K, Umemura K, Ogasawara Y, Ishii K, Kimura H (2009) A source of hydrogen sulfide and a mechanism of its release in the brain. Antioxid Redox Signal 11(2):205–214

Ishii I, Akahoshi N, Yamada H, Nakano S, Izumi T, Suematsu M (2010) Cystathionine gamma-Lyase-deficient mice require dietary cysteine to protect against acute lethal myopathy and oxidative injury. J Biol Chem 285(34):26358–26368

Islam MZ, Shen X, Pardue S, Kevil CG, Shackelford RE (2022) The ataxia-telangiectasia mutated gene product regulates the cellular acid-labile sulfide fraction. DNA Repair (Amst) 116:103344

Jackson MR, Melideo SL, Jorns MS (2012) Human sulfide:quinone oxidoreductase catalyzes the first step in hydrogen sulfide metabolism and produces a sulfane sulfur metabolite. Biochemistry 51(34):6804–6815

Jafarinia H, van der Giessen E, Onck PR (2024) C9orf72 polyPR directly binds to various nuclear transport components. elife 12:12

Jensen BK, Schuldi MH, McAvoy K, Russell KA, Boehringer A, Curran BM et al (2020) Synaptic dysfunction induced by glycine-alanine dipeptides in C9orf72-ALS/FTD is rescued by SV2 replenishment. EMBO Mol Med 12(5):e10722

Jiang W, Zou W, Hu M, Tian Q, Xiao F, Li M et al (2022) Hydrogen sulphide attenuates neuronal apoptosis of substantia nigra by re-establishing autophagic flux via promoting leptin signalling in a 6-hydroxydopamine rat model of Parkinson's disease. Clin Exp Pharmacol Physiol 49(1): 122–133

Kaczmarek AT, Bender D, Gehling T, Kohl JB, Daimaguler HS, Santamaria-Araujo JA et al (2022) A defect in molybdenum cofactor binding causes an attenuated form of sulfite oxidase deficiency. J Inherit Metab Dis 45(2):169–182

Kamoun P (2001) Mental retardation in down syndrome: a hydrogen sulfide hpothesis. Med Hypotheses 57(3):389–392

Kamsler A, Daily D, Hochman A, Stern N, Shiloh Y, Rotman G et al (2001) Increased oxidative stress in ataxia telangiectasia evidenced by alterations in redox state of brains from Atm-deficient mice. Cancer Res 61(5):1849–1854

Kanemaru E, Ichinose F (2024) Essential role of sulfide oxidation in brain health and neurological disorders. Pharmacol Ther 266:108787

Kanemaru E, Shimoda K, Marutani E, Morita M, Miranda M, Miyazaki Y et al (2024) Exclusion of sulfide:quinone oxidoreductase from mitochondria causes Leigh-like disease in mice by impairing sulfide metabolism. J Clin Invest 134(15)

Kawaguchi Y, Okamoto T, Taniwaki M, Aizawa M, Inoue M, Katayama S et al (1994) CAG expansions in a novel gene for Machado-Joseph disease at chromosome 14q32.1. Nat Genet 8(3):221–228

Keum JW, Shin A, Gillis T, Mysore JS, Abu Elneel K, Lucente D et al (2016) The HTT CAG-expansion mutation determines age at death but not disease duration in Huntington disease. Am J Hum Genet 98(2):287–298

Kida K, Yamada M, Tokuda K, Marutani E, Kakinohana M, Kaneki M et al (2011) Inhaled hydrogen sulfide prevents neurodegeneration and movement disorder in a mouse model of Parkinson's disease. Antioxid Redox Signal 15(2):343–352

Kimura Y, Goto Y, Kimura H (2010) Hydrogen sulfide increases glutathione production and suppresses oxidative stress in mitochondria. Antioxid Redox Signal 12(1):1–13

Kirkinezos IG, Bacman SR, Hernandez D, Oca-Cossio J, Arias LJ, Perez-Pinzon MA et al (2005) Cytochrome c association with the inner mitochondrial membrane is impaired in the CNS of G93A-SOD1 mice. J Neurosci 25(1):164–172

Klein C, Lohmann K (2009) Parkinson disease(s): is "Parkin disease" a distinct clinical entity? Neurology 72(2):106–107

Kleiner G, Barca E, Ziosi M, Emmanuele V, Xu Y, Hidalgo-Gutierrez A et al (2018) CoQ (10) supplementation rescues nephrotic syndrome through normalization of H(2)S oxidation pathway. Biochim Biophys Acta Mol basis Dis 1864(11):3708–3722

Klockgether T, Paulson H (2011) Milestones in ataxia. Mov Disord 26(6):1134–1141

Klockgether T, Mariotti C, Paulson HL (2019) Spinocerebellar ataxia. Nat Rev Dis Primers 5(1):24

Kruger WD (2017) Cystathionine beta-synthase deficiency: of mice and men. Mol Genet Metab 121(3):199–205

Kuo SM, Lea TC, Stipanuk MH (1983) Developmental pattern, tissue distribution, and subcellular distribution of cysteine: alpha-ketoglutarate aminotransferase and 3-mercaptopyruvate sulfurtransferase activities in the rat. Biol Neonate 43(1–2):23–32

Lake NJ, Compton AG, Rahman S, Thorburn DR (2016) Leigh syndrome: one disorder, more than 75 monogenic causes. Ann Neurol 79(2):190–203

Lambert TW, Goodwin VM, Stefani D, Strosher L (2006) Hydrogen sulfide (H2S) and sour gas effects on the eye. A historical perspective. Sci Total Environ 367(1):1–22

Landry AP, Ballou DP, Banerjee R (2021) Hydrogen sulfide oxidation by sulfide Quinone oxidoreductase. Chembiochem 22(6):949–960

Lee JH, Paull TT (2007) Activation and regulation of ATM kinase activity in response to DNA double-strand breaks. Oncogene 26(56):7741–7748

Lee M, Tazzari V, Giustarini D, Rossi R, Sparatore A, Del Soldato P et al (2010) Effects of hydrogen sulfide-releasing L-DOPA derivatives on glial activation: potential for treating Parkinson disease. J Biol Chem 285(23):17318–17328

Leigh D (1951) Subacute necrotizing encephalomyelopathy in an infant. J Neurol Neurosurg Psychiatry 14(3):216–221

Li X, Valencia A, Sapp E, Masso N, Alexander J, Reeves P et al (2010) Aberrant Rab11-dependent trafficking of the neuronal glutamate transporter EAAC1 causes oxidative stress and cell death in Huntington's disease. J Neurosci 30(13):4552–4561

Li L, Liu H, Dong P, Li D, Legant WR, Grimm JB et al (2016) Real-time imaging of Huntingtin aggregates diverting target search and gene transcription. elife:5

Li H, Shi G, Zha H, Zheng L, Luo Z, Wang Y (2024) Inhibition of histone deacetylase promotes a neuroprotective mechanism in an experimental model of Parkinson's disease. Arch Med Sci 20(2):664–674

Libiad M, Sriraman A, Banerjee R (2015) Polymorphic variants of human Rhodanese exhibit differences in thermal stability and sulfur transfer kinetics. J Biol Chem 290(39):23579–23588

Linden DR, Sha L, Mazzone A, Stoltz GJ, Bernard CE, Furne JK et al (2008) Production of the gaseous signal molecule hydrogen sulfide in mouse tissues. J Neurochem 106(4):1577–1585

Linden DR, Furne J, Stoltz GJ, Abdel-Rehim MS, Levitt MD, Szurszewski JH (2012) Sulphide quinone reductase contributes to hydrogen sulphide metabolism in murine peripheral tissues but not in the CNS. Br J Pharmacol 165(7):2178–2190

Liu TW, Chen CM, Chang KH (2022) Biomarker of Neuroinflammation in Parkinson's disease. Int J Mol Sci 23(8)

Liu F, Tian Q, Tang HL, Cheng X, Zou W, Zhang P (2024) Hydrogen sulfide attenuates depression-like behaviours in Parkinson's disease model rats by improving synaptic plasticity in a hippocampal Warburg effect-dependent manner. Pharmacol Biochem Behav 234:173677

Luna-Sanchez M, Hidalgo-Gutierrez A, Hildebrandt TM, Chaves-Serrano J, Barriocanal-Casado E, Santos-Fandila A et al (2017) CoQ deficiency causes disruption of mitochondrial sulfide oxidation, a new pathomechanism associated with this syndrome. EMBO Mol Med 9(1):78–95

Luo H, Todi SV, Paulson HL, Costa MDC (2023a) Regional and age-dependent changes in ubiquitination in cellular and mouse models of spinocerebellar ataxia type 3. Front Mol Neurosci 16:1154203

Luo Y, Chatre L, Melhem S, Al-Dahmani ZM, Homer NZM, Miedema A et al (2023b) Thiosulfate sulfurtransferase deficiency promotes oxidative distress and aberrant NRF2 function in the brain. Redox Biol 68:102965

Mani S, Yang G, Wang R (2011) A critical life-supporting role for cystathionine gamma-lyase in the absence of dietary cysteine supply. Free Radic Biol Med 50(10):1280–1287

Marangi G, Traynor BJ (2015) Genetic causes of amyotrophic lateral sclerosis: new genetic analysis methodologies entailing new opportunities and challenges. Brain Res 1607:75–93

Martelli A, Testai L, Citi V, Marino A, Bellagambi FG, Ghimenti S et al (2014) Pharmacological characterization of the vascular effects of aryl isothiocyanates: is hydrogen sulfide the real player? Vasc Pharmacol 60(1):32–41

Marutani E, Kosugi S, Tokuda K, Khatri A, Nguyen R, Atochin DN et al (2012) A novel hydrogen sulfide-releasing N-methyl-D-aspartate receptor antagonist prevents ischemic neuronal death. J Biol Chem 287(38):32124–32135

Marutani E, Morita M, Hirai S, Kai S, Grange RMH, Miyazaki Y et al (2021) Sulfide catabolism ameliorates hypoxic brain injury. Nat Commun 12(1):3108

Matos CA, de Almeida LP, Nobrega C (2019) Machado-Joseph disease/spinocerebellar ataxia type 3: lessons from disease pathogenesis and clues into therapy. J Neurochem 148(1):8–28

McDonald CJ, Ostini L, Wallace DF, John AN, Watters DJ, Subramaniam VN (2011) Iron loading and oxidative stress in the Atm-/- mouse liver. Am J Physiol Gastrointest Liver Physiol 300(4): G554–G560

Meredith MJ, Dodson ML (1987) Imparied glutathione biosynthesis in cultured human ataxia-telangiectasia cells. Cancer Res 47(17):4576–4581

Mironov A, Seregina T, Nagornykh M, Luhachack LG, Korolkova N, Lopes LE et al (2017) Mechanism of H(2)S-mediated protection against oxidative stress in Escherichia coli. Proc Natl Acad Sci USA 114(23):6022–6027

Mitchell CW, Davenport SJ (1924) Hydrogen sulphide literature. Public Health Rep 39:1–13

Modis K, Bos EM, Calzia E, van Goor H, Coletta C, Papapetropoulos A et al (2014) Regulation of mitochondrial bioenergetic function by hydrogen sulfide. Part II. Pathophysiological and therapeutic aspects. Br J Pharmacol 171(8):2123–2146

Modis K, Ju Y, Ahmad A, Untereiner AA, Altaany Z, Wu L et al (2016) S-Sulfhydration of ATP synthase by hydrogen sulfide stimulates mitochondrial bioenergetics. Pharmacol Res 113 (Pt A):116–124

Morikawa T, Kajimura M, Nakamura T, Hishiki T, Nakanishi T, Yukutake Y et al (2012) Hypoxic regulation of the cerebral microcirculation is mediated by a carbon monoxide-sensitive hydrogen sulfide pathway. Proc Natl Acad Sci USA 109(4):1293–1298

Morton NM, Beltram J, Carter RN, Michailidou Z, Gorjanc G, McFadden C et al (2016) Genetic identification of thiosulfate sulfurtransferase as an adipocyte-expressed antidiabetic target in mice selected for leanness. Nat Med 22(7):771–779

Mustafa AK, Gadalla MM, Sen N, Kim S, Mu W, Gazi SK et al (2009) H2S signals through protein S-sulfhydration. Sci Signal 2(96):ra72

Mustafa AK, Sikka G, Gazi SK, Steppan J, Jung SM, Bhunia AK et al (2011) Hydrogen sulfide as endothelium-derived hyperpolarizing factor sulfhydrates potassium channels. Circ Res 109(11): 1259–1268

Myers JT (1920) The production of hydrogen Sulphide by bacteria. J Bacteriol 5(3):231–252

Nagashima F, Miyazaki Y, Kanemaru E, Ezaka M, Hara H, Sugiura K et al (2023) Sulfide:quinone oxidoreductase ameliorates neurodegeneration in a murine model of Parkinson's disease. Redox Biol 59:102562

Neumann M, Sampathu DM, Kwong LK, Truax AC, Micsenyi MC, Chou TT et al (2006) Ubiquitinated TDP-43 in frontotemporal lobar degeneration and amyotrophic lateral sclerosis. Science 314(5796):130–133

Oh YM, Lee SW, Kim WK, Chen S, Church VA, Cates K et al (2022) Age-related Huntington's disease progression modeled in directly reprogrammed patient-derived striatal neurons highlights impaired autophagy. Nat Neurosci 25(11):1420–1433

Panagaki T, Janickova L, Petrovic D, Zuhra K, Ditroi T, Juranyi EP et al (2024) Neurobehavioral dysfunction in a mouse model of Down syndrome: upregulation of cystathionine beta-synthase, H(2)S overproduction, altered protein persulfidation, synaptic dysfunction, endoplasmic reticulum stress, and autophagy. Geroscience 46(5):4275–4314

Paul BD (2022) Cysteine metabolism and hydrogen sulfide signaling in Huntington's disease. Free Radic Biol Med 186:93–98

Paul BD, Snyder SH (2012) H(2)S signalling through protein sulfhydration and beyond. Nat Rev Mol Cell Biol 13(8):499–507

Paul BD, Snyder SH (2014) Neurodegeneration in Huntington's disease involves loss of cystathionine gamma-lyase. Cell Cycle 13(16):2491–2493

Paul BD, Snyder SH (2015) H2S: a novel Gasotransmitter that signals by Sulfhydration. Trends Biochem Sci 40(11):687–700
Paul BD, Snyder SH (2019) Impaired redox signaling in Huntington's disease: therapeutic implications. Front Mol Neurosci 12:68
Paul BD, Sbodio JI, Xu R, Vandiver MS, Cha JY, Snowman AM et al (2014) Cystathionine gamma-lyase deficiency mediates neurodegeneration in Huntington's disease. Nature 509(7498):96–100
Paul BD, Sbodio JI, Snyder SH (2018) Cysteine metabolism in neuronal redox homeostasis. Trends Pharmacol Sci 39(5):513–524
Paul BD, Snyder SH, Kashfi K (2021) Effects of hydrogen sulfide on mitochondrial function and cellular bioenergetics. Redox Biol 38:101772
Paul BD, Sbodio JI, Snyder SH (2022) Mutant huntingtin derails cysteine metabolism in Huntington's disease at both transcriptional and post-translational levels. Antioxidants (Basel) 11(8)
Paull TT, Woolley PR (2024) A-T neurodegeneration and DNA damage-induced transcriptional stress. DNA Repair (Amst) 135:103647
Pengo M, Squitieri F (2024) Beyond CAG repeats: the multifaceted role of genetics in Huntington disease. Genes (Basel) 15(6)
Phukan J, Elamin M, Bede P, Jordan N, Gallagher L, Byrne S et al (2012) The syndrome of cognitive impairment in amyotrophic lateral sclerosis: a population-based study. J Neurol Neurosurg Psychiatry 83(1):102–108
Poewe W, Seppi K, Tanner CM, Halliday GM, Brundin P, Volkmann J et al (2017) Parkinson disease. Nat Rev Dis Primers 3:17013
Quick KL, Dugan LL (2001) Superoxide stress identifies neurons at risk in a model of ataxia-telangiectasia. Ann Neurol 49(5):627–635
Quinet G, Paz-Cabrera MC, Freire R (2024) Biochemical analysis to study wild-type and polyglutamine-expanded ATXN3 species. PLoS One 19(12):e0315868
Rahman S, Blok RB, Dahl HH, Danks DM, Kirby DM, Chow CW et al (1996) Leigh syndrome: clinical features and biochemical and DNA abnormalities. Ann Neurol 39(3):343–351
Ramazzini B (1700) De Morbis Artificum Diatriba. Antonii Capponi, Mutinae (Modena)
Ravikumar B, Duden R, Rubinsztein DC (2002) Aggregate-prone proteins with polyglutamine and polyalanine expansions are degraded by autophagy. Hum Mol Genet 11(9):1107–1117
Reed BR, Crane J, Garrett N, Woods DL, Bates MN (2014) Chronic ambient hydrogen sulfide exposure and cognitive function. Neurotoxicol Teratol 42:68–76
Rizzo AA (1967) The possible role of hydrogen sulfide in human periodontal disease. I. Hydrogen sulfide production in periodontal pockets. Periodontics 5(5):233–236
Rosen DR, Siddique T, Patterson D, Figlewicz DA, Sapp P, Hentati A et al (1993) Mutations in Cu/Zn superoxide dismutase gene are associated with familial amyotrophic lateral sclerosis. Nature 362(6415):59–62
Rothblum-Oviatt C, Wright J, Lefton-Greif MA, McGrath-Morrow SA, Crawford TO, Lederman HM (2016) Ataxia telangiectasia: a review. Orphanet J Rare Dis 11(1):159
Rowland LP, Shneider NA (2001) Amyotrophic lateral sclerosis. N Engl J Med 344(22):1688–1700
Savitsky K, Bar-Shira A, Gilad S, Rotman G, Ziv Y, Vanagaite L et al (1995) A single ataxia telangiectasia gene with a product similar to PI-3 kinase. Science 268(5218):1749–1753
Sawant N, Morton H, Kshirsagar S, Reddy AP, Reddy PH (2021) Mitochondrial abnormalities and synaptic damage in Huntington's disease: a focus on defective Mitophagy and mitochondria-targeted therapeutics. Mol Neurobiol 58(12):6350–6377
Sbodio JI, Snyder SH, Paul BD (2016) Transcriptional control of amino acid homeostasis is disrupted in Huntington's disease. Proc Natl Acad Sci USA 113(31):8843–8848
Sbodio JI, Snyder SH, Paul BD (2018) Golgi stress response reprograms cysteine metabolism to confer cytoprotection in Huntington's disease. Proc Natl Acad Sci USA 115(4):780–785
Sbodio JI, Snyder SH, Paul BD (2019) Redox mechanisms in neurodegeneration: from disease outcomes to therapeutic opportunities. Antioxid Redox Signal 30(11):1450–1499

Schwarz G (2016) Molybdenum cofactor and human disease. Curr Opin Chem Biol 31:179–187
Scott SSO, Pedroso JL, Barsottini OGP, Franca-Junior MC, Braga-Neto P (2020) Natural history and epidemiology of the spinocerebellar ataxias: insights from the first description to nowadays. J Neurol Sci 417:117082
Scrivner O, Kumar MR, Sorokolet K, Wong A, Kebaara B, Farmer PJ (2021) Characterization of endogenous and extruded H(2)S and small oxoacids of sulfur (SOS) in cell cultures. ACS Chem Biol 16(8):1413–1424
Sen N, Paul BD, Gadalla MM, Mustafa AK, Sen T, Xu R et al (2012) Hydrogen sulfide-linked sulfhydration of NF-kappaB mediates its antiapoptotic actions. Mol Cell 45(1):13–24
Sestito S, Daniele S, Pietrobono D, Citi V, Bellusci L, Chiellini G et al (2019) Memantine prodrug as a new agent for Alzheimer's disease. Sci Rep 9(1):4612
Shatalin K, Shatalina E, Mironov A, Nudler E (2011) H2S: a universal defense against antibiotics in bacteria. Science 334(6058):986–990
Shen X, Peter EA, Bir S, Wang R, Kevil CG (2012) Analytical measurement of discrete hydrogen sulfide pools in biological specimens. Free Radic Biol Med 52(11–12):2276–2283
Shen X, Carlstrom M, Borniquel S, Jadert C, Kevil CG, Lundberg JO (2013) Microbial regulation of host hydrogen sulfide bioavailability and metabolism. Free Radic Biol Med 60:195–200
Shibuya N, Tanaka M, Yoshida M, Ogasawara Y, Togawa T, Ishii K et al (2009) 3-Mercaptopyruvate sulfurtransferase produces hydrogen sulfide and bound sulfane sulfur in the brain. Antioxid Redox Signal 11(4):703–714
Singh SB, Lin HC (2015) Hydrogen sulfide in physiology and diseases of the digestive tract. Microorganisms 3(4):866–889
Snijder PM, Baratashvili M, Grzeschik NA, Leuvenink HGD, Kuijpers L, Huitema S et al (2016) Overexpression of cystathionine gamma-Lyase suppresses detrimental effects of spinocerebellar ataxia type 3. Mol Med 21(1):758–768
Spalloni A, Greco V, Ciriminna G, Corasolla Carregari V, Marini F, Pieroni L et al (2019) Impact of pharmacological inhibition of hydrogen Sulphide production in the SOD1G93A-ALS mouse model. Int J Mol Sci 20(10)
Spencer JP, Jenner P, Daniel SE, Lees AJ, Marsden DC, Halliwell B (1998) Conjugates of catecholamines with cysteine and GSH in Parkinson's disease: possible mechanisms of formation involving reactive oxygen species. J Neurochem 71(5):2112–2122
Stipanuk MH (2020) Metabolism of sulfur-containing amino acids: how the body copes with excess methionine, cysteine, and sulfide. J Nutr 150(Suppl 1):2494S–2505S
Stipanuk MH, Beck PW (1982) Characterization of the enzymic capacity for cysteine desulphhydration in liver and kidney of the rat. Biochem J 206(2):267–277
Sultana J, Ragagnin AMG, Parakh S, Saravanabavan S, Soo KY, Vidal M et al (2024) C9orf72-associated dipeptide repeat expansions perturb ER-Golgi vesicular trafficking, inducing Golgi fragmentation and ER stress, in ALS/FTD. Mol Neurobiol 61(12):10318–10338
Sun M, Latourelle JC, Wooten GF, Lew MF, Klein C, Shill HA et al (2006) Influence of heterozygosity for parkin mutation on onset age in familial Parkinson disease: the GenePD study. Arch Neurol 63(6):826–832
Sun Y, Lu F, Yu X, Wang B, Chen J, Lu F et al (2020) Exogenous H(2)S promoted USP8 Sulfhydration to regulate Mitophagy in the hearts of db/db mice. Aging Dis 11(2):269–285
Sun Y, Li D, Su Y, Zhao H, Pang W, Zhao W et al (2023) Protective effect of hydrogen sulfide is mediated by negative regulation of epigenetic histone acetylation in Parkinson's disease. Arch Med Sci 19(4):1124–1135
Szabo C (2020) The re-emerging pathophysiological role of the cystathionine-beta-synthase – hydrogen sulfide system in Down syndrome. FEBS J 287(15):3150–3160
Szabo C, Ransy C, Modis K, Andriamihaja M, Murghes B, Coletta C et al (2014) Regulation of mitochondrial bioenergetic function by hydrogen sulfide. Part I. Biochemical and physiological mechanisms. Br J Pharmacol 171(8):2099–2122

Tang G, Wu L, Liang W, Wang R (2005) Direct stimulation of K(ATP) channels by exogenous and endogenous hydrogen sulfide in vascular smooth muscle cells. Mol Pharmacol 68(6): 1757–1764

Tang G, Yang G, Jiang B, Ju Y, Wu L, Wang R (2013) H(2)S is an endothelium-derived hyperpolarizing factor. Antioxid Redox Signal 19(14):1634–1646

The Huntington's Disease Collaborative Research Group (1993) A novel gene containing a trinucleotide repeat that is expanded and unstable on Huntington's disease chromosomes. Cell 72(6):971–983

Tiranti V, D'Adamo P, Briem E, Ferrari G, Mineri R, Lamantea E et al (2004) Ethylmalonic encephalopathy is caused by mutations in ETHE1, a gene encoding a mitochondrial matrix protein. Am J Hum Genet 74(2):239–252

Tripathi SJ, Chakraborty S, Miller E, Pieper AA, Paul BD (2023) Hydrogen sulfide signalling in neurodegenerative diseases. Br J Pharmacol. https://doi.org/10.1111/bph.16170

Vandiver MS, Paul BD, Xu R, Karuppagounder S, Rao F, Snowman AM et al (2013) Sulfhydration mediates neuroprotective actions of parkin. Nat Commun 4:1626

Villarejo M, Westley J (1963) Rhodanese-catalyzed reduction of thiosulfate by reduced lipoic acid. J Biol Chem 238:1185–1186

Wang R (2009) Hydrogen sulfide: a new EDRF. Kidney Int 76(7):700–704

Wang Y, Clark LN, Louis ED, Mejia-Santana H, Harris J, Cote LJ et al (2008) Risk of Parkinson disease in carriers of parkin mutations: estimation using the kin-cohort method. Arch Neurol 65(4):467–474

Warenycia MW, Goodwin LR, Benishin CG, Reiffenstein RJ, Francom DM, Taylor JD et al (1989) Acute hydrogen sulfide poisoning. Demonstration of selective uptake of sulfide by the brainstem by measurement of brain sulfide levels. Biochem Pharmacol 38(6):973–981

Wiatr K, Marczak L, Perot JB, Brouillet E, Flament J, Figiel M (2021) Broad influence of mutant Ataxin-3 on the proteome of the adult brain, young neurons, and axons reveals central molecular processes and biomarkers in SCA3/MJD using Knock-in mouse model. Front Mol Neurosci 14: 658339

Wright DJ, Renoir T, Smith ZM, Frazier AE, Francis PS, Thorburn DR et al (2015) N-Acetylcysteine improves mitochondrial function and ameliorates behavioral deficits in the R6/1 mouse model of Huntington's disease. Transl Psychiatry 5:e492

Wright DJ, Gray LJ, Finkelstein DI, Crouch PJ, Pow D, Pang TY et al (2016) N-acetylcysteine modulates glutamatergic dysfunction and depressive behavior in Huntington's disease. Hum Mol Genet 25(14):2923–2933

Wright GEB, Collins JA, Kay C, McDonald C, Dolzhenko E, Xia Q et al (2019) Length of uninterrupted CAG, independent of Polyglutamine size, results in increased somatic instability, hastening onset of Huntington disease. Am J Hum Genet 104(6):1116–1126

Xie L, Hu LF, Teo XQ, Tiong CX, Tazzari V, Sparatore A et al (2013) Therapeutic effect of hydrogen sulfide-releasing L-Dopa derivative ACS84 on 6-OHDA-induced Parkinson's disease rat model. PLoS One 8(4):e60200

Xu S, Shieh M, Paul BD, Xian M (2023) Hydrogen sulfide: recent development of its dual donors and hybrid drugs. Br J Pharmacol. https://doi.org/10.1111/bph.16211

Yagensky O, Kohansal-Nodehi M, Gunaseelan S, Rabe T, Zafar S, Zerr I et al (2019) Increased expression of heme-binding protein 1 early in Alzheimer's disease is linked to neurotoxicity. elife 8:8

Yang G, Wu L, Jiang B, Yang W, Qi J, Cao K et al (2008) H2S as a physiologic vasorelaxant: hypertension in mice with deletion of cystathionine gamma-lyase. Science 322(5901):587–590

Yang J, Minkler P, Grove D, Wang R, Willard B, Dweik R et al (2019) Non-enzymatic hydrogen sulfide production from cysteine in blood is catalyzed by iron and vitamin B(6). Commun Biol 2:194

Yin WL, Yin WG, Huang BS, Wu LX (2017) Neuroprotective effects of lentivirus-mediated cystathionine-beta-synthase overexpression against 6-OHDA-induced parkinson's disease rats. Neurosci Lett 657:45–52

Yu YC, Kuo CL, Cheng WL, Liu CS, Hsieh M (2009) Decreased antioxidant enzyme activity and increased mitochondrial DNA damage in cellular models of Machado-Joseph disease. J Neurosci Res 87(8):1884–1891

Zafeiriou DI, Augoustides-Savvopoulou P, Haas D, Smet J, Triantafyllou P, Vargiami E et al (2007) Ethylmalonic encephalopathy: clinical and biochemical observations. Neuropediatrics 38(2): 78–82

Zeng L, Zhang D, McLoughlin HS, Zalon AJ, Aravind L, Paulson HL (2018) Loss of the spinocerebellar ataxia type 3 disease protein ATXN3 alters transcription of multiple signal transduction pathways. PLoS One 13(9):e0204438

Zhang K, Donnelly CJ, Haeusler AR, Grima JC, Machamer JB, Steinwald P et al (2015) The C9orf72 repeat expansion disrupts nucleocytoplasmic transport. Nature 525(7567):56–61

Zhang L, Wei K, Xu J, Yang D, Zhang C, Wang Z et al (2016) Belamcanda chinensis (L.) DC-An ethnopharmacological, phytochemical and pharmacological review. J Ethnopharmacol 186:1–13

Zhao W, Zhang J, Lu Y, Wang R (2001) The vasorelaxant effect of H(2)S as a novel endogenous gaseous K(ATP) channel opener. EMBO J 20(21):6008–6016

Ziosi M, Di Meo I, Kleiner G, Gao XH, Barca E, Sanchez-Quintero MJ et al (2017) Coenzyme Q deficiency causes impairment of the sulfide oxidation pathway. EMBO Mol Med 9(1):96–111

Zivanovic J, Kouroussis E, Kohl JB, Adhikari B, Bursac B, Schott-Roux S et al (2019) Selective Persulfide detection reveals evolutionarily conserved antiaging effects of S-Sulfhydration. Cell Metab 30(6):1152–70 e13

The Role of H_2S in Atherosclerosis and Associated Cardiometabolic Comorbidities

Odysseia Savvoulidou, Turtushikh Damba, Daniel F. J. Ketelhuth, and Maria Peleli

Contents

The original version of this chapter was previously published without open access. A correction to this chapter is available at https://doi.org/10.1007/164_2026_795

Authors Odysseia Savvoulidou and Turtushikh Damba have equally contributed to this chapter.

O. Savvoulidou · T. Damba · D. F. J. Ketelhuth · M. Peleli (✉)
Cardiovascular and Renal Research Unit, Department of Molecular Medicine, University of Southern Denmark, Odense, Denmark
e-mail: mpeleli@sdu.dk

A. Papapetropoulos, C. Szabo (eds.), *Hydrogen Sulfide: A Toxicant with Healing Potential*, Handbook of Experimental Pharmacology 292,
https://doi.org/10.1007/164_2025_760

Abstract

Hydrogen sulfide (H_2S) is an important gasotransmitter with multiple roles and is involved in several pathophysiological processes, including atherosclerosis and associated cardiometabolic comorbidities. This chapter examines the molecular mechanisms by which H_2S can influence the development and progression of atherosclerosis, including its effects on vascular tone, angiogenesis, oxidative stress, and inflammation, which can also affect atherosclerotic plaque stability. Moreover, we describe how H_2S affects the outcomes of cardiometabolic comorbidities associated with atherosclerosis, such as diabetes, hypertension, cardiac ischemia-reperfusion injury (IRI), stroke, metabolic-associated fatty liver disease (MAFLD), and kidney disease. Finally, we discuss how H_2S levels could potentially serve as biomarkers and the potential of increasing H_2S levels (both donors and metabolic pathway modulators) as promising therapeutic agents for improving vascular function, reducing plaque formation, and mitigating cardiovascular disease risk.

Keywords

Atherosclerosis · Cardiometabolic comorbidities · Hydrogen sulfide · Inflammation · Oxidative stress · Plaque stabilization

Abbreviations

ABCA1	ATP-binding cassette transporter A1
AOAA	Aminooxyacetic acid
Bax/Bcl-2	Bcl-2-associated X protein, B-cell lymphoma 2
CAD	Carotid artery disease
CAT	Cysteine aminotransferase
CBS	Cystathionine β-synthase
CKD	Chronic kidney disease
CO	Carbon monoxide
CSE	Cystathionine γ-lyase
CVD	Cardiovascular diseases
Cys-SSH	Cysteine-persulfide
EC	Endothelial Cells
ED	Endothelial Dysfunction
EDRF	Endothelium-derived relaxing factor
eNOS	Endothelial nitric oxide synthase
ER	Endoplasmic reticulum
ERK1/2	Extracellular signal-regulated kinase
ESRD	End-stage renal disease
GSSH	Glutathione persulfide
H_2S	Hydrogen sulfide
H_2S_2	Hydrogen persulfide

H_2S_3	Hydrogen trisulfide
HFD	High-fat diet
HHcy	Hyperhomocysteinemia
Homocys	Homocysteine
HuR	Human antigen R
ICAM-1	Intercellular adhesion molecule-1
IL-10	Interleukin-10
IL-1β	Interleukin 1 beta
IL-6	Interleukin-6
iNOS	Inducible nitric oxide synthase
IRI	Ischemia-reperfusion injury
K_{ATP}	ATP-sensitive potassium channels
KCNQ	Potassium voltage-gated channel subfamily Q member 1
KO	Knockout
L-Cys	L-Cysteine
LDL	Low-density lipoprotein
LOX-1	Lectin-like oxidized low-density lipoprotein receptor-1
MAFLD	Metabolic-associated fatty liver disease
MAPK	Mitogen-activated protein kinase
Na_2S	Sodium sulfide
NaHS	Sodium hydrosulfide
NASH	Non-alcoholic steatohepatitis
NF-κB	Nuclear factor kappa-light-chain-enhancer of activated B cells
NLRP3	Nucleotide-binding domain, leucine-rich–containing family, pyrin domain–containing-3
NO	Nitric oxide
OA	Osteoarthritis,
oxLDL	Oxidized LDL
PAD	Periphery artery disease
PAG	D, L-Propargylglycine
PDI	Protein disulfide isomerase
PI3K	Phosphatidylinositol-3-kinase
PPARα	Peroxisome proliferator-activated receptor alpha
PPARγ	Peroxisome proliferator-activated receptor gamma
RNS	Reactive nitrogen species
ROS	Reactive oxygen species
shRNA	Short hairpin RNA
Sirt1	Sirtuin 1
TFEB	Transcription factor EB
TNF	Tumor necrosis factor
Trx	Thioredoxin
VEGFR2	Vascular endothelial growth factor receptor-2
VSMC	Vascular smooth muscle cells
3-MP	3-Mercaptopyruvate
3-MST	3-Mercaptopyruvate sulfurtransferase

1 The Physiology of Hydrogen Sulfide (H_2S) Production

By the late 1990s, H_2S was recognized as an important gasotransmitter alongside nitric oxide (NO) and carbon monoxide (CO), whose production and metabolism are predominantly regulated by enzymes (Giuffre and Vicente 2018; Murphy et al. 2019).

H_2S is produced through nonenzymatic and enzymatic pathways and can either exert immediate biological effects or be stored for later release. Mammalian cells contain two sulfur stores, acid-labile sulfur and sulfane sulfur (Wang 2012). The levels, both in free and bound forms, typically range from nanomolar to micromolar concentrations (>100 μM), exhibiting substantial variability depending on the tissue type and measurement method, such as the notably higher concentrations found in the aorta (Cirino et al. 2023; Kabil and Banerjee 2014).

H_2S production can be upregulated in response to physiological stimuli such as shear stress and hypoxia, and its bioavailability and metabolism can be influenced by microbiota, and dysregulation of H_2S levels is linked to numerous diseases, including cardiovascular and metabolic disorders, and neurodegenerative diseases (Predmore et al. 2012; Shen et al. 2013). Furthermore, H_2S can modulate the activity of ion channels, transcription factors, and enzymes involved in cellular metabolism, and it also affects mitochondrial respiration, apoptosis, biogenesis, and morphology (Murphy et al. 2019; Predmore et al. 2012).

1.1 Synthetic Pathways of H_2S

Yang et al. (2019) indicated that the nonenzymatic generation of H_2S in most human tissues utilizes iron and vitamin B6, playing key role in the process by using cysteine as a substrate. Additionally, nonenzymatic processes involve dietary compounds, such as diallyl disulfide and sulforaphane (Benavides et al. 2007; Kashfi and Olson 2013), as well as contributions from gut microbiota (Tomasova et al. 2016). These factors collectively influence H_2S levels through various biochemical pathways, with over 65% of plasma H_2S and sulfane sulfur being derived from gut microbiota (Shen et al. 2013).

Nonenzymatic H_2S – produced by dietary sulfur compounds, chemical reactions, and especially gut microbiota – is an underrecognized modulator of cardiovascular health. The gut microbiota is the body's largest source of H_2S, which is increasingly linked to atherosclerosis through mechanisms involving inflammation, barrier dysfunction, and microbial translocation (Gui et al. 2021). Although the precise cardiovascular effects of colon-derived H_2S remain incompletely defined, targeting microbial sulfur metabolism offers promising therapeutic potential (Cirino et al. 2023; Dilek et al. 2020).

Conversely, enzymatic H_2S production is well established and occurs in various mammalian organs, including the liver, kidney, brain, and heart. This process primarily involves three key enzymes: cystathionine β-synthase (CBS), cystathionine γ-lyase (CSE), and 3-mercaptopyruvate sulfurtransferase (3-MST) (Cirino et al. 2023; Yang et al. 2019). 3-MST is present in both mitochondria and the cytosol. In contrast, CSE is found exclusively in the cytosol, however studies have shown that an increase in intracellular calcium can induce its translocation to the mitochondria (Cirino et al. 2023; Kanagy et al. 2017; Nagahara et al. 1998). CBS is also found to a large extent in the cytosol, but under certain conditions, such as hypoxia, an increase in CBS in the mitochondrial fraction has also been reported (Teng et al. 2013). Elevated levels of H_2S are observed in many types of cancer (Augsburger and Szabo 2020; Szabo 2016, 2021), while reduced levels are commonly associated with cardiovascular diseases (CVD) (Mani et al. 2013; Shen et al. 2015).

CBS and CSE are involved in H_2S generation, cysteine production, and catabolism through the transsulfuration pathway, which responds to various stimuli, such as hormonal changes and conditions that increase cyclic AMP levels (Kabil and Banerjee 2014; Murphy et al. 2019). 3-MST catalyzes the production of H_2S through a two-step pathway (Nagahara 2018). Initially, cysteine aminotransferase (CAT) converts cysteine and α-ketoglutarate into 3-mercaptopyruvate (3-MP), and subsequently, 3-MST transfers the sulfur group from 3-MP to a cysteine residue, forming a pyruvate and a persulfide, which can react with thiols or be reduced by thioredoxin (Trx) to produce H_2S (Kanagy et al. 2017; Kimura 2011; Mikami et al. 2011; Murphy et al. 2019; Nagahara 2018).

3-MST is also involved in the production of several persulfide species, such as cysteine-persulfide (Cys-SSH) and glutathione persulfide (GSSH) in the cytosol, through the utilization of hydrogen persulfide (H_2S_2) and hydrogen trisulfide (H_2S_3) (Kimura et al. 2017), which are essential for bioenergetics by generating intramitochondrial H_2S production, which helps maintain mitochondrial electron flow and supports cellular energy production (Abdollahi Govar et al. 2020; Modis et al. 2013b). Studies have shown that deficiency of this enzyme leads to reduced cellular antioxidants and a corresponding increase in reactive oxygen species (ROS) (Maria Peleli et al. 2020; Williams et al. 2003), and during oxidative stress, the 3-MST-derived intramitochondrial H_2S is diminished, resulting in compromised bioenergetics and cytoprotection (Modis et al. 2013a).

It is worth noting that in biological systems, there are three main biochemical pathways for direct H_2S signaling. H_2S can interact with ROS and reactive nitrogen species (RNS) directly, bind to metal centers, participate in electron transfer, or influence protein function through a process known as S-sulfhydration or persulfidation, where it modifies cysteine residues (Filipovic 2015). S-sulfhydration is pivotal as it supports metabolic functions by modulating enzyme activities, contributes to mitochondrial respiration and overall bioenergetics, and plays a crucial role in protein quality control, which is essential for maintaining cellular homeostasis and extending longevity (Nishimura et al. 2024). Zivanovic J. et al. revealed that S-sulfhydration exerts antiaging effects by mitigating age-related cellular damage (Zivanovic et al. 2019), and many studies highlight the key role of CSE in inducing S-sulfhydration, which contributes to the preservation of endothelial cell function (Bibli et al. 2019).

2 H_2S, Vascular Reactivity, and Angiogenesis

In the cardiovascular system, H_2S generally induces vasodilation and angiogenesis, reduces blood pressure and vascular inflammation, and protects against myocardial IRI (Shibuya et al. 2009; Zhao and Wang 2002). However, one should emphasize, that this is a rather simplified statement that has to be interpreted cautiously since exceptions apply both for the enzymatically (endogenously) derived H_2S and for the H_2S derived from chemical donors.

Solid evidence confirms that CSE, CBS, and 3-MST are all expressed in the vascular wall, although their expression levels and cellular localization vary considerably depending on the vascular bed and specific vessel type examined (Mendiola et al. 2021). However, the endogenously produced H_2S has complex effects in the vasculature, with CSE-derived H_2S potentially causing either vasodilation or vasoconstriction (Cirino et al. 2023; Peleli et al. 2023; Szabo 2017). On the other hand, 3-MST-derived H_2S has a minimal impact on vascular reactivity, playing a limited role in blood vessel dilation and constriction (Peleli et al. 2020).

The endothelium plays a vital role in protecting the vasculature from inflammatory insult, regulating blood volume and electrolyte homeostasis, and mediating angiogenesis and vascular remodeling. Numerous studies have established that H_2S is an endothelium-derived relaxing factor (EDRF) produced in the endothelium via CSE and 3-MST pathways and released to hyperpolarize and relax vascular smooth muscle cells (VSMC) (Bibli et al. 2019; Shibuya et al. 2009; Wang et al. 2015). H_2S often enhances the NO/cGMP/sGC/PKG pathway in the EC, which is essential for the execution of vasodilation. The interaction between H_2S and NO is particularly important, as these molecules can potentiate each other's effects to different extents (Papapetropoulos et al. 2009; Szabo 2017). Another hallmark study by Coletta et al. showed that H_2S and NO are mutually dependent in promoting endothelium-dependent vasorelaxation. Their interaction converges at the cGMP–PKG pathway, highlighting a cooperative mechanism essential for vascular homeostasis (Coletta et al. 2012). Last but not least, it is important to mention that this positive feedback loop between H_2S and endothelial NO synthase (eNOS)-derived NO is not always the case, since some studies have indicated the opposite; that H_2S attenuates eNOS and NO bioavailability, resulting therefore in vasoconstriction (Peleli et al. 2023; Prieto-Lloret and Aaronson 2015; Szijarto et al. 2018).

In VSMC, H_2S has been shown to induce relaxation of blood vessels by opening adenosine triphosphate-sensitive potassium (K_ATP) channels, reducing intracellular calcium levels, and decreasing vascular tone (Bir et al. 2012). This effect helps lower blood pressure and prevent vascular hypertrophy (Kimura 2011). However, other studies have shown that H_2S can also elicit vasoconstriction under specific conditions. For example, studies have shown that the H_2S donor NaHS reverses β-adrenoceptor–induced relaxation by downregulating cAMP in rat aortic rings (Lim et al. 2008) whereas another study showed that the same H_2S donor causes constriction of cerebral arteries via inhibition of adenylyl cyclase/cAMP signaling (Li et al. 2015). Similar findings in human and mouse cerebral arterioles have also been recently reported by us, suggesting that CSE triggers vasoconstriction, which is significantly attenuated by the H_2S donor sodium hydrosulfide (NaHS) (Peleli et al. 2023). Additionally, in rat aorta, high H_2S concentrations reduced ATP and limited K_ATP-mediated relaxation, instead triggering metabolic inhibition-linked constriction (Kiss et al. 2008), suggesting that the net vascular effect of H_2S depends critically on its concentration and the balance between K^+ channel activation and cAMP/ATP signaling.

In addition to the spatial differences in expression across vascular beds, several physiological stimuli can dynamically regulate the expression of H_2S-producing enzymes. Laminar shear flow and ROS are known to boost CSE and 3-MST expression. Meanwhile, calcium levels and Sp1 enhance CSE expression in smooth muscle cells (Kanagy et al. 2017). CSE helps regulate SMC proliferation by modulating key signaling pathways involved in cell growth and division. CBS performs a similar function to CSE in VSMC, contributing to the formation of new blood vessels through the production of H_2S (Predmore et al. 2012).

In relation to angiogenesis, which is the process of new blood vessel formation from pre-existing vessels, H_2S acts as a significant endogenous stimulator by promoting the proliferation, survival, migration, and tube formation of EC, partly by enhancing vascular NO signaling and activating endothelial NO synthase (eNOS) (Kimura 2011; Papapetropoulos et al. 2009; Szabo 2017; Wang et al. 2015). In particular, Papapetropoulos et al. demonstrated that a key mechanism of H_2S-induced angiogenesis involves the upregulation and phosphorylation of the vascular endothelial growth factor receptor-2 (VEGFR2), which amplifies the cellular response to VEGF, a crucial growth factor in blood vessel formation (Papapetropoulos et al. 2009). H_2S also activates the phosphatidylinositol-3-kinase (PI3K)/Akt pathway, leading to increased endothelial cell proliferation and survival and stimulating the extracellular signal-regulated kinase (ERK1/2) pathway, which facilitates cell migration and proliferation (Papapetropoulos et al. 2009).

In conclusion, H_2S is a vital gasotransmitter with diverse physiological roles in the vascular system, including the modulation of vascular tone, promotion of angiogenesis, and exertion of anti-inflammatory effects (Papapetropoulos et al. 2009; Wang 2012). As described in detail in the following paragraphs, H_2S also plays a critical role in atherosclerosis by influencing various cell types within atherosclerotic plaques, such as the endothelium, VSMC, and inflammatory cells (particularly macrophages). Understanding the synthesis pathways and specific functions of H_2S in different cell types of the vascular wall, both under physiological conditions and within atherosclerotic lesions, is crucial for harnessing its therapeutic potential. Continued research in this area promises to uncover new insights and innovative treatments for CVD, including atherosclerosis.

3 Effects of H_2S on the Different Cell Types of the Atherosclerotic Plaque

3.1 H_2S and Endothelial Cells (EC)

Bibli et al. have demonstrated the importance of CSE in preventing endothelial dysfunction (ED) during atherosclerosis and identified human antigen R (HuR) as a key molecular target of endogenous H_2S (Bibli et al. 2019). Hyperhomocysteinemia (HHcy) is another independent risk factor for atherosclerosis and related CVD, causing vascular damage, but NaHS was found to reduce Hcy levels, decrease oxidative damage, and protect against HHcy-induced cellular toxicity, by attenuating endothelial endoplasmic reticulum (ER) stress (Jiang et al. 2021; Mani et al. 2014). Oxidative stress, which contributes to atherosclerosis by causing ED and foam cell formation, is countered by H_2S through enhanced antioxidant levels, ROS scavenging, and nuclear factor kappa-light-chain-enhancer of activated B cells (NF-κB) and nucleotide-binding domain, leucine-rich–containing family, pyrin domain–containing-3 (NLRP3) inflammasome inhibition (Liu et al. 2013; Mani et al. 2014; Sun et al. 2019). Further mechanistic insight was provided by a recent study showing that H_2S promotes sulfhydration of protein disulfide isomerase (PDI), which alleviates homocysteine-induced ER stress in EC and suppresses atherogenesis (Jiang et al. 2021).

3.2 H_2S and VSMC

H_2S inhibits VSMC proliferation and migration, induces VSMC apoptosis, and interacts with the MAPK pathway to regulate vascular remodeling (Lv et al. 2021; Yang et al. 2006). Moreover, H_2S protects against vascular calcification, a frequent complication of atherosclerosis, by inhibiting osteoblastic differentiation and calcification, and potentially reducing atherosclerosis development by upregulating inducible NO synthase (iNOS) expression, NO generation, and protein S-nitrosylation (Lin et al. 2016; S. Mani et al. 2014). Notably, H_2S was identified as a potent inhibitor of phosphate-induced vascular calcification in VSMC, preventing calcification by maintaining the contractile phenotype and suppressing osteogenic gene expression (Zavaczki et al. 2011).

VSMC-derived H_2S activates the transcription factor EB (TFEB), which mediates autophagy, helping to maintain plaque integrity and preventing its progression (Chen et al. 2022). Recent evidence shows that VSMC-derived H_2S activates the transcription factor EB (TFEB) via sulfhydration, which in turn triggers autophagy, enhances collagen production, reduces apoptosis, and ultimately promotes atherosclerotic plaque stabilization, thus lowering the risk of acute cardiovascular events, such as myocardial infarction (Chen et al. 2022).

H_2S regulates vascular tone through both endothelium-independent and -dependent mechanisms, inducing vasodilation by activating VSMC ATP-sensitive potassium (KATP) channels and, to a lesser extent, potassium voltage-gated channel subfamily Q member 1 (KCNQ) channels, and has been shown to inhibit the proliferation and migration of VSMC cells, regulate inflammation, protect the vascular endothelium, and modulate blood pressure, making it a key player in the prevention and management of atherosclerosis (Sun et al. 2019). It is also important to note that CBS is a key modulator of mitochondrial behavior, crucial for maintaining endothelial cell function, and that loss of its function is associated with cardiovascular disorders (Rao et al. 2020).

3.3 H_2S and Inflammatory Cells

H_2S has emerged as a complex but crucial mediator of inflammation, comparable to NO and CO (Wallace et al. 2012). At low concentrations, H_2S exerts anti-inflammatory effects by reducing oxidative stress and inhibiting pro-inflammatory pathways such as NF-κB, while promoting interleukin-10 (IL-10) production (Longen et al. 2016). However, H_2S signaling is dose- and context-dependent – prolonged or excessive H_2S can promote inflammasome activation or macrophage apoptosis. This dual role highlights the need for balanced H_2S regulation to control macrophage-driven inflammation in diseases like atherosclerosis (Sun et al. 2021).

The downregulation of the CSE/H_2S pathway plays a significant role, as studies have shown that ox-LDL disrupts CSE/H_2S homeostasis, leading to ox-LDL-induced inflammation, and that CSE-deficient mice develop more severe atherosclerosis (Wang et al. 2013). The CSE/H_2S system is regulated by miR-216a, which suppresses cholesterol efflux in THP-1 macrophage-derived foam cells by targeting CSE and downregulating the PI3K/AKT/ ATP-binding cassette transporter A1 (ABCA1) pathway (Gong et al. 2016). This led to reduced ABCA1 expression, lower cholesterol efflux, and increased intracellular cholesterol content, reinforcing the role of CSE/H_2S in lipid homeostasis.

Supporting these findings, Du et al. showed that in ApoE KO mice on a high-fat diet, treatment with H_2S donors (NaHS or GYY4137) directly sulfhydrated Sirt1, enhancing its deacetylase activity, which in turn lowered macrophage foam cell formation, reduced plaque burden, decreased macrophage infiltration, and reduced aortic inflammation (Du et al. 2019). Moreover, in hepatocytes and in livers from ApoE KO mice, H_2S has been shown to enhance ABCA1 expression by promoting nuclear translocation of PPARα, increasing reverse cholesterol transport, lowering serum lipids, and reducing atherosclerotic plaque size (Li et al. 2016). In adipocytes, H_2S sulfhydrates PPARγ, thereby enhancing its nuclear translocation and transcriptional activity (Cai et al. 2016), suggesting that a similar mechanism operates in macrophages to promote cholesterol efflux and impede foam cell formation. Notably, both PPARα and PPARγ exert anti-inflammatory, endothelial-protective, and VSMC-stabilizing effects, contributing to plaque stabilization and slowing atherosclerosis progression (Duval et al. 2002; Luan et al. 2023).

Beyond CSE, other enzymes such as 3-MST and CBS also contribute to endogenous H_2S production, though their specific roles in vascular pathology are less well understood. A recent study by Zampas et al. highlighted the protective role of

3-MST against metabolic syndrome and vascular inflammation. In this study, 8-week-old mice were subjected to a high-fat diet (HFD) for 15 weeks, followed by eNOS inhibition to induce hypertension and full metabolic syndrome. 3-MST KO mice showed worsened glucose intolerance, increased body weight, and immune activation in the vasculature, indicated by T-cell infiltration. These changes were associated with ED and reduced aortic H_2S levels. Thus, pharmacological restoration of H_2S with GYY4137 improved glucose handling, reduced blood pressure, and cardiac fibrosis, and enhanced vascular relaxation (Zampas et al. 2025).

The role of CBS, the third major H_2S-producing enzyme, is less well understood in relation to immune cell activation and atherosclerosis. However, current published data indicate that CBS-derived H_2S plays an anti-inflammatory role in immune cells. In a mouse model of cerebral ischemia, reduced CBS expression in microglia promoted M1 polarization, whereas restoring CBS or treating with H_2S donors shifted microglia toward an M2 phenotype via AMPK activation (Du et al. 2014). Another study (Lendoiro-Cino et al. 2023) aimed to mimic the inflammatory environment of osteoarthritis (OA) in patients with type 2 diabetes (T2D), a condition characterized by chronic metabolic stress and low-grade inflammation. Under high-glucose conditions, THP-1-derived macrophages showed reduced expression of H_2S-producing enzymes, including CBS, and adopted a pro-inflammatory M1 phenotype. Overall, these findings indicate that CBS may support anti-inflammatory macrophage programming and limit immune-driven vascular damage. This suggests a potential atheroprotective role of CBS, although further studies are needed to confirm its relevance in vascular inflammation and atherosclerosis.

In relation to different cell targets that are crucial for mediating the anti-inflammatory effects of H_2S, it has been shown that H_2S inhibits leukocyte adherence to the vascular endothelium, a key process in inflammation, by downregulating adhesion molecules like intercellular adhesion molecule-1 (ICAM-1) suppresses inflammatory responses through the anti-inflammatory protein annexin-1, which is crucial for resolving inflammation (Gemici and Wallace 2015; Wallace et al. 2012; Zanardo et al. 2006). H_2S also inhibits phosphodiesterase activity, suppresses NF-κB, and scavenges oxidants and peroxynitrite, which may contribute to its anti-inflammatory effects (Wallace et al. 2012). Emerging evidence suggests that H_2S plays a crucial role in adipocyte function by promoting adipogenesis and influencing inflammation, with its effects including the promotion of adipocyte differentiation and the regulation of macrophage polarization between anti-inflammatory M2 and pro-inflammatory M1 phenotypes, thus presenting a potential for targeted therapeutic interventions in metabolic disorders (Comas et al. 2019).

A summary of the role of H_2S in atherosclerosis in relation to the molecular mechanisms taking place in the different vascular wall cell types is shown in Fig. 1.

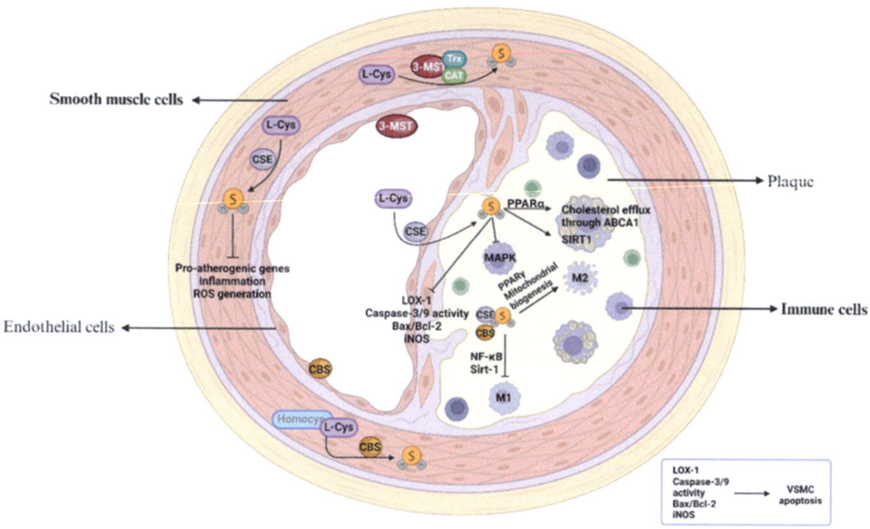

Fig. 1 Suggested mechanisms by which H_2S-producing enzymes modulate vascular wall cell function in atherosclerosis. In EC, CBS, CSE, and 3-MST generate sulfur species that suppress pro-atherogenic gene expression, inflammation, and ROS production. In smooth muscle cells, the same enzymes inhibit MAPK-mediated pathways that lead to vascular calcifications and promote apoptotic signaling via LOX-1, Bax/Bcl-2, and caspase-3/9, contributing to VSMC apoptosis and plaque remodeling. In macrophages, H_2S produced by CBS and CSE inhibits NF-κB–driven inflammation and enhances Sirt1 activity in M1 cells. In M2 macrophages, H_2S promotes anti-inflammatory signaling and cholesterol efflux through ABCA1 expression. Although not yet directly demonstrated in plaque macrophages, studies in hepatocytes and adipocytes suggest that H_2S may induce PPARα nuclear translocation and PPARγ sulfhydration, and the same transcription factors have been shown to play a critical role in plaque macrophages regulating cholesterol efflux, mitochondrial biogenesis, and foam cell formation. (The figure was created using BioRender (https://biorender.com))

4 H_2S in Cardiometabolic Diseases: Beyond the Vascular Role

4.1 H_2S and Cardiometabolic Diseases

Cardiometabolic diseases encompass a range of disorders that affect the cardiovascular system and metabolic processes, including diabetes, obesity, hypertension, cardiac IRI, stroke, metabolic-associated fatty liver disease (MAFLD), and different types of kidney diseases. These conditions are significant risk factors for the development of atherosclerosis (Targher et al. 2021). The protective effects of H_2S in these diseases have been extensively studied and are schematically summarized in Fig. 2.

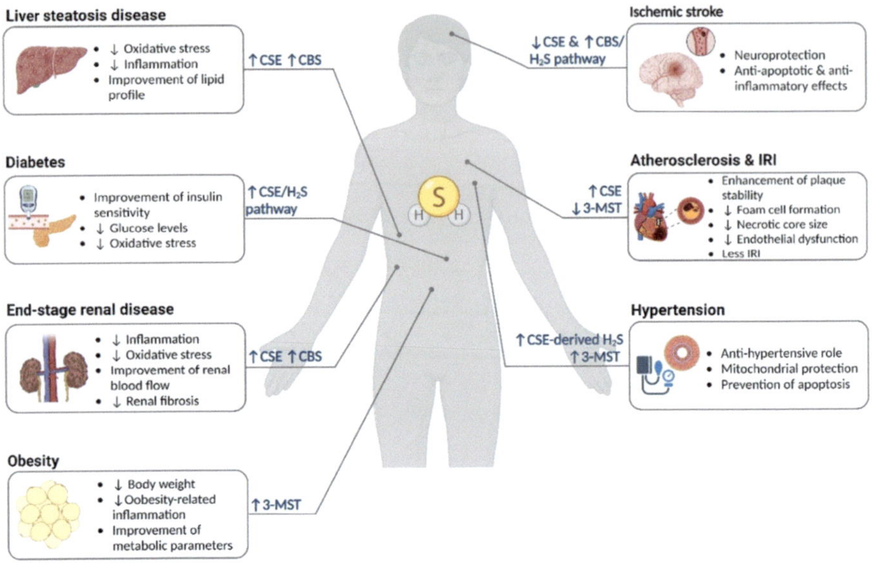

Fig. 2 H_2S, atherosclerosis, and other cardiometabolic comorbidities. Enzymatically produced H_2S exerts pleiotropic protective effects through multiple mechanisms, including reducing oxidative stress, inflammation, and fibrosis while enhancing vasodilation and metabolic regulation. Modulation of H_2S levels presents a promising therapeutic strategy for a wide range of cardiometabolic conditions. (The figure was created by using the Biorender software https://biorender.com)

4.2 Obesity and Diabetes

Obesity is associated with chronic inflammation and oxidative stress, contributing to the development of atherosclerosis and type 2 diabetes (Wang 2012). H_2S has been shown to exert anti-inflammatory and antioxidant effects in obese and diabetic models, improving vascular function and reducing the risk of atherosclerosis (Du et al. 2019; Xie et al. 2016). Enhanced activity and expression of 3-mercaptopyruvate sulfur transferase (3-MST) can reduce body weight and improve metabolic parameters in obese mice, suggesting that H_2S could be beneficial in managing obesity-related vascular complications (Katsouda et al. 2022). Similarly, Coletta et al. demonstrated that in a streptozotocin (STZ)-induced diabetes model, hyperglycemia impaired 3-MST-dependent H_2S production in EC, leading to defective angiogenesis and mitochondrial dysfunction. These impairments were rescued by treatment with 3-mercaptopyruvate and lipoic acid, highlighting the therapeutic potential of restoring 3-MST/H_2S signaling in diabetic vascular complications (Coletta et al. 2015).

In diabetes, the dysregulation of glucose and lipid metabolism leads to ED, inflammation, and oxidative stress, all of which contribute to the development of atherosclerosis (Sena et al. 2018). Regulation (most commonly upregulation, but sometimes also pharmacological inhibition) of CSE-derived H_2S has been shown to improve insulin sensitivity, reduce blood glucose levels, and decrease oxidative stress in diabetic models, thereby mitigating the risk of atherosclerosis (El-Sayed et al. 2016; Geng et al. 2013; Yang et al. 2013). In agreement with these findings, a previous study in 2011 (Suzuki et al. 2011) demonstrated that hyperglycemia increases mitochondrial ROS production in EC, leading to enhanced H_2S catabolism and ED. H_2S replacement therapy or CSE overexpression preserved mitochondrial function, reduced oxidative stress and DNA damage, and restored endothelium-dependent vasodilation, highlighting the protective role of H_2S in hyperglycemia-induced vascular injury. Recently, Sun et al. demonstrated that H_2S donors could prevent ED and reduce vascular inflammation in diabetic rats, highlighting the therapeutic potential of H_2S in diabetes-related vascular complications (Sun et al. 2023).

4.3 Hypertension and Cardiac IRI

Hypertension is a major risk factor for atherosclerosis (Mani et al. 2014). H_2S has been shown to exert vasodilatory effects, reduce blood pressure, and improve vascular function in hypertensive models (Altaany et al. 2013; Szabo and Papapetropoulos 2017). Recent studies have shown that H_2S donors can significantly lower blood pressure and improve vascular compliance (Li et al. 2018a). When it comes to the specific biological role of the two main H_2S-producing enzymes in the cardiovascular system, CSE-derived H_2S has anti-hypertensive, anti-atherosclerotic, and cardioprotective effects in CSE KO mice, whereas the role of 3-MST is more complicated (Elrod et al. 2007; Mani et al. 2013; Wang et al. 2013; Yang et al. 2008).

We have recently shown that aged 3-MST KO mice are characterized by significantly elevated blood pressure, cardiomegaly, and obesity, but at a young age, they are actually protected from cardiac (IRI) (Katsouda et al. 2022; Maria Peleli et al. 2020). At the molecular level, a hallmark mechanism leading to atherosclerosis and cardiac IRI is the intracellular elevation of reactive oxygen species (ROS) derived mainly from mitochondria and NADPH oxidases (Batty et al. 2022; Peoples et al. 2019). We and others have shown that both CSE and 3-MST play a significant antioxidant role in various cardiovascular organs and organisms ranging from mice to zebrafish (Barrow et al. 2022; Katsouda et al. 2020; Mani et al. 2013; Maria Peleli et al. 2020).

4.4 Stroke

Stroke remains a leading cause of mortality worldwide. Ischemic stroke – caused by cerebral artery occlusion – accounts for approximately 87% of cases, while hemorrhagic stroke, due to blood vessel rupture, comprises the remaining 13% (Feigin et al. 2022). Atherosclerosis is a key underlying cause of ischemic stroke. Several controllable risk factors contribute to both atherosclerosis and ischemic stroke, including obesity, diabetes, hypertension, and hyperlipidemia. Uncontrollable factors, such as age, gender, ethnicity, and family history, also influence stroke risk (Gopalakrishnan et al. 2019).

H_2S plays a dual role in ischemic stroke: therapeutically, it has been proposed to exert neuroprotective effects by reducing oxidative stress, inflammation, and neuronal apoptosis, and additionally, it mitigates excitotoxicity by influencing glutamate signaling and maintaining ion homeostasis (Gopalakrishnan et al. 2019). Conversely, excessive or dysregulated H_2S production can exacerbate stroke injury, contributing to inflammation and neuronal damage, and potentially aggravating stroke outcomes (Gopalakrishnan et al. 2019). Many studies have highlighted the roles of CBS and H_2S in various physiological and pathological conditions, particularly during hypoxia, IRI, and cerebral inflammation (Omorou et al. 2022). CBS/H_2S has been shown to exert neuroprotective, anti-apoptotic, and anti-inflammatory effects in the brain and other tissues, suggesting that maintaining or enhancing CBS activity could be beneficial in mitigating the damage caused by hypoxia, ischemic stroke, and other related conditions (Omorou et al. 2022).

4.5 Metabolic-Associated Fatty Liver Disease (MAFLD)

MAFLD is a strong predictor of atherosclerosis (Tsutsumi et al. 2021). Hence, non-alcoholic steatohepatitis (NASH), an earlier stage of MAFLD, has also been proposed to contribute to the development of atherosclerosis due to increased secretion of atherogenic and thrombotic agents (Stols-Gonçalves et al. 2019). The newly formed term MAFLD (metabolic-associated fatty liver disease) is a strong predictor of atherosclerosis (Tsutsumi et al. 2021). NASH, an inflammatory stage of MAFLD, contributes to the development of atherosclerosis due to the promoted secretion of atherogenic and thrombotic agents (Stols-Gonçalves et al. 2019).

H_2S appears to play a protective role in liver steatosis by mitigating oxidative stress and inflammation, both of which are key contributors to MAFLD pathogenesis (Wu et al. 2020). Furthermore, H_2S donors have been shown to improve lipid profiles, reduce hepatic steatosis, and decrease plaque formation in animal models (Paul and Snyder 2015).

The H_2S-producing enzymes CSE, CBS, and 3-MST play critical and opposing roles in MAFLD development. H_2S appears to play a protective role in liver steatosis by mitigating oxidative stress and inflammation, both of which are key contributors to the pathogenesis of MAFLD (Wu et al. 2020). Furthermore, H_2S donors have been shown to improve lipid profiles, reduce hepatic steatosis, and decrease plaque formation in animal models (Paul and Snyder 2015). The H_2S-producing enzymes CSE, CBS, and 3-MST play critical and opposing roles for the development of MAFLD. CBS plays a clear protective role and attenuates the progression of MAFLD, whereas the role of CSE and 3-MST appears to be more complicated. Studies have shown that CBS upregulation ameliorates liver metabolic damage and improves overall liver function and metabolism (Dong et al. 2023). In contrast, CBS knockout (KO) mice exhibit higher levels of oxidative stress, fibrosis, and steatosis in their liver, further highlighting the protective function of CBS in maintaining liver health (Robert et al. 2005).

Recent studies from the Papapetropoulos group have shed more light on the role of 3-MST and CSE in MAFLD by showing that global 3-MST KO mice are characterized by steatotic livers upon HFD or aging, whereas double global 3-MST/CSE KO develop spontaneous liver steatosis (Katsouda et al. 2020, 2025). Interestingly, these results contrast with earlier findings from a study using partial 3-MST inhibition via heterozygous KO or short hairpin RNA (shRNA), which suggested that 3-MST promotes the progression of MAFLD (Li et al. 2018b). This discrepancy underscores how the choice of genetic model can significantly influence our understanding of an enzyme's role in disease.

Similarly, contradictory findings exist for CSE depending on the cellular origin in the liver. In particular, a recent study in Hepatology showed that hepatocyte-specific CSE KO mice have an ameliorated MAFLD phenotype (Xu et al. 2022) whereas other cellular studies focusing on hepatic stellate cells show that CSE inhibition has a clear anti-fibrotic effect (Damba et al. 2019, 2024), which, however, remains to be confirmed in in vivo animal models.

4.6 Kidney Disease

Atherosclerosis can lead to various kidney pathologies due to reduced blood flow and subsequent damage to renal tissues (Chade et al. 2005). Interestingly, a reverse correlation also exists, meaning that serious kidney pathologies such as chronic kidney disease, renal IRI, and end-stage renal disease can accelerate the development of atherosclerotic lesions in the cardiovascular system and increase morbidity and mortality (Valdivielso et al. 2019). H_2S is known to exert several beneficial effects, including anti-inflammatory, antioxidant, and vasodilatory actions, which could mitigate the impact of atherosclerosis on the kidneys (Peleli et al. 2022). H_2S has been shown to reduce inflammation and oxidative stress in kidney tissues, which are key contributors to chronic kidney disease (CKD) progression (Aminzadeh and Vaziri 2012; Shirazi et al. 2019).

Another intriguing aspect of H_2S biology in inflammation is its role in the kidney-brain axis, an emerging concept describing the bidirectional crosstalk between renal and central nervous system functions. In this context, H_2S has been shown to attenuate the production of pro-inflammatory cytokines, such as interleukin-1 beta (IL-1β), tumor necrosis factor (TNF), and interleukin-6 (IL-6), as well as neurotoxins, thereby limiting the parallel progression of kidney and brain disorders (Calabrese et al. 2020). Moreover, H_2S may improve renal blood flow and protect against ROS-induced ischemic damage (Bos et al. 2013). Finally, research suggests that H_2S can attenuate renal fibrosis and improve kidney function, potentially delaying progression to end-stage renal disease (ESRD) (Song et al. 2014).

5 Role of H_2S as Biomarker and Pharmacological Potential of H_2S Donors and Inhibitors in Atherosclerosis and Associated Cardiometabolic Diseases

5.1 H_2S as a Biomarker

The role of H_2S as a biomarker extends beyond that of CVD. Its levels has been correlated with various metabolic conditions, such as diabetes and obesity, which are known risk factors for atherosclerosis (Yang et al. 2013). Monitoring the H_2S levels in patients with these conditions could provide valuable insights into their cardiovascular risk profiles and guide therapeutic interventions (Byrne and Targher 2015; Wu et al. 2020).

To evaluate whether H_2S can serve as a reliable biomarker for atherosclerosis, several critical factors must be considered – particularly, which form (species) of H_2S is being measured, the method used for its quantification, and patient-specific variables such as ethnicity and sex differences.

For example, a study published in 2018 demonstrated that African American patients with periphery or carotid artery disease (PAD or CAD) had lower plasma levels of H_2S, regardless of sex, whereas significant sex specific differences existed in a matching Caucasian group of patients (Rajpal et al. 2018).

The choice of analytical tool used to detect and quantify different types of H_2S species could also play a crucial role. One study examined patients with PAD undergoing surgical revascularization and used two methods: one measured free sulfide concentrations, and the other assessed the H_2S-producing capacity by incubating plasma with substrates and cofactors to stimulate enzymatic activity. Only the H_2S-producing capacity, and not the free sulfide levels, was inversely associated with mortality risk over 36 months (Longchamp et al. 2021). In contrast, another study measured only free plasma sulfide in patients with various vascular diseases (CAD, PAD, etc.) and found surprisingly higher levels than in controls without vascular disease (Peter et al. 2013). These contrasting findings may stem from differences in analytical techniques, the different types of measured H_2S species, or varying inclusion criteria for defining vascular disease. Moreover, the elevated plasma H_2S levels observed in vascular patients could reflect compensatory systemic responses, such as increased non-vascular H_2S production or impaired tissue utilization.

In conclusion, plasma H_2S levels may not directly mirror changes within the vascular system and their interpretation as biomarkers should be approached with caution.

5.2 Pharmacological Potential of H_2S Donors

Pharmacologically, H_2S donors and inhibitors are being actively investigated for their therapeutic potential. H_2S donors, such as sodium hydrosulfide (NaHS) and GYY4137, have shown promise in preclinical studies for their ability to release H_2S

slowly and sustainably, thus exerting protective effects against atherosclerosis (Altaany et al. 2013; Szabó and Papapetropoulos 2011). These donors have been demonstrated to inhibit endothelial cell apoptosis, reduce oxidative stress, and suppress inflammatory cytokine production, all of which are crucial for the prevention of plaque formation and progression in atherosclerosis (Kolluru et al. 2013; Mani et al. 2014). Pharmacologically, H_2S donors and inhibitors are being actively investigated for their therapeutic potential. H_2S donors, such as sodium hydrosulfide (NaHS) and GYY4137, have shown promise in preclinical studies for their ability to release H_2S slowly and sustainably, thus exerting protective effects against atherosclerosis (Altaany et al. 2013; Szabó and Papapetropoulos 2011). These donors have been demonstrated to inhibit endothelial cell apoptosis, reduce oxidative stress, and suppress inflammatory cytokine production, all of which are crucial in the prevention of plaque formation and progression in atherosclerosis (Kolluru et al. 2013; S. Mani et al. 2014).

GYY4137, a slow-releasing H_2S donor, has been shown to reduce hypertension and improve vascular function in animal models (Papapetropoulos et al. 2009). Another study demonstrated that GYY4137 could prevent the formation of atherosclerotic lesions in mice by inhibiting the proliferation and migration of VSMC (Liu et al. 2013). These findings highlight the therapeutic potential of H_2S donors in managing atherosclerosis and associated cardiometabolic diseases (Li et al. 2018a; Wang 2012). Recent advancements in drug delivery systems have also enhanced the pharmacological potential of H_2S donors. Nanoparticle-based delivery systems, for example, allow for the targeted and controlled release of H_2S, improving the efficacy and reducing the potential side effects of these compounds (Chen et al. 2019). These innovations have opened new avenues for the development of H_2S-based therapeutics for managing atherosclerosis and associated cardiometabolic conditions (Kimura 2015; Li et al. 2018a).

5.3 Pharmacological Potential of H_2S Inhibitors

Selective inhibition of the H_2S producing enzymes could help delineate the specific pathways through which H_2S contributes to disease processes (Paul and Snyder 2015). Genetic or pharmacological inhibition of CSE can exacerbate hypertension and vascular dysfunction in animal models, suggesting a protective role for endogenous H_2S in cardiovascular health (Papapetropoulos et al. 2009).

In the context of cardiometabolic diseases, modulating H_2S levels through these pharmacological agents holds the potential for therapeutic intervention, aiming to restore homeostasis and prevent disease progression (Gadalla and Snyder 2010; Wang 2012). Recent studies have shown that inhibition of CBS can reduce the progression of NAFLD and improve insulin sensitivity in animal models, indicating that targeted modulation of H_2S-producing enzymes could be beneficial in managing metabolic diseases (Yang et al. 2013). In the context of cardiometabolic diseases, modulating H_2S levels through these pharmacological agents holds potential for therapeutic intervention, aiming to restore homeostasis and prevent disease progression (Gadalla and Snyder 2010; Wang 2012). Recent studies have shown that inhibiting CBS can reduce the progression of NAFLD and improve insulin sensitivity in animal models, indicating that targeted modulation of H_2S-producing enzymes could be beneficial in managing metabolic diseases (Yang et al. 2013).

5.4 Therapeutic Applications of H_2S in CVD

The therapeutic potential of H_2S donors have been explored in the context of myocardial IRI injury, a condition that is often associated with atherosclerosis. H_2S donors have been shown to reduce infarct size, improve cardiac function, and enhance angiogenesis in animal models of myocardial infarction (Polhemus et al. 2015; Szabó and Papapetropoulos 2011). In addition, studies have demonstrated that in CSE KO models, the infarct size is larger, whereas in 3-MST KO models, the infarct size is smaller, highlighting the complex role of these enzymes in cardiovascular health (Maria Peleli et al. 2020; Zhu et al. 2020). The therapeutic potential of H_2S donors have been explored in the context of myocardial IRI, a condition often associated with atherosclerosis. H_2S donors have been shown to reduce infarct size, improve cardiac function, and enhance angiogenesis in animal models of myocardial infarction (Polhemus et al. 2015; Szabó and Papapetropoulos 2011). In addition, studies have demonstrated that in CSE KO models, the infarct size is higher, whereas in 3-MST KO models, the infarct size is smaller, highlighting the complex role of these enzymes in cardiovascular health (Maria Peleli et al. 2020; Zhu et al. 2020).

For instance, treatment with H_2S donors have been shown to improve mitochondrial function and reduce oxidative stress in cardiomyocytes, thereby preserving cardiac function after ischemic events (Papapetropoulos et al. 2009). These protective effects are mediated through various mechanisms, including the activation of ATP-sensitive potassium channels, inhibition of the renin-angiotensin system, and modulation of oxidative stress and inflammation (Kolluru et al. 2013; Mani et al. 2014).

The pharmacological modulation of H_2S levels also holds promise for the treatment of other cardiometabolic diseases such as hypertension and heart failure. H_2S has been shown to exert vasodilatory effects, reduce blood pressure, and improve cardiac function in preclinical studies (Altaany et al. 2013; Szabó and Papapetropoulos 2011).

Moreover, H_2S donors have shown potential for improving metabolic parameters in animal models of diabetes and obesity. For instance, treatment with H_2S donors have been shown to improve insulin sensitivity, reduce blood glucose levels, and decrease body weight in obese and diabetic mice (Yang et al. 2013). These findings suggest that H_2S-based therapies could be beneficial in managing metabolic disorders, which are risk factors for atherosclerosis.

In humans, the only H_2S donor that has been successfully tested in clinical trials is the slow-releasing donor SG1002. This drug has successfully passed phase I clinical trials and has shown a safe profile and the ability to increase plasma H_2S and NO levels in patients with heart failure and associated cardiometabolic complications

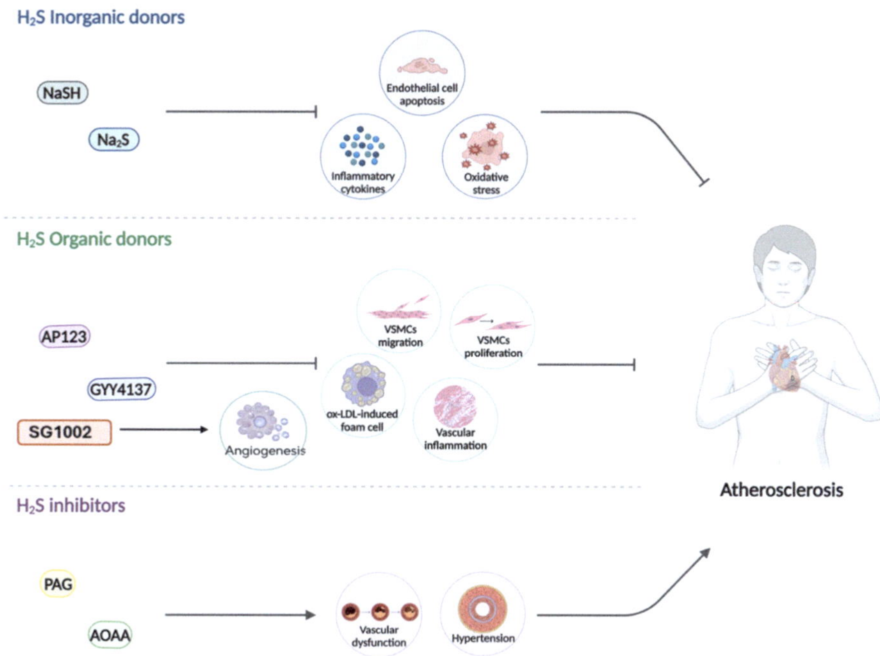

Fig. 3 Summary of the role of pharmacological H_2S donors and inhibitors in atherosclerosis. H_2S organic and inorganic donors show promising potential for treating atherosclerosis via mechanisms that involve the attenuation of inflammation, foam cell formation, oxidative stress, endothelial apoptosis, and VSMC migration and proliferation. Although not directly tested in atherosclerosis, SG1002, an oral H_2S prodrug that has passed phase I clinical trials, has shown in preclinical models vascular protective effects, including enhanced angiogenesis and preservation of vessel density in ischemic models – effects that may offer therapeutic benefit in atherosclerotic disease. In contrast, CSE or CSE/CBS inhibitors, such as D, L-propargylglycine (PAG) and aminooxyacetic acid (AOAA), could have a proatherogenic role by supporting vascular dysfunction and hypertension. (The figure was created by using the Biorender software https://biorender.com)

such as T2D, dyslipidemia, hypertension, and ischemic heart disease (Polhemus et al. 2015).

A summary of the pharmacological properties and effects of H_2S donors and inhibitors on atherosclerosis is shown in Fig. 3.

6 Conclusions and Future Perspectives

Investigating the molecular mechanisms underlying the protective effects of H_2S could provide new insights into its role in cardiovascular health. Understanding how H_2S interacts with other signaling pathways and identifying its downstream targets will be the key to developing targeted therapies. Additionally, exploring the genetic regulation of H_2S-producing enzymes and their role in disease pathogenesis could reveal novel therapeutic targets (Paul and Snyder 2015).

The integration of H_2S-based therapies with existing treatment modalities, such as lifestyle interventions and pharmacological agents could enhance their efficacy and provides a comprehensive approach for managing cardiometabolic diseases. Personalized medical approaches, taking into account individual differences in H_2S metabolism and response to therapy could further optimize treatment outcomes (Gadalla and Snyder 2010; Yang et al. 2013).

In conclusion, H_2S represents a promising biomarker and therapeutic target for atherosclerosis and its associated cardiometabolic diseases. Continued research and innovations in this field have the potential to significantly advance our understanding and treatment of these complex conditions, ultimately improving patient outcomes and reducing the burden of cardiovascular disease (Szabó and Papapetropoulos 2011; Wang 2012).

Conflicts of Interest None.

Authorship Statement **Conceptualization:** M.P, O.S, T.D, **Investigation:** O.S, T.D, M.P, **Writing – Original Draft:** O.S, T.D, D.F.J.K, M.P, **Writing – Review and Editing:** T.D, D.F.J. K, M.P, **Funding acquisition:** D.F.J.K, M.P, **Supervision:** D.F.J.K, M.P.

Funding Maria Peleli is supported by funds from the Novo Nordisk Foundation (NNF-Emerging Investigator Grant in Endocrinology & Metabolism NNF22OC0072951), Daniel F.J. Ketelhuth's research is supported by grants from the Novo Nordisk Foundation (0064142; 0075258); Independent Research Fund Denmark (2034-00136B); Simon Fougner Hartmanns Familiefond (2023-0066); and the University of Southern Denmark.

References

Abdollahi Govar A, Toro G, Szaniszlo P, Pavlidou A, Bibli SI, Thanki K, Resto VA, Chao C, Hellmich MR, Szabo C, Papapetropoulos A, Modis K (2020) 3-Mercaptopyruvate sulfurtransferase supports endothelial cell angiogenesis and bioenergetics. Br J Pharmacol 177(4):866–883. https://doi.org/10.1111/bph.14574

Altaany Z, Yang G, Wang R (2013) Crosstalk between hydrogen sulfide and nitric oxide in endothelial cells. J Cell Mol Med 17(7):879–888

Aminzadeh MA, Vaziri ND (2012) Downregulation of the renal and hepatic hydrogen sulfide (H2S)-producing enzymes and capacity in chronic kidney disease. Nephrol Dial Transplant 27(2):498–504

Augsburger F, Szabo C (2020) Potential role of the 3-mercaptopyruvate sulfurtransferase (3-MST)-hydrogen sulfide (H(2)S) pathway in cancer cells. Pharmacol Res 154:104083. https://doi.org/10.1016/j.phrs.2018.11.034

Barrow K, Wang Y, Yu R, Zhu J, Yang G (2022) H2S protects from oxidative stress-driven ACE2 expression and cardiac aging. Mol Cell Biochem 477(5):1393–1403

Batty M, Bennett MR, Yu E (2022) The role of oxidative stress in atherosclerosis. Cells 11(23):3843

Benavides GA, Squadrito GL, Mills RW, Patel HD, Isbell TS, Patel RP, Darley-Usmar VM, Doeller JE, Kraus DW (2007) Hydrogen sulfide mediates the vasoactivity of garlic. Proc Natl Acad Sci USA 104(46):17977–17982. https://doi.org/10.1073/pnas.0705710104

Bibli SI, Hu J, Sigala F, Wittig I, Heidler J, Zukunft S, Tsilimigras DI, Randriamboavonjy V, Wittig J, Kojonazarov B, Schurmann C, Siragusa M, Siuda D, Luck B, Abdel Malik R, Filis KA, Zografos G, Chen C, Wang DW, Pfeilschifter J, Brandes RP, Szabo C, Papapetropoulos A,

Fleming I (2019) Cystathionine gamma Lyase Sulfhydrates the RNA binding protein human antigen R to preserve endothelial cell function and delay Atherogenesis. Circulation 139(1): 101–114. https://doi.org/10.1161/CIRCULATIONAHA.118.034757

Bir SC, Kolluru GK, McCarthy P, Shen X, Pardue S, Pattillo CB, Kevil CG (2012) Hydrogen sulfide stimulates ischemic vascular remodeling through nitric oxide synthase and nitrite reduction activity regulating hypoxia-inducible factor-1alpha and vascular endothelial growth factor-dependent angiogenesis. J Am Heart Assoc 1(5):e004093. https://doi.org/10.1161/JAHA.112.004093

Bos EM, Wang R, Snijder PM, Boersema M, Damman J, Fu M, Moser J, Hillebrands J-L, Ploeg RJ, Yang G (2013) Cystathionine γ-lyase protects against renal ischemia/reperfusion by modulating oxidative stress. J Am Soc Nephrol 24(5):759–770

Byrne CD, Targher G (2015) NAFLD: a multisystem disease. J Hepatol 62(1):S47–S64

Cai J, Shi X, Wang H, Fan J, Feng Y, Lin X, Yang J, Cui Q, Tang C, Xu G (2016) Cystathionine γ lyase–hydrogen sulfide increases peroxisome proliferator-activated receptor γ activity by sulfhydration at C139 site thereby promoting glucose uptake and lipid storage in adipocytes. Biochim Biophys Acta (BBA)-Mol Cell Biol Lipids 1861(5):419–429

Calabrese V, Scuto M, Salinaro AT, Dionisio G, Modafferi S, Ontario ML, Greco V, Sciuto S, Schmitt CP, Calabrese EJ (2020) Hydrogen sulfide and carnosine: modulation of oxidative stress and inflammation in kidney and brain axis. Antioxidants 9(12):1303

Chade AR, Lerman A, Lerman LO (2005) Kidney in early atherosclerosis. Hypertension 45(6): 1042–1049

Chen W, Ni D, Rosenkrans ZT, Cao T, Cai W (2019) Smart H2S-triggered/therapeutic system (SHTS)-based Nanomedicine. Adv Sci 6(22):1901724

Chen Z, Ouyang C, Zhang H, Gu Y, Deng Y, Du C, Cui C, Li S, Wang W, Kong W (2022) Vascular smooth muscle cell-derived hydrogen sulfide promotes atherosclerotic plaque stability via TFEB (transcription factor EB)-mediated autophagy. Autophagy 18(10):2270–2287

Cirino G, Szabo C, Papapetropoulos A (2023) Physiological roles of hydrogen sulfide in mammalian cells, tissues, and organs. Physiol Rev 103(1):31–276. https://doi.org/10.1152/physrev.00028.2021

Coletta C, Papapetropoulos A, Erdelyi K, Olah G, Modis K, Panopoulos P, Asimakopoulou A, Gero D, Sharina I, Martin E, Szabo C (2012) Hydrogen sulfide and nitric oxide are mutually dependent in the regulation of angiogenesis and endothelium-dependent vasorelaxation. Proc Natl Acad Sci USA 109(23):9161–9166. https://doi.org/10.1073/pnas.1202916109

Coletta C, Módis K, Szczesny B, Brunyánszki A, Oláh G, Rios EC, Yanagi K, Ahmad A, Papapetropoulos A, Szabo C (2015) Regulation of vascular tone, angiogenesis and cellular bioenergetics by the 3-mercaptopyruvate sulfurtransferase/H 2 S pathway: functional impairment by hyperglycemia and restoration by DL-α-lipoic acid. Mol Med 21:1–14

Comas F, Latorre J, Cusso O, Ortega F, Lluch A, Sabater M, Castells-Nobau A, Ricart W, Ribas X, Costas M, Fernandez-Real JM, Moreno-Navarrete JM (2019) Hydrogen sulfide impacts on inflammation-induced adipocyte dysfunction. Food Chem Toxicol 131:110543. https://doi.org/10.1016/j.fct.2019.05.051

Damba T, Zhang M, Buist-Homan M, van Goor H, Faber KN, Moshage H (2019) Hydrogen sulfide stimulates activation of hepatic stellate cells through increased cellular bio-energetics. Nitric Oxide 92:26–33

Damba T, Zhang M, Serna Salas SA, Wu Z, van Goor H, Arenas AF, Muñoz-Ortega MH, Ventura-Juárez J, Buist-Homan M, Moshage H (2024) Inhibition of endogenous hydrogen sulfide production reduces activation of hepatic stellate cells via the induction of cellular senescence. Cell Cycle 23(6):629–644

Dilek N, Papapetropoulos A, Toliver-Kinsky T, Szabo C (2020) Hydrogen sulfide: an endogenous regulator of the immune system. Pharmacol Res 161:105119. https://doi.org/10.1016/j.phrs.2020.105119

Dong B, Sun Y, Cheng B, Xue Y, Li W, Sun X (2023) Activating transcription factor (ATF) 6 upregulates cystathionine β synthetase (CBS) expression and hydrogen sulfide (H2S) synthesis to ameliorate liver metabolic damage. Eur J Med Res 28(1):540

Du C, Jin M, Hong Y, Li Q, Wang X-H, Xu J-M, Wang F, Zhang Y, Jia J, Liu C-F (2014) Downregulation of cystathionine β-synthase/hydrogen sulfide contributes to rotenone-induced microglia polarization toward M1 type. Biochem Biophys Res Commun 451(2):239–245

Du C, Lin X, Xu W, Zheng F, Cai J, Yang J, Cui Q, Tang C, Cai J, Xu G (2019) Sulfhydrated sirtuin-1 increasing its deacetylation activity is an essential epigenetics mechanism of anti-atherogenesis by hydrogen sulfide. Antioxid Redox Signal 30(2):184–197

Duval C, Chinetti G, Trottein F, Fruchart J-C, Staels B (2002) The role of PPARs in atherosclerosis. Trends Mol Med 8(9):422–430

Elrod JW, Calvert JW, Morrison J, Doeller JE, Kraus DW, Tao L, Jiao X, Scalia R, Kiss L, Szabo C (2007) Hydrogen sulfide attenuates myocardial ischemia-reperfusion injury by preservation of mitochondrial function. Proc Natl Acad Sci 104(39):15560–15565

El-Sayed SS, Zakaria MN, Abdel-Ghany RH, Abdel-Rahman AA (2016) Cystathionine-γ lyase-derived hydrogen sulfide mediates the cardiovascular protective effects of moxonidine in diabetic rats. Eur J Pharmacol 783:73–84

Feigin VL, Brainin M, Norrving B, Martins S, Sacco RL, Hacke W, Fisher M, Pandian J, Lindsay P (2022) World stroke organization (WSO): global stroke fact sheet 2022. Int J Stroke 17(1): 18–29. https://doi.org/10.1177/17474930211065917

Filipovic MR (2015) Persulfidation (S-sulfhydration) and H2S. Handb Exp Pharmacol 230:29–59. https://doi.org/10.1007/978-3-319-18144-8_2

Gadalla MM, Snyder SH (2010) Hydrogen sulfide as a gasotransmitter. J Neurochem 113(1):14–26

Gemici B, Wallace JL (2015) Anti-inflammatory and cytoprotective properties of hydrogen sulfide. Methods Enzymol 555:169–193. https://doi.org/10.1016/bs.mie.2014.11.034

Geng B, Cai B, Liao F, Zheng Y, Zeng Q, Fan X, Gong Y, Yang J, Cui QH, Tang C (2013) Increase or decrease hydrogen sulfide exert opposite lipolysis, but reduce global insulin resistance in high fatty diet induced obese mice. PLoS One 8(9):e73892

Giuffre A, Vicente JB (2018) Hydrogen sulfide biochemistry and interplay with other gaseous mediators in mammalian physiology. Oxidative Med Cell Longev 2018:6290931. https://doi.org/10.1155/2018/6290931

Gong D, Cheng H-p, Xie W, Zhang M, Liu D, Lan G, Huang C, Zhao Z-w, Chen L-y, Yao F (2016) Cystathionine γ-lyase (CSE)/hydrogen sulfide system is regulated by miR-216a and influences cholesterol efflux in macrophages via the PI3K/AKT/ABCA1 pathway. Biochem Biophys Res Commun 470(1):107–116

Gopalakrishnan P, Shrestha B, Kaskas AM, Green J, Alexander JS, Pattillo CB (2019) Hydrogen sulfide: therapeutic or injurious in ischemic stroke? Pathophysiology 26(1):1–10. https://doi.org/10.1016/j.pathophys.2018.10.005

Gui DD, Luo W, Yan BJ, Ren Z, Tang ZH, Liu LS, Zhang JF, Jiang ZS (2021) Effects of gut microbiota on atherosclerosis through hydrogen sulfide. Eur J Pharmacol 896:173916. https://doi.org/10.1016/j.ejphar.2021.173916

Jiang S, Xu W, Chen Z, Cui C, Fan X, Cai J, Gong Y, Geng B (2021) Hydrogen sulphide reduces hyperhomocysteinaemia-induced endothelial ER stress by sulfhydrating protein disulphide isomerase to attenuate atherosclerosis. J Cell Mol Med 25(7):3437–3448. https://doi.org/10.1111/jcmm.16423

Kabil O, Banerjee R (2014) Enzymology of H2S biogenesis, decay and signaling. Antioxid Redox Signal 20(5):770–782. https://doi.org/10.1089/ars.2013.5339

Kanagy NL, Szabo C, Papapetropoulos A (2017) Vascular biology of hydrogen sulfide. Am J Physiol Cell Physiol 312(5):C537–C549. https://doi.org/10.1152/ajpcell.00329.2016

Kashfi K, Olson KR (2013) Biology and therapeutic potential of hydrogen sulfide and hydrogen sulfide-releasing chimeras. Biochem Pharmacol 85(5):689–703. https://doi.org/10.1016/j.bcp.2012.10.019

Katsouda A, Peleli M, Asimakopoulou A, Papapetropoulos A, Beis D (2020) Generation and characterization of a CRISPR/Cas9—induced 3-mst deficient zebrafish. Biomolecules 10(2): 317

Katsouda A, Valakos D, Dionellis VS, Bibli S-I, Akoumianakis I, Karaliota S, Zuhra K, Fleming I, Nagahara N, Havaki S (2022) MPST sulfurtransferase maintains mitochondrial protein import and cellular bioenergetics to attenuate obesity. J Exp Med 219(7):e20211894

Katsouda A, Markou M, Valakos D, Theodorou I, Papapetropoulos A (2025) Cth/Mpst double ablation results in early onset fatty liver disease in lean mice. Redox Biol 83:103641

Kimura H (2011) Hydrogen sulfide: its production, release and functions. Amino Acids 41(1): 113–121. https://doi.org/10.1007/s00726-010-0510-x

Kimura H (2015) Physiological roles of hydrogen sulfide and polysulfides. In: Chemistry, biochemistry and pharmacology of hydrogen sulfide, pp 61–81

Kimura Y, Koike S, Shibuya N, Lefer D, Ogasawara Y, Kimura H (2017) 3-Mercaptopyruvate sulfurtransferase produces potential redox regulators cysteine- and glutathione-persulfide (Cys-SSH and GSSH) together with signaling molecules H(2)S(2), H(2)S(3) and H(2)S. Sci Rep 7(1):10459. https://doi.org/10.1038/s41598-017-11004-7

Kiss L, Deitch EA, Szabó C (2008) Hydrogen sulfide decreases adenosine triphosphate levels in aortic rings and leads to vasorelaxation via metabolic inhibition. Life Sci 83(17–18):589–594

Kolluru GK, Shen X, Bir SC, Kevil CG (2013) Hydrogen sulfide chemical biology: pathophysiological roles and detection. Nitric Oxide 35:5–20

Lendoiro-Cino N, Rodríguez-Coello A, Saborido A, F-Burguera E, Fernández-Rodríguez JA, Meijide-Faílde R, Blanco FJ, Vaamonde-García C (2023) Study of hydrogen sulfide biosynthesis in synovial tissue from diabetes-associated osteoarthritis and its influence on macrophage phenotype and abundance. J Physiol Biochem 79(3):653–667

Li S, Ping N-n, Cao L, Mi Y-n, Cao Y-x (2015) H2S induces vasoconstriction of rat cerebral arteries via cAMP/adenylyl cyclase pathway. Toxicol Appl Pharmacol 289(3):389–396

Li D, Xiong Q, Peng J, Hu B, Li W, Zhu Y, Shen X (2016) Hydrogen sulfide up-regulates the expression of ATP-binding cassette transporter A1 via promoting nuclear translocation of PPARα. Int J Mol Sci 17(5):635

Li Z, Polhemus DJ, Lefer DJ (2018a) Evolution of hydrogen sulfide therapeutics to treat cardiovascular disease. Circ Res 123(5):590–600

Li M, Xu C, Shi J, Ding J, Wan X, Chen D, Gao J, Li C, Zhang J, Lin Y (2018b) Fatty acids promote fatty liver disease via the dysregulation of 3-mercaptopyruvate sulfurtransferase/hydrogen sulfide pathway. Gut 67(12):2169–2180

Lim JJ, Liu Y-H, Khin ESW, Bian J-S (2008) Vasoconstrictive effect of hydrogen sulfide involves downregulation of cAMP in vascular smooth muscle cells. Am J Phys Cell Phys 295(5):C1261–C1270

Lin Y, Chen Y, Zhu N, Zhao S, Fan J, Liu E (2016) Hydrogen sulfide inhibits development of atherosclerosis through up-regulating protein S-nitrosylation. Biomed Pharmacother 83:466–476. https://doi.org/10.1016/j.biopha.2016.07.003

Liu Z, Han Y, Li L, Lu H, Meng G, Li X, Shirhan M, Peh MT, Xie L, Zhou S (2013) The hydrogen sulfide donor, GYY 4137, exhibits anti-atherosclerotic activity in high fat fed apolipoprotein E−/− mice. Br J Pharmacol 169(8):1795–1809

Longchamp A, MacArthur MR, Trocha K, Ganahl J, Mann CG, Kip P, King WW, Sharma G, Tao M, Mitchell SJ (2021) Plasma hydrogen sulfide is positively associated with post-operative survival in patients undergoing surgical revascularization. Front Cardiovasc Med 8:750926

Longen S, Beck KF, Pfeilschifter J (2016) H2S-induced thiol-based redox switches: biochemistry and functional relevance for inflammatory diseases. Pharmacol Res 111:642–651. https://doi.org/10.1016/j.phrs.2016.07.026

Luan J, Ji X, Liu L (2023) PPARγ in atherosclerotic endothelial dysfunction: regulatory compounds and PTMs. Int J Mol Sci 24(19):14494

Lv B, Chen S, Tang C, Jin H, Du J, Huang Y (2021) Hydrogen sulfide and vascular regulation – an update. J Adv Res 27:85–97. https://doi.org/10.1016/j.jare.2020.05.007

Mani S, Li H, Untereiner A, Wu L, Yang G, Austin RC, Dickhout JG, Lhoták Š, Meng QH, Wang R (2013) Decreased endogenous production of hydrogen sulfide accelerates atherosclerosis. Circulation 127(25):2523–2534

Mani S, Untereiner A, Wu L, Wang R (2014) Hydrogen sulfide and the pathogenesis of atherosclerosis. Antioxid Redox Signal 20(5):805–817

Mendiola PJ, Naik JS, Gonzalez Bosc LV, Gardiner AS, Birg A, Kanagy NL (2021) Hydrogen Sulfide Actions in the Vasculature. Compr Physiol 11(4):2467–2488. https://doi.org/10.1002/cphy.c200036. PMID: 34558672; PMCID: PMC11758848

Mikami Y, Shibuya N, Kimura Y, Nagahara N, Ogasawara Y, Kimura H (2011) Thioredoxin and dihydrolipoic acid are required for 3-mercaptopyruvate sulfurtransferase to produce hydrogen sulfide. Biochem J 439(3):479–485. https://doi.org/10.1042/BJ20110841

Modis K, Asimakopoulou A, Coletta C, Papapetropoulos A, Szabo C (2013a) Oxidative stress suppresses the cellular bioenergetic effect of the 3-mercaptopyruvate sulfurtransferase/hydrogen sulfide pathway. Biochem Biophys Res Commun 433(4):401–407. https://doi.org/10.1016/j.bbrc.2013.02.131

Modis K, Coletta C, Erdelyi K, Papapetropoulos A, Szabo C (2013b) Intramitochondrial hydrogen sulfide production by 3-mercaptopyruvate sulfurtransferase maintains mitochondrial electron flow and supports cellular bioenergetics. FASEB J 27(2):601–611. https://doi.org/10.1096/fj.12-216507

Murphy B, Bhattacharya R, Mukherjee P (2019) Hydrogen sulfide signaling in mitochondria and disease. FASEB J 33(12):13098–13125. https://doi.org/10.1096/fj.201901304R

Nagahara N (2018) Multiple role of 3-mercaptopyruvate sulfurtransferase: antioxidative function, H (2) S and polysulfide production and possible SO(x) production. Br J Pharmacol 175(4):577–589. https://doi.org/10.1111/bph.14100

Nagahara N, Ito T, Kitamura H, Nishino T (1998) Tissue and subcellular distribution of mercaptopyruvate sulfurtransferase in the rat: confocal laser fluorescence and immunoelectron microscopic studies combined with biochemical analysis. Histochem Cell Biol 110(3):243–250. https://doi.org/10.1007/s004180050286

Nishimura A, Yoon S, Matsunaga T, Ida T, Jung M, Ogata S, Morita M, Yoshitake J, Unno Y, Barayeu U, Takata T, Takagi H, Motohashi H, van der Vliet A, Akaike T (2024) Longevity control by supersulfide-mediated mitochondrial respiration and regulation of protein quality. Redox Biol 69:103018. https://doi.org/10.1016/j.redox.2023.103018

Omorou M, Liu N, Huang Y, Al-Ward H, Gao M, Mu C, Zhang L, Hui X (2022) Cystathionine beta-synthase in hypoxia and ischemia/reperfusion: A current overview. Arch Biochem Biophys 718:109149. https://doi.org/10.1016/j.abb.2022.109149

Papapetropoulos A, Pyriochou A, Altaany Z, Yang G, Marazioti A, Zhou Z, Jeschke MG, Branski LK, Herndon DN, Wang R (2009) Hydrogen sulfide is an endogenous stimulator of angiogenesis. Proc Natl Acad Sci 106(51):21972–21977

Paul BD, Snyder SH (2015) H2S: a novel gasotransmitter that signals by sulfhydration. Trends Biochem Sci 40(11):687–700

Peleli M, Bibli SI, Li Z, Chatzianastasiou A, Varela A, Katsouda A, Zukunft S, Bucci M, Vellecco V, Davos CH, Nagahara N, Cirino G, Fleming I, Lefer DJ, Papapetropoulos A (2020) Cardiovascular phenotype of mice lacking 3-mercaptopyruvate sulfurtransferase. Biochem Pharmacol 176:113833. https://doi.org/10.1016/j.bcp.2020.113833

Peleli M, Zampas P, Papapetropoulos A (2022) Hydrogen sulfide and the kidney: physiological roles, contribution to pathophysiology, and therapeutic potential. Antioxid Redox Signal 36(4–6):220–243

Peleli M, Lyngso KS, Poulsen FR, Hansen PBL, Papapetropoulos A, Stubbe J (2023) Inhibition of cystathionine-gamma lyase dampens vasoconstriction in mouse and human intracerebral arterioles. Acta Physiol (Oxf) 239(1):e14021. https://doi.org/10.1111/apha.14021

Peoples JN, Saraf A, Ghazal N, Pham TT, Kwong JQ (2019) Mitochondrial dysfunction and oxidative stress in heart disease. Exp Mol Med 51(12):1–13

Peter EA, Shen X, Shah SH, Pardue S, Glawe JD, Zhang WW, Reddy P, Akkus NI, Varma J, Kevil CG (2013) Plasma free H2S levels are elevated in patients with cardiovascular disease. J Am Heart Assoc 2(5):e000387

Polhemus DJ, Li Z, Pattillo CB, Gojon G Sr, Gojon G Jr, Giordano T, Krum H (2015) A novel hydrogen sulfide prodrug, SG 1002, promotes hydrogen sulfide and nitric oxide bioavailability in heart failure patients. Cardiovasc Ther 33(4):216–226

Predmore BL, Lefer DJ, Gojon G (2012) Hydrogen sulfide in biochemistry and medicine. Antioxid Redox Signal 17(1):119–140. https://doi.org/10.1089/ars.2012.4612

Prieto-Lloret J, Aaronson PI (2015) Potentiation of hypoxic pulmonary vasoconstriction by hydrogen sulfide precursors 3-Mercaptopyruvate and D-cysteine is blocked by the cystathionine gamma Lyase inhibitor Propargylglycine. Adv Exp Med Biol 860:81–87. https://doi.org/10.1007/978-3-319-18440-1_10

Rajpal S, Katikaneni P, Deshotels M, Pardue S, Glawe J, Shen X, Akkus N, Modi K, Bhandari R, Dominic P (2018) Total sulfane sulfur bioavailability reflects ethnic and gender disparities in cardiovascular disease. Redox Biol 15:480–489

Rao G, Murphy B, Dey A, Dwivedi SKD, Zhang Y, Roy RV, Chakraborty P, Bhattacharya R, Mukherjee P (2020) Cystathionine beta synthase regulates mitochondrial dynamics and function in endothelial cells. FASEB J 34(7):9372–9392. https://doi.org/10.1096/fj.202000173R

Robert K, Nehmé J, Bourdon E, Pivert G, Friguet B, Delcayre C, Delabar JM, Janel N (2005) Cystathionine β synthase deficiency promotes oxidative stress, fibrosis, and steatosis in mice liver. Gastroenterology 128(5):1405–1415

Sena CM, Carrilho F, Seiça RM (2018) Endothelial dysfunction in type 2 diabetes: targeting inflammation. Endothelial Dysfunction: Old Concepts and New Challenges 24:23110

Shen X, Carlstrom M, Borniquel S, Jadert C, Kevil CG, Lundberg JO (2013) Microbial regulation of host hydrogen sulfide bioavailability and metabolism. Free Radic Biol Med 60:195–200. https://doi.org/10.1016/j.freeradbiomed.2013.02.024

Shen Y, Shen Z, Miao L, Xin X, Lin S, Zhu Y, Guo W, Zhu YZ (2015) miRNA-30 family inhibition protects against cardiac ischemic injury by regulating cystathionine-gamma-lyase expression. Antioxid Redox Signal 22(3):224–240. https://doi.org/10.1089/ars.2014.5909

Shibuya N, Mikami Y, Kimura Y, Nagahara N, Kimura H (2009) Vascular endothelium expresses 3-mercaptopyruvate sulfurtransferase and produces hydrogen sulfide. J Biochem 146(5):623–626. https://doi.org/10.1093/jb/mvp111

Shirazi MK, Azarnezhad A, Abazari MF, Poorebrahim M, Ghoraeian P, Sanadgol N, Bokharaie H, Heydari S, Abbasi A, Kabiri S (2019) The role of nitric oxide signaling in renoprotective effects of hydrogen sulfide against chronic kidney disease in rats: involvement of oxidative stress, autophagy and apoptosis. J Cell Physiol 234(7):11411–11423

Song K, Wang F, Li Q, Shi Y-B, Zheng H-F, Peng H, Shen H-Y, Liu C-F, Hu L-F (2014) Hydrogen sulfide inhibits the renal fibrosis of obstructive nephropathy. Kidney Int 85(6):1318–1329

Stols-Gonçalves D, Hovingh GK, Nieuwdorp M, Holleboom AG (2019) NAFLD and atherosclerosis: two sides of the same dysmetabolic coin? Trends Endocrinol Metab 30(12):891–902

Sun HJ, Wu ZY, Nie XW, Bian JS (2019) Role of endothelial dysfunction in cardiovascular diseases: the link between inflammation and hydrogen sulfide. Front Pharmacol 10:1568. https://doi.org/10.3389/fphar.2019.01568

Sun F, Luo JH, Yue TT, Wang FX, Yang CL, Zhang S, Wang XQ, Wang CY (2021) The role of hydrogen sulphide signalling in macrophage activation. Immunology 162(1):3–10

Sun H-J, Xiong S-P, Wang Z-C, Nie X-W, Bian J-S (2023) Hydrogen sulfide in diabetic complications revisited: the state of the art, challenges, and future directions. Antioxid Redox Signal 38(1–3):18–44

Suzuki K, Olah G, Modis K, Coletta C, Kulp G, Gerö D, Szoleczky P, Chang T, Zhou Z, Wu L (2011) Hydrogen sulfide replacement therapy protects the vascular endothelium in hyperglycemia by preserving mitochondrial function. Proc Natl Acad Sci 108(33):13829–13834

Szabo C (2016) Gasotransmitters in cancer: from pathophysiology to experimental therapy. Nat Rev Drug Discov 15(3):185–203. https://doi.org/10.1038/nrd.2015.1

Szabo C (2017) Hydrogen sulfide, an enhancer of vascular nitric oxide signaling: mechanisms and implications. Am J Physiol Cell Physiol 312(1):C3–C15. https://doi.org/10.1152/ajpcell.00282.2016

Szabo C (2021) Hydrogen sulfide, an endogenous stimulator of mitochondrial function in cancer cells. Cells 10(2). https://doi.org/10.3390/cells10020220

Szabó C, Papapetropoulos A (2011) Hydrogen sulphide and angiogenesis: mechanisms and applications. Br J Pharmacol 164(3):853–865

Szabo C, Papapetropoulos A (2017) International union of basic and clinical pharmacology. CII: pharmacological modulation of H2S levels: H2S donors and H2S biosynthesis inhibitors. Pharmacol Rev 69(4):497–564

Szijarto IA, Marko L, Filipovic MR, Miljkovic JL, Tabeling C, Tsvetkov D, Wang N, Rabelo LA, Witzenrath M, Diedrich A, Tank J, Akahoshi N, Kamata S, Ishii I, Gollasch M (2018) Cystathionine gamma-Lyase-produced hydrogen sulfide controls endothelial NO bioavailability and blood pressure. Hypertension 71(6):1210–1217. https://doi.org/10.1161/HYPERTENSIONAHA.117.10562

Targher G, Tilg H, Byrne CD (2021) Non-alcoholic fatty liver disease: a multisystem disease requiring a multidisciplinary and holistic approach. Lancet Gastroenterol Hepatol 6(7):578–588

Teng H, Wu B, Zhao K, Yang G, Wu L, Wang R (2013) Oxygen-sensitive mitochondrial accumulation of cystathionine beta-synthase mediated by Lon protease. Proc Natl Acad Sci USA 110(31):12679–12684. https://doi.org/10.1073/pnas.1308487110

Tomasova L, Konopelski P, Ufnal M (2016) Gut bacteria and hydrogen sulfide: the new old players in circulatory system homeostasis. Molecules 21(11). https://doi.org/10.3390/molecules21111558

Tsutsumi T, Eslam M, Kawaguchi T, Yamamura S, Kawaguchi A, Nakano D, Koseki M, Yoshinaga S, Takahashi H, Anzai K (2021) MAFLD better predicts the progression of atherosclerotic cardiovascular risk than NAFLD: generalized estimating equation approach. Hepatol Res 51(11):1115–1128. https://onlinelibrary.wiley.com/doi/10.1111/hepr.13685

Valdivielso JM, Rodríguez-Puyol D, Pascual J, Barrios C, Bermúdez-López M, Sánchez-Niño MD, Pérez-Fernández M, Ortiz A (2019) Atherosclerosis in chronic kidney disease: more, less, or just different? Arterioscler Thromb Vasc Biol 39(10):1938–1966

Wallace JL, Ferraz JG, Muscara MN (2012) Hydrogen sulfide: an endogenous mediator of resolution of inflammation and injury. Antioxid Redox Signal 17(1):58–67. https://doi.org/10.1089/ars.2011.4351

Wang R (2012) Physiological implications of hydrogen sulfide: a whiff exploration that blossomed. Physiol Rev 92(2):791–896

Wang X-H, Wang F, You S-J, Cao Y-J, Cao L-D, Han Q, Liu C-F, Hu L-F (2013) Dysregulation of cystathionine γ-lyase (CSE)/hydrogen sulfide pathway contributes to ox-LDL-induced inflammation in macrophage. Cell Signal 25(11):2255–2226

Wang R, Szabo C, Ichinose F, Ahmed A, Whiteman M, Papapetropoulos A (2015) The role of H2S bioavailability in endothelial dysfunction. Trends Pharmacol Sci 36(9):568–578. https://doi.org/10.1016/j.tips.2015.05.007

Williams RA, Kelly SM, Mottram JC, Coombs GH (2003) 3-Mercaptopyruvate sulfurtransferase of Leishmania contains an unusual C-terminal extension and is involved in thioredoxin and antioxidant metabolism. J Biol Chem 278(3):1480–1486. https://doi.org/10.1074/jbc.M209395200

Wu D, Zhong P, Wang Y, Zhang Q, Li J, Liu Z, Ji A, Li Y (2020) Hydrogen sulfide attenuates high-fat diet-induced non-alcoholic fatty liver disease by inhibiting apoptosis and promoting autophagy via reactive oxygen species/phosphatidylinositol 3-kinase/AKT/mammalian target of rapamycin signaling pathway. Front Pharmacol 11:585860

Xie L, Gu Y, Wen M, Zhao S, Wang W, Ma Y, Meng G, Han Y, Wang Y, Liu G (2016) Hydrogen sulfide induces Keap1 S-sulfhydration and suppresses diabetes-accelerated atherosclerosis via Nrf2 activation. Diabetes 65(10):3171–3184

Xu W, Cui C, Cui C, Chen Z, Zhang H, Cui Q, Xu G, Fan J, Han Y, Tang L (2022) Hepatocellular cystathionine γ lyase/hydrogen sulfide attenuates nonalcoholic fatty liver disease by activating farnesoid X receptor. Hepatology 76(6):1794–1810

Yang G, Wu L, Wang R (2006) Pro-apoptotic effect of endogenous H2S on human aorta smooth muscle cells. FASEB J 20(3):553–555. https://doi.org/10.1096/fj.05-4712fje

Yang G, Wu L, Jiang B, Yang W, Qi J, Cao K, Meng Q, Mustafa AK, Mu W, Zhang S (2008) H2S as a physiologic vasorelaxant: hypertension in mice with deletion of cystathionine γ-lyase. Science 322(5901):587–590

Yang G, Zhao K, Ju Y, Mani S, Cao Q, Puukila S, Khaper N, Wu L, Wang R (2013) Hydrogen sulfide protects against cellular senescence via S-sulfhydration of Keap1 and activation of Nrf2. Antioxid Redox Signal 18(15):1906–1919

Yang J, Minkler P, Grove D, Wang R, Willard B, Dweik R, Hine C (2019) Non-enzymatic hydrogen sulfide production from cysteine in blood is catalyzed by iron and vitamin B(6). Commun Biol 2:194. https://doi.org/10.1038/s42003-019-0431-5

Zampas P, Li Z, Katsouda A, Varela A, Psarras S, Davos CH, Lefer DJ, Papapetropoulos A (2025) Protective role of 3-mercaptopyruvate sulfurtransferase (MPST) in the development of metabolic syndrome and vascular inflammation. Pharmacol Res 211:107542

Zanardo RC, Brancaleone V, Distrutti E, Fiorucci S, Cirino G, Wallace JL (2006) Hydrogen sulfide is an endogenous modulator of leukocyte-mediated inflammation. FASEB J 20(12):2118–2120. https://doi.org/10.1096/fj.06-6270fje

Zavaczki E, Jeney V, Agarwal A, Zarjou A, Oros M, Katko M, Varga Z, Balla G, Balla J (2011) Hydrogen sulfide inhibits the calcification and osteoblastic differentiation of vascular smooth muscle cells. Kidney Int 80(7):731–739

Zhao W, Wang R (2002) H(2)S-induced vasorelaxation and underlying cellular and molecular mechanisms. Am J Physiol Heart Circ Physiol 283(2):H474–H480. https://doi.org/10.1152/ajpheart.00013.2002

Zhu L, Duan W, Wu G, Zhang D, Wang L, Chen D, Chen Z, Yang B (2020) Protective effect of hydrogen sulfide on endothelial cells through Sirt1-FoxO1-mediated autophagy. Ann Transl Med 8(23):1586

Zivanovic J, Kouroussis E, Kohl JB, Adhikari B, Bursac B, Schott-Roux S, Petrovic D, Miljkovic JL, Thomas-Lopez D, Jung Y, Miler M, Mitchell S, Milosevic V, Gomes JE, Benhar M, Gonzalez-Zorn B, Ivanovic-Burmazovic I, Torregrossa R, Mitchell JR, Whiteman M, Schwarz G, Snyder SH, Paul BD, Carroll KS, Filipovic MR (2019) Selective persulfide detection reveals evolutionarily conserved antiaging effects of S-Sulfhydration. Cell Metab 30(6):1152–1170 e1113. https://doi.org/10.1016/j.cmet.2019.10.007

The Role of Hydrogen Sulfide-Derived Perivascular Adipose Tissue in Vascular Diseases

Emma Mitidieri, Chiara Indolfi, Vincenzo Brancaleone, Raffaella Sorrentino, and Roberta d'Emmanuele di Villa Bianca

Contents

Abstract

Perivascular adipose tissue (PVAT) is a metabolically active, endocrine organ that plays a crucial role in regulating blood vessel tone, endothelial function, vascular smooth muscle cell growth, and proliferation and contributes significantly to the onset and progression of cardiovascular diseases. In a healthy state, PVAT displays anticontractile, anti-inflammatory, and antioxidative properties, which are critical for maintaining vascular homeostasis. However, under certain pathophysiological conditions, PVAT exerts pro-contractile effects by decreasing the production of anticontractile and/or increasing that of pro-contractile factors.

E. Mitidieri · C. Indolfi · R. Sorrentino (✉) · R. d'Emmanuele di Villa Bianca
Department of Pharmacy, School of Medicine and Surgery, Naples, Italy
e-mail: rafsorre@unina.it

V. Brancaleone
Department of Health Science, University of Basilicata, Potenza, Italy

A. Papapetropoulos, C. Szabo (eds.), *Hydrogen Sulfide: A Toxicant with Healing Potential*, Handbook of Experimental Pharmacology 292,
https://doi.org/10.1007/164_2025_755

In this context, recent studies have identified hydrogen sulfide (H_2S) as a key vascular anti-contractile factor released from PVAT. The enzymes responsible for H_2S biosynthesis are differentially expressed in PVAT, depending on the vascular bed and species, and their function can be altered by metabolic and cardiovascular diseases. These alterations can influence H_2S signalling, further contributing to vascular dysfunction. PVAT-derived H_2S may have particular importance in obesity-related vascular disease, hypertension, and diabetes as it has direct paracrine effects on the vasculature. Understanding the role of PVAT-derived H_2S in both healthy and diseased states may provide new insights into preventing vascular dysfunction associated with PVAT changes. The dissection of the specific contributions of each enzyme involved in PVAT-derived H_2S biosynthesis could be relevant to fully understanding the complex role of H_2S in vascular health vs vascular disease. Further research into modulating PVAT-derived H_2S provides an exciting avenue to explore novel pharmacological targets against vascular disease pathogenesis.

Keywords

Cardiovascular disease · Hydrogen sulfide · Perivascular adipose tissue

1 Introduction

Adipose tissue is considered a central metabolic organ closely involved in the systemic regulation of energy metabolism via nutrient exchange and secretion of adipose-derived hormones and cytokines, together termed adipokines (Stern et al. 2016). Adipose tissue differentiates into white (WAT) and brown (BAT). WAT is the most abundant fat depot in mammals and comprises two main subdivisions, subcutaneous adipose tissue and visceral adipose tissue. WAT also acts as an endocrine organ, secreting hormones and proteins that play a role in metabolism, homeostasis, and inflammation (Hagberg et al. 2024). BAT represents a small proportion of the total fat mass in adult humans (0.2–3.0%) and is predominantly located in the neck, supraclavicular, perirenal/adrenal, and vertebral fat depots. WAT could undergo a phenotypic switch under certain conditions to become 'browned' (beige) taking on a thermogenic brown-like phenotype. However, in non-pathological states, this is believed to represent a very small fraction of WAT adipocytes in humans (Song et al. 2020; Herz et al. 2019). Perivascular adipose tissue (PVAT) is a particular form of adipose tissue that shields the blood vessels. PVAT consists of both WAT and BAT and the different phenotypes seem to be largely dependent on their anatomical locations. As an example, while the abdominal aorta PVAT shows a WAT-like phenotype, the thoracic aorta is surrounded by BAT-like phenotype. The close connection to the vasculature makes the role of PVAT unique; it not only protects the vessels from external stimuli but also, more relevantly, represents a metabolically active organ, releasing a wide variety of molecules, recognized as PVAT-derived relaxing factors (PDRFs), which exert a paracrine effect on the vasculature. These substances are generally referred to as adipokines that diffuse into the vasculature contributing to the vascular tone homeostasis.

2 Perivascular Adipose Tissue

2.1 Anatomical Localization, Composition, and Role

PVAT has been recognized as an endocrine organ with important physiological and pathological effects in vasculature (Ramirez et al. 2017; Chang et al. 2020). As well known, the endothelium is the main component of the tunica intima; it is a thin layer of squamous cells lining the internal surface of the entire cardiovascular system. It contributes to a fine regulation of the vascular tone with a well-combined system of autocrine and paracrine interplay. This mechanism is based on the release of local factors that influence not only the vasculature tone but also organs within structure and function. The release of these factors is dependent on changes in local perfusion pressure (shear stress), and this interplay contributes to controlling the vascular homeostasis. In this regard, it has been demonstrated that PVAT contributes as well to the vascular homeostasis. Capacitance vessels are shielded by PVAT, as well as small and resistant vessels such as mesentery, skeletal muscle micro-vessels, coronary arteries, and subcutaneous arteries (Löhn et al. 2002; Verlohren et al. 2004; Weston et al. 2013; Zavaritskaya et al. 2013). On the other hand, microcirculation and cerebral vasculature do not contain PVAT (Li et al. 2021). As previously reported, PVAT differs from the other adipose tissues distributed in the body (subcutaneous, omental, or visceral) (Miao et al. 2012; Gil-Ortega et al. 2015). As stated before, PVAT is composed of WAT and BAT; WAT serves as a lipid storage, and BAT is associated with thermogenesis and is a reservoir of active substances due to its difference in the phenotype (Harms and Seale 2013). Of note, PVAT contains a plethora of different types of cells (i.e. adipocytes, neuronal cells (autonomic nerves), immune cells, and the stromal vascular fraction that consists of pre-adipocytes, fibroblasts, endothelial cells, and mesenchymal stem cells derived from the local microcirculation) with different roles. Thus, it suggests an intriguing role. Concerning the immune cells, PVAT contains a community of innate and adaptive immune cells, consisting of lymphocytes (T and B cells), macrophages, eosinophils, NK-cells, and mast cells. In the healthy state, macrophages and T-cells form the dominant immune-cell types in PVAT that together with the eosinophils contribute to the anti-inflammatory and anti-constrictive effects of PVAT on the vessel wall (Gil-Ortega et al. 2015; Nosalski et al. 2017).

Soltis and Cassis in 1991 clearly demonstrated that PVAT serves more than structural support; it is a working part of the blood vessel (Soltis et al. 1991). After this evidence, several groups have produced evidence that one of the functions of PVAT in health conditions is to release anticontractile factors firstly named ADRF i.e., "*adventitium*-derived relaxing factor" (Lohn et al. 2002). Later, these active molecules have been recognized as adiponectin, nitric oxide (NO), hydrogen sulfide (H_2S), and so on. The anti-contractile functions of PVAT have been observed in humans and other species, as well as in multiple types of blood vessels (Szasz et al. 2012; Verlohren et al. 2004; Antoniades et al. 2023). They found that phenylephrine, angiotensin II (AngII), or serotonin contractions in aortic rings were weaker when PVAT was preserved. This anticontractile effect was also observed in vessels without PVAT by using medium obtained from PVAT-incubated specimens, suggesting that a release of active substances in the medium occurs (Löhn et al. 2002, Verlohren et al. 2004; Gollasch et al. 2012; Szasz et al. 2012). In particular, the interaction between PVAT and the vascular wall, which renders PVAT a potential therapeutic target in cardiovascular diseases, has been investigated. PVAT could represent a clinically relevant source of diagnostic and prognostic biomarkers of vascular function, which may guide precision medicine in atherosclerosis, hypertension, heart failure, and other cardiovascular diseases. More recently, a role as a 'biosensor' of vascular inflammation has been highlighted since imaging technologies that visualize PVAT in clinical practice allow non-invasive quantification of coronary inflammation and the related residual cardiovascular inflammatory risk, guiding deployment of therapeutic interventions (Antoniades et al. 2023).

2.2 Perivascular Adipose Tissue-Releasing Factors and Their Function

PVAT releases bioactive molecules such as adipokines, anti-inflammatory, and pro-inflammatory factors as cytokines/chemokines (interleukin (IL)-6, tumour necrosis factor (TNF)-α, interferon (INF)-γ, monocyte chemoattractant protein, miRNAs, prostacyclin, NO, H_2S, angiotensin (Ang)-1–7, AngII, endothelin 1 (ET-1), palmitic acid methyl ester (PAME), and reactive oxygen species (ROS) such as hydrogen peroxide, named PVAT-derived factor. All these molecules contribute to the regulation/dysregulation of vascular homeostasis in health and cardiometabolic diseases (Fig. 1) (Malinowski et al. 2008; Almabrouk et al. 2014; Gollasch et al. 2012; Lee et al. 2011; Nòbrega et al. 2019; Szasz et al. 2012; Mitidieri et al. 2022, Tano et al. 2014; Szasz et al. 2013; Gao et al. 2007).

More interestingly, it should be underlined that a bi-directional communication between vascular cells and PVAT exists, and this contributes to the regulating vascular function. A better understanding of PVAT-specific secretome changes in response to vascular signals, including adipokines, cytokines, and lipid peroxidation products, is essential to identify new diagnostic biomarkers and therapeutic targets in cardiovascular medicine.

Under physiological conditions, PVAT exerts a net vasodilatory, antioxidant, and anti-inflammatory effect on the vasculature by releasing a plethora of active molecules (Kim et al. 2020). PVAT-derived factors are released in a paracrine (i.e. by diffusion) or vasocrine (i.e. mediated by the local microcirculation) way (Antoniades et al. 2023).

The imbalance between vasoactive molecules (vasorelaxants and vasoconstrictors) together with anti/pro-inflammatory stimulus (anti/pro-oxidant)

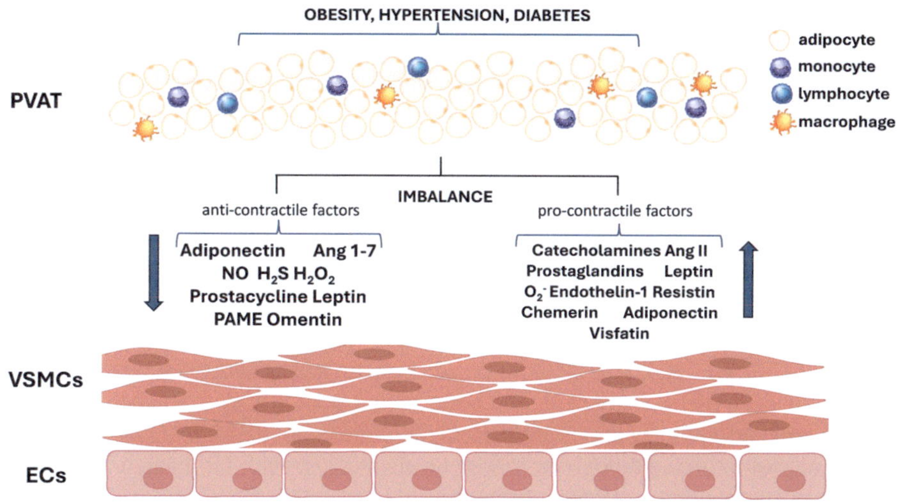

Fig. 1 PVAT released factors in the control of vascular tone in health and disease. PVAT is composed of adipocytes and immune cells that modulate vascular tone through the secretion of vasoactive factors (adipokines). Adipokines and other soluble factors can either reduce (anti-contractile) or enhance (pro-contractile) VSMC contractility. Ang 1–7, Angiotensin 1–7; Ang II, Angiotensin II; H_2S, hydrogen sulfide; NO, nitric oxide; PAME, palmitic acid methyl ester; PVAT, perivascular adipose tissue; VMSCs, vascular smooth muscle cells; ECs, endothelial cells

can affect the pathological process of vascular remodelling, participating in the development of hypertension, atherosclerosis, and cardiometabolic diseases.

The vasodilator effect is due to the secretion of adipocyte-derived relaxing factors as adiponectin, apelin, leptin, and omentin, which induce vasodilatation either by acting directly on vascular smooth muscle cells (VSMCs) (e.g. adipocyte-derived relaxing factor, leptin, and adiponectin) and/or indirectly by increasing the generation of endothelium-derived vasodilators such as nitric oxide (NO) and endothelium-derived hyperpolarizing factor (e.g. apelin, omentin, leptin, and adiponectin).

Leptin displays a dual effect: i) In physiological conditions, it exerts protective effects such as endothelium- and NO-dependent vasorelaxation, increase in angiogenesis, and decrease in lipid accumulation and lipotoxicity. ii) In cardiometabolic disease, leptin has detrimental effects such as sympathetic activation, insulin resistance, and enhanced platelet aggregation, where a selective leptin resistance has been suggested, especially in obese individuals (Wolk et al. 2006; Li et al. 2014; Tahergorbi et al. 2015).

Adiponectin explicates the action by the activation of adiponectin receptors AdipoR1 and AdipoR2 (Yamauchi et al. 2014). This hormone is a local vasodilator; it has been observed that PVAT anti-contractile effect is lost in eosinophil-deficient mice and is restored after eosinophil reconstitution. This anticontraction effect is mediated by the release of adiponectin and NO upon activation of adipocyte 3-adrenoceptors by eosinophil-derived catecholamines. In PVAT-intact mouse mesenteric arteries, 3-adrenoceptors stimulation causes the release of a hyperpolarizing factor, probably adiponectin, capable of hyperpolarizing the underlying VSMCs via the opening of BK_{Ca} channels and NO release from adipocytes (Weston et al. 2013). In obese mice, the anti-contractile effect of femoral artery PVAT is markedly reduced, as an impairment of the acetylcholine-induced relaxation associated with a reduction of p-eNOSSer1177 and Cu/Zn-SOD expression was found. Besides, a decrease in adiponectin levels and AdipoR1 expression along with an increase in leptin and ROS production occurred. Interestingly, the aerobic activity reversed this impairment (Sousa et al. 2021). The adiponectin-induced vasodilation acts by increasing not only the eNOS activity but also the substrate tetrahydrobiopterin in endothelial cells and stimulates VSMCs directly by activating BKCa 1.1 or Kv channels or indirectly through the inhibition of VSMC NADPH oxidase (Ahmed et al. 2023). Moreover, adiponectin exerts anti-inflammatory (Van de Voorde et al. 2014; Zang et al. 2009; Hulsmans et al. 2012), antidiabetic (Chang et al. 2010) and vascular protective effects (Withers et al. 2014).

In contrast to other adipokines, whose levels increase with fat mass, adiponectin decreases in individuals suffering from obesity (Arita et al. 1999) or type 2 diabetes showing a strong negative correlation with multiple indices of insulin resistance (Hotta et al. 2000; Weyer et al. 2001). Numerous studies demonstrated that adiponectin plasma levels are inversely associated with hypertension (Pery-Okonny et al. 2016) or cardiovascular disease (Xu et al. 2012; Sarvottam et al. 2013). Therefore, the vasodilating and protective effects of adiponectin are probably mediated by multiple effects on the endothelium, including AMP-activated protein kinase-dependent endothelial NO activation and NO production (Cheng et al. 2007) with an improvement of the redox state in vessels (Margaritis et al. 2013).

Omentin exists in two isoforms, omentin-1 and omentin-2, and the former is predominant in human plasma (Shuldiner et al. 2006). Omentin, which exerts anti-inflammatory, antioxidant, anti-atherogenic, and cardioprotective functions, is expressed in PVAT. Direct evidence demonstrating the involvement of omentin in PVAT anti-contractile effect is still lacking, as PVAT-specific omentin knockout animal models are not available yet. Nevertheless, human recombinant omentin relaxes noradrenaline-induced contraction of both rat aorta and mesenteric arteries in an endothelium-dependent manner via eNOSSer1177 phosphorylation and independently of Akt or tyrosine kinase activation (Yamawaky et al. 2010). This protein is downregulated in patients with carotid plaque (Liu et al. 2011) and coronary arterial dysfunction (Shibata et al. 2011). Omentin derived from adipose tissue inhibits the growth of VSMCs (Uemura et al. 2015; Kazama et al. 2014a). In addition, the administration of omentin in rats inhibits vascular remodelling and abnormal contractile response and reduces blood pressure, suggesting this protein as a drug candidate for the prevention or treatment of vascular diseases (Kazama et al. 2013, 2014b). However, the contribution of omentin to PVAT anti-contractile effect awaits more experimental evidence, particularly that involving animal models with omentin gene-specific deletion in PVAT.

Concerning the role of PVAT-derived PAME, there are only a few studies available; it is a potent vasodilator released from the sympathetic ganglion (Lee et al. 2019). PAME was first identified as PDRF in aortic PVAT of Wistar–Kyoto rats, where it induces vasodilation by the opening of KV channels (Lee et al. 2011). Noticeably, the levels of the enzyme involved in the PAME synthesis, the catechol-O-methyltransferase, are significantly reduced in SHR, suggesting a role for PAME in hypertension (Liu et al. 2020). Due to the limited evidence, the role of PAME in PVAT regulation of vascular tone requires further studies, particularly in small blood vessels.

Other adipokines such as chemerin, visfatin, and resistin contribute to the increase in vascular tone together with pro-inflammatory effect, proliferative response of VSMCs (Zavaritskaia et al. 2013), and impairment of insulin-stimulated glucose uptake.

Chemerin is a multifunctional inflammatory adipokine, which exerts its effect by binding to and activating the chemokine-like receptor 1. Chemerin and its active fragment chemerin-9 contract different vascular beds in a concentration-dependent manner. It causes an increased L type Ca2+ channels opening and stimulation of Src kinase and Rho kinase activity. PVAT-derived chemerin also contributes to obesity-induced endothelial dysfunction and hypertension (Ahmed et al. 2023). In this regard, other studies are needed to well define chemerin's contribution to vascular diseases to understand whether chemerin receptor antagonists might be valuable therapeutic weapons.

Resistin bounds to receptors such as Toll-Like Receptor 4 trigger various intracellular signalling cascades leading to vascular inflammation, lipid accumulation, oxidative stress, and cardiovascular diseases (Kavanami et al. 2004), endothelial dysfunction (Chen et al. 2010), angiogenesis (Mu et al. 2006), and VSMCs dysfunction (Filkovà et al. 2009). PVAT-derived resistin increases the susceptibility to hypertension in males compared to female rats (Small et al. 2019). Moreover, it is implicated in the development of insulin resistance, which is associated with obesity and diabetes; it attenuates insulin-induced vasodilation by inhibiting tyrosine/serine phosphorylation of insulin receptor substrate-1 and its sequential interaction with phosphatidylinositol 3-kinase, thus impairing Akt and eNOS phosphorylation (Steppan et al. 2001; Luo et al. 2018).

Interestingly, resistin has been considered a predictive factor for coronary atherosclerosis in humans and also correlates with markers of inflammation such as TNF-α, soluble TNF-α receptor-2, or IL-6 (Yaturu et al. 2006; Burnet et al. 2005; Reilly et al. 2005). Moreover, this adipokine positively correlates with the levels of biomarkers for cardiac injury, suggesting resistin as a marker of the severity of myocardium ischaemic injury (Chu et al. 2008).

Visfatin is a multifunctional adipokine, which can activate insulin receptors and toll-Like receptors and is associated with several diseases such as hypertension, diabetes, obesity, and atherosclerosis (Auguet et al. 2016; Yu et al. 2019). PVAT-derived visfatin can induce endothelial dysfunction and VSMC proliferation (Wang et al. 2009; Ahmed et al. 2023).

The renin-angiotensin system (RAS) represents another source of PVAT-derived factors. Several studies indicate Ang1–7 (Lee et al. 2009) and Ang II (Galvez-Prieto et al. 2008) as potential candidates. Ang1–7 produced by PVAT binds to its endothelial receptor, causing vasodilation indirectly through the endothelial NO production (Lee et al. 2009; Gil-Ortega et al. 2015; Nòbrega et al. 2019). In parallel, PVAT also produces the vasoconstrictor Ang II (Huang et al. 2010; Lu et al. 2010), a powerful growth factor stimulating VSMC growth and migration (Kim et al. 2000). PVAT-specific RAS activation has been implicated in the development of atherosclerosis associated with chronic kidney disease (CKD) and may represent a therapeutic target for preventing CKD-associated atherogenesis (Kawahito et al. 2013).

Beyond the aforementioned mediators, gaseous molecules such as NO and H_2S are feasible candidates for PVRFs as well, and this chapter will discuss the role of PVAT-derived H_2S in the control of vascular homeostasis in physiological and pathological conditions.

3 Perivascular Adipose Tissue-Derived Hydrogen Sulfide

3.1 Perivascular Adipose Tissue-Derived Hydrogen Sulfide in Physiological Conditions

Emerging evidence recognizes H_2S as a vascular anti-contractile factor released from PVAT. The initial evidence supporting this hypothesis was presented by Fang et al. (2009), who first demonstrated that PVAT has the endogenous ability to produce H_2S (Fig. 2) (Fang et al. 2009). The authors showed that PVAT-derived H_2S is mainly produced by CSE acting as ADRF. In detail, the inhibition of CSE by using both reversible (beta-cyano-L-alanine or BCA) and irreversible (PAG) inhibitors resulted in reduced H_2S production. This finding underscored the role of CSE as a key enzyme involved in PVAT-derived H_2S synthesis and its subsequent effects on vascular tone. Furthermore, the authors used an experimental model of rat aortic rings demonstrating that PVAT significantly attenuated the contractile response to vasoactive agents like phenylephrine, serotonin, and Ang II, compared to rings deprived of PVAT. This suggests that H_2S released from PVAT exerts an anti-contractile effect on the vascular smooth muscle. Notably, the inhibition of CSE in the presence of PVAT increases the contraction, further confirming the critical role of PVAT-derived H_2S in regulating vascular tone. These findings provide compelling evidence for the role of PVAT as a modulator of vascular reactivity through its production of H_2S, which acts as an anti-contractile mediator. Interestingly, in the same study, in a hypertensive rat model induced by abdominal aortic banding, the expression of CSE and H_2S production was upregulated in PVAT but not in the aortic tissue itself. This suggests that the increase in PVAT-derived H_2S may be a compensatory response to elevated blood pressure, contributing to vascular protection in the context of hypertension. The authors further demonstrated that transplanting PVAT into the peri-adventitia of stenotic aortas reduced elevated arterial blood pressure and lowered Ang II levels, suggesting a protective role for PVAT/CSE-derived H_2S in mitigating hypertension-induced vascular changes (Fang et al. 2009).

Kohn and co-workers (2012) confirmed the identity of PVAT-derived H_2S as ADRF in rat aortas and demonstrated the involvement of XE991-sensitive KCNQ

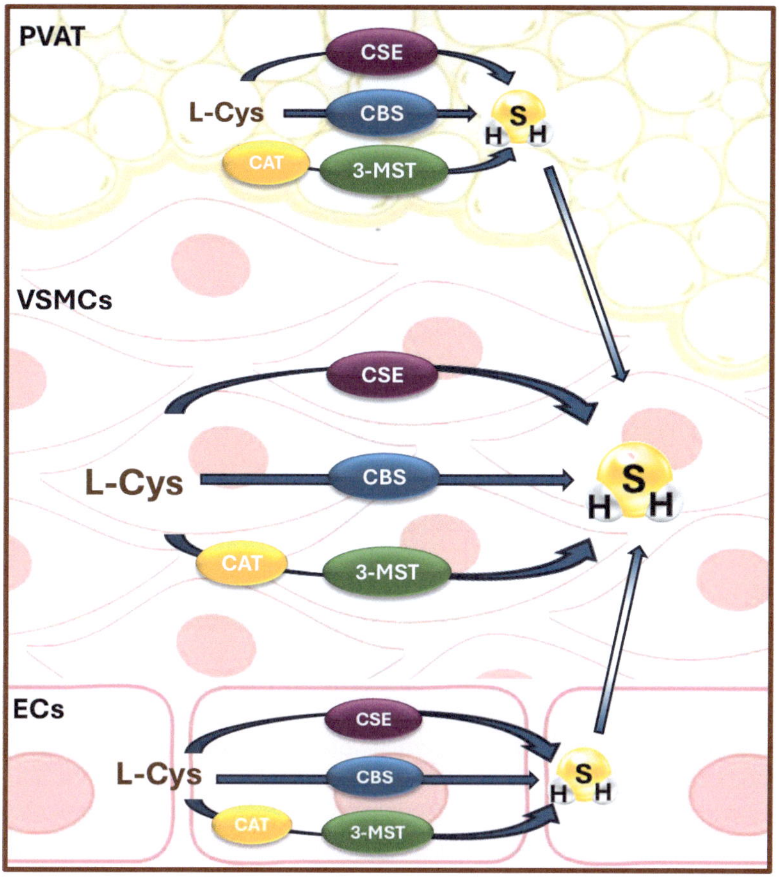

Fig. 2 Biosynthetic pathway of H_2S in PVAT, VSMCs, and ECs. The enzymes CBS, CSE, and 3-MST are expressed in PVAT, SMCs, and ECs, with their expression pattern varying across species and vascular regions. These enzymes contribute to the generation of H_2S, which modulates vascular tone by contributing to the regulation of the whole vascular function. 3-MST, 3-mercaptosulfurtransferase; CAT, cysteine aminotransferase; CBS, cystathionine-β-synthase; CSE, cystathionine-γ-lyase; PVAT, perivascular adipose tissue; VMSCs, vascular smooth muscle cells; ECs, endothelial cells

channels in mediating the anti-contractile effect of PVAT/CSE-derived H_2S (Kohn et al. 2012). Their findings showed differences between the aortas of rats and mice. Specifically, the serotonin-induced contraction was less pronounced in aortic tissue with intact PVAT compared to that denuded of PVAT, and this effect was reversed by KCNQ channel inhibition using XE991 in both species. Furthermore, blocking CBS with AOAA did not alter the anticontractile effects of PVAT in both rats and mice, suggesting that CBS-derived H_2S does not contribute significantly to this effect. However, the inhibition of CSE with PAG reduced the anticontractile effect of PVAT in rats but not in mice, indicating that CSE-derived H_2S modulates the effect in rats rather than acting as a direct candidate ADRF in mouse aortas (Kohn et al. 2012).

These findings support the idea that the mechanisms by which PVAT regulates vascular tone via H_2S may differ across species and locations, emphasizing the complexity of PVAT's role in vascular biology. Indeed, the study highlights significant interspecies differences in the response of aortic tissue to PVAT-derived H_2S, with neither CSE- nor CBS-derived H_2S playing a role in the anticontractile effects on mouse aortas. This observation could explain the divergent effects of CSE- and CBS-derived H_2S in modulating the anticontractile action of PVAT.

Similar findings were observed in mouse mesenteric arteries, where the presence of PVAT attenuated the contraction induced by agents such as serotonin and KCl. This effect was reversed by PAG or BCA, inhibitors of CSE, as well as with XE991, an inhibitor of KCNQ channels, underscoring the role of both CSE-derived H_2S and KCNQ channels in the paracrine regulation of vascular tone by PVAT. Furthermore, the authors demonstrated that these paracrine effects could be replicated by synthetic KCNQ channel openers, suggesting that the opening of KCNQ channels plays a crucial role in the paracrine control of vascular tone by PVAT, with the involvement of H_2S as a key mediator. These results indicate the crucial role of KCNQ channel opening in the paracrine control of vascular tone by PVAT, which is at least in part mediated or modulated by H_2S (Schleifenbaum et al. 2010).

However, concerning the cited studies, it's relevant to consider that the inhibitors used are not fully selective (Asimakopolou et al. 2013). Specifically, AOAA, which was used to block CBS, has been shown to inhibit CSE more effectively than CBS. This raises concerns regarding the specificity of the conclusions drawn by Kohn and colleagues, as the effects observed in their study could be attributed to the inhibition of CSE rather than CBS. Therefore, the interpretation of the results, particularly the role of CBS in PVAT-derived H_2S production, may need to be revisited. To address this uncertainty, further investigations are essential to definitively identify whether CSE or CBS is the primary enzymatic contributor to H_2S production in PVAT. These additional studies are crucial for clarifying the precise mechanisms involved and for accurately interpreting the role of H_2S in regulating vascular tone.

Nevertheless, it is evident that the contribution of PVAT-derived H_2S to vascular tone regulation varies significantly across different vessels and species. Specifically, coronary artery tone is influenced by the surrounding PVAT, which releases diffusible vasoactive factors that modulate arterial tone through a combination of anti-contractile and anti-relaxant mechanisms. These mechanisms appear to be dependent on the quantity of PVAT, regardless of the vascular region or species involved.

Although PVAT slightly increases resting arterial tension, it has been demonstrated to significantly attenuate the contractile response to vasoactive agents such as serotonin, stable thromboxane analogues, and phenylephrine in coronary arteries. Notably, this modulatory effect is absent when vessels are contracted using depolarizing agents like KCl, which directly alter the membrane potential. These findings suggest that PVAT does not influence contractile responses mediated by membrane depolarization. Therefore, the observed discrepancies in contractile responses between arteries with and without PVAT are likely not attributable to mechanical interference during vascular constriction.

The anticontractile factors released by PVAT play a crucial role in regulating coronary artery tone, not only controlling basal coronary tone but also modulating agonist-induced contractions. This occurs through a mechanism involving the reduction of the sensitivity of Rho-kinase-dependent smooth muscle Ca^{2+} channels (Aalbaek et al. 2015). The Rho-kinase pathway is an important regulator of smooth muscle contraction, and its inhibition by PVAT-derived factors suggests a protective mechanism against excessive vasoconstriction in coronary arteries. This implies that PVAT acts against excessive narrowing of coronary vessels, potentially protecting the myocardium from ischaemia. Interestingly, exogenous sources of NO or H_2S produce similar degrees of vascular relaxation in arteries, regardless of the presence of PVAT. This finding raises the question of whether these molecules, which are also involved in PVAT's anticontractile effect, may operate via a different mechanism when introduced exogenously compared to their endogenous release by PVAT.

Moreover, the role of the endothelium in this process is of particular interest. Indeed, removal of the endothelium enhanced the anticontractile effect of PVAT in response to agonist-induced contraction. This suggests that endothelial cells may normally modulate the effects of PVAT, perhaps by releasing additional vasoactive substances such as NO or by influencing the signalling pathways involved in smooth muscle relaxation and contraction. In the absence of endothelium, PVAT's effects may be more pronounced, underscoring the complex interplay between endothelial and PVAT-derived factors in regulating vascular tone. To further explore this interaction, arteries were exposed to acetylcholine or methacholine, which stimulates endothelial cells to release NO and induce vasodilation. The relaxation response to cholinergic stimulation was significantly reduced in the presence of PVAT, suggesting that PVAT can inhibit endothelial function or reduce endothelial-dependent relaxation. This phenomenon likely involves a balance between the pro-relaxant effects of endothelial NO and the anti-relaxant effects of PVAT-derived factors like H_2S.

Notably, the use of DL-propargylglycine, an inhibitor of CSE, significantly attenuated the observed difference in vasorelaxation between arteries with and without PVAT. This result provides compelling evidence that H_2S, derived from PVAT, is an important regulator of vascular tone. Furthermore, the fact that this difference in relaxation was abolished when Kv7 channels, important targets for H_2S, were inhibited by XE991, further strengthens the argument that PVAT-derived H_2S acts through specific ion channels to modulate smooth muscle function. Kv7 channels are known to contribute to vasorelaxation by facilitating the hyperpolarization of smooth muscle cells, thereby reducing their contractility. The fact that blocking these channels negates the effect of PVAT on vasorelaxation underscores their crucial role in the PVAT-H_2S signalling pathway.

Taken together, these results strongly suggest that H_2S, released by PVAT, serves as a critical anticontractile factor in the regulation of coronary artery tone, and that the KCNQ (Kv7) channels play a pivotal role in this process. PVAT appears to exert dual regulation over coronary artery tone, modulating both vasoconstriction and vasorelaxation. This dual control provides a dynamic mechanism by which PVAT can finely regulate vascular resistance, ensuring adequate blood flow to the myocardium under varying physiological conditions. Such regulation may be particularly important in the context of pathological conditions such as atherosclerosis, where PVAT's ability to buffer excessive vasoconstriction and promote vasodilation could help preserve coronary function and prevent ischemia.

However, it is important to recognize certain limitations in the studies reported here. A significant limitation is the use of non-selective inhibitors to identify the enzymes responsible for the generation of PVAT-derived H_2S (Asimakopolou et al. 2013). These inhibitors may not be highly specific and could have off-target effects, potentially confounding the interpretation of results. For example, other enzymes or signalling pathways might also contribute to PVAT's anticontractile effects, and these contributions may be overlooked if the inhibitors do not precisely target the relevant enzymes. Therefore, further studies using more selective and specific inhibitors, once available, are essential to definitively characterize the enzymes and molecular pathways involved in PVAT-derived H_2S production.

In conclusion, the dual role of PVAT in regulating coronary artery tone through both vasoconstriction and vasorelaxation highlights the importance of PVAT as a dynamic and fine-tuned regulator of vascular resistance. Understanding the molecular mechanisms by which PVAT exerts these effects, including the contribution of H_2S and Kv7 channels, will be crucial for developing therapeutic strategies aimed at improving coronary function in pathological conditions. Further research using selective inhibitors and more advanced techniques will be necessary to fully elucidate the complex role of PVAT in vascular health.

3.2 Perivascular Adipose Tissue-Derived Hydrogen Sulfide in Pathological Conditions

The influence of PVAT on blood vessel tone varies with metabolic state, inflammation, and clinical risk factors. In healthy conditions, PVAT's protective and vasorelaxant properties are predominant, whereas in pathological conditions, its effects appear more detrimental (Guzik et al. 2007). In this scenario, accumulating evidence shows a key role of PVAT-derived H_2S in cardiovascular diseases (Bełtowski et al. 2013; Mitidieri et al. 2022).

Several studies have shown that obesity is associated with a reduced anticontractile effect or even a paradoxical pro-contractile response of PVAT in both animals and humans (Ketonen et al. 2010; Ma et al. 2010; Lynch et al. 2013; Greenstein et al. 2009; Owen et al. 2013). In this regard, the role of H_2S in PVAT dysfunction was examined in an experimental model of obesity induced by a high-calorie diet on the anticontractile effect of PVAT in aortic rings in rats (Bełtowski et al. 2012, 2013). More in detail, the contractility of endothelium-denuded aortic rings without PVAT was similar to the control, 1-month obese, and 3-month obese rats. However, obesity had a time-dependent effect on the phenylephrine-induced contraction of aortic rings with intact PVAT since the anticontractile effect of PVAT was increased in the 1-month obese rats but was reduced in the 3-month obese group, by comparison with the control animals. In line, H_2S production in isolated PVAT slices was higher in the 1-month obese, and lower in the 3-month obese, than in the control rats. H_2S production in stimulated condition (i.e. by L-cysteine) as well as CSE activity toward cystathionine in these tissues were not altered in the 1-month obese group but were reduced in the 3-month obese group, coupled with a reduction in CSE expression as both mRNA and protein levels. Besides, the anticontractile effect of PVAT in all 3 groups was markedly attenuated by either propargylglycine or glibenclamide, indicating that the CSE/H_2S/K_{ATP} pathway is involved. Thus, a down-regulation of CSE-derived H_2S in PVAT occurs in obesity with increasing aging. Interestingly, the treatment of 3-month obese rats with rosiglitazone, a peroxisome PPAR-λ agonist, increased CSE expression and activity as well as H_2S production by isolated PVAT slices improving the anticontractile effect of PVAT on aortic rings (Bełtowski 2016).

Basically, it has been reported that H_2S signalling in PVAT is determined by the rate of its synthesis by CSE and its mitochondrial oxidation. Indeed, obesity induced by short-term feeding with a high-fat diet (4 weeks) increased H_2S-derived PVAT by suppressing mitochondrial biogenesis and decreasing mitochondrial H_2S oxidation. These effects were reversed by rimonabant, CB1antagonist, which had, however, no effect in rats fed standard chow. Thus, an interplay between H_2S-derived PVAT and the cannabinoid system occurs in obesity with an initial protective effect in the short term (Beltowski et al. 2015). Indeed, the treatment with arachidonyl-2-chloroethanolamide, a CB1 agonist, increased H_2S production by isolated PVAT and the anticontractile effect of PVAT coupled to a reduction of mitochondrial H_2S oxidation (Tedesco et al. 2010). Therefore, PVAT-derived H_2S is upregulated by short-term obesity or brief cannabinoid treatment, which may serve as a compensatory mechanism to maintain vascular homeostasis despite endothelial dysfunction. On the other hand, prolonged obesity can lead to H_2S deficiency in PVAT, potentially exacerbating harmful vascular complications. Nevertheless, oxygen pressure was markedly reduced within the PVAT in obese animals, consistently with many studies indicating that obesity is associated with adipose tissue hypoxia due to greater diameter of adipocytes and reduced relative density of capillaries (Trayhurn et al. 2013). Particularly, in the context of hypoxia, H_2S plays a significant role in cellular metabolism and mitochondrial function. Under normal oxygen conditions, mitochondria are responsible for the efficient production of ATP through oxidative phosphorylation. However, in hypoxic conditions, where oxygen availability is limited, mitochondrial function is impaired, and cells must adapt to sustain energy production.

H_2S is involved in mitochondrial bioenergetics, acting both as a signalling molecule and a modulator of mitochondrial function. At low concentrations, H_2S can serve as an endogenous regulator of mitochondrial activity by influencing several key mitochondrial processes. Specifically, H_2S has been shown to affect mitochondrial respiration by modulating the activity of cytochrome c oxidase in the electron transport chain. In the presence of hypoxia, H_2S can enhance mitochondrial efficiency by acting as an alternative electron donor, potentially compensating for reduced oxygen levels.

Additionally, H_2S has been implicated in protecting mitochondrial integrity under conditions of oxidative stress, which is often associated with hypoxia. It exerts antioxidative effects by scavenging ROS and preventing mitochondrial damage. This protective role helps to preserve mitochondrial function and cellular viability during hypoxic stress (Szabo et al. 2014; Guo et al. 2012; Huang et al. 2023).

Based on this evidence, the hypoxia occurring in PVAT could be responsible for the increase in H_2S signalling in short-lasting obesity (1-month obese rats). In line, it has also been reported that PVAT enhances the hypoxic relaxation response in porcine coronary arteries through mechanisms involving H_2S (Donovan et al. 2018). All three H_2S-synthesizing enzymes have been detected in the artery and myocardium, but only CBS and 3-mercaptopyruvate sulphurtransferase (3-MST) were detected in PVAT. Specifically, coronary arteries with intact PVAT exhibited reduced hypoxic contraction and increased relaxation, indicating a distinct in situ effect of PVAT and the involvement of diffusible factors, identifying it as H_2S. In fact, by using a classical pharmacological approach on arteries with intact PVAT, the enhancement of hypoxia-induced relaxation was reversed by inhibiting CBS but not CSE, where CBS is the primary enzyme responsible for H_2S generation during hypoxic relaxation in coronary arteries; the augmentation of hypoxic relaxation was not reversed by aspartate as inhibitor of 3-MST pathway (Donovan et al. 2017, 2018). Finally, the potential of sulfur-containing AMP and GMP analogues, AMPS and GMPS, as H_2S donors in PVAT from obese rats has been explored. AMPS and GMPS can be hydrolyzed to H_2S by PVAT upon activation of P2X7 receptors; both AMPS and GMPS can restore the anticontractile effects of PVAT found in obese animals without P2X7 stimulation (Beltowski et al. 2015).

Similarly to what occurs in obesity, an impairment in the aorta anticontractile effect of PVAT was found in spontaneously hypertensive rats (SHR) in which the density and the total number of brown adipocytes were greater than in normotensive rats (WKY) while the cross-section area of PVAT was similar between them. The functional impairment observed in SHR-PVAT may be related to changes in adipocyte composition but not to reduced PVAT mass in SHR (Lu et al. 2011). In the experimental model of hypertension induced by abdominal aortic banding in rats, the H_2S synthesis and CSE expression in the aorta without PVAT was unchanged compared with the normotensive animals, but of note, it was increased in isolated PVAT (Fang et al. 2009). Transplanting PVAT into peri-adventitia of stenotic aortas ameliorated the increase in blood pressure and decreased angiotensin II levels in the aorta. Therefore, the CSE-derived H_2S in PVAT could be a backup vasodilatory mechanism that is up-regulated in response to both the acute effect of vasoconstrictors and chronic hypertension. In addition, H_2S production by PVAT isolated from ageing (12-month-old) rats was lower than in 6-month-old animals by 35%, and lower than in 1-month-old rats by 65%, despite the age-related increase in CSE expression, suggesting that the anticontractile effect of H_2S-derived PVAT decreases with age, even if the mechanism of the dissociation between CSE expression and H_2S production in ageing animals is unclear (Fang et al. 2009). However, firstly, the authors did not take into consideration CBS and 3-MST in this study; secondly, this effect could be due to an impaired CSE activity rather than a CSE deficiency and/or enhanced H_2S scavenging or metabolism.

The anticontractile effect of PVAT was impaired in SHR compared to WKY rats and also on the mesenteric artery (Li et al. 2013). In this case, the difference between strains was eliminated by XE991, a blocker of Kv7 (KCNQ) voltage-dependent potassium channels. To date, Kv7 represents one of the major targets for H_2S-mediated relaxation in vessels; thus, an impairment in PVAT-derived H_2S can occur in hypertension (Abramavicius et al. 2021; Hedegaard et al. 2016; Testai et al. 2023).

Similarly, Galvez et al. (2006) noted that SHR exhibits a reduced PVAT in the mesenteric bed, suggesting that changes in PVAT mass and function may contribute to increased vascular resistance in hypertension (Galvez et al. 2006). More recently, it has also been found that the aorta of SHR showed an increased contractile response to noradrenaline in the presence or absence of PVAT compared to WKY. CSE inhibition reduced basal tension in the presence of PVAT in both groups, where a more pronounced effect was evident in WKYs compared with SHRs. In the absence of PVAT, CSE inhibition markedly increased basal tension in WKYs compared with SHR, suggesting the pro-relaxing effect of basally released H_2S plays a major role within the vasculature. Indeed, also a significant impairment in Ach-induced vasorelaxation was found in the aorta of SHR in the absence of PVAT. Thus, PVAT plays a role in the regulation of vascular tone under hypertensive conditions (Golas et al. 2022). In SHR, unlike normotensive rats, PVAT displayed pro-contractile effects in mesenteric arteries following perivascular nerve stimulation together with an impairment in basal H_2S production. Both PVAT and endogenous H_2S showed pro- and anti-contractile properties in mesenteric arteries, depending on the pathological condition and activated signalling pathways. However, nerve stimulation could also be associated with excessive production of pro-contractile factors under pathological conditions. In mesenteric arteries of normotensive rats, the presence of PVAT coupled with CSE inhibition increased basal tone while reducing contractions induced by exogenous noradrenaline. Conversely, in SHR, PVAT significantly reduced noradrenaline-induced contractions, unaffected by CSE inhibition, suggesting the release of a different regulatory factor, unlike from H_2S. Nevertheless, in SHR arteries with PVAT, the potentiation of exogenous H_2S-induced vasorelaxation was noted. Further investigation is needed to understand the mechanisms and potential interactions between the anti-contractile properties of PVAT and the upregulation of H_2S activity (Cacanyiova et al. 2019). A cooperative role of both PVAT and H_2S has also been observed in the aorta of hypertensive pregnant rats, where the anti-contractile effect mediated by K_{ATP} channel activation in PVAT was partially lost, but a key role for the PVAT-derived H_2S/K_{ATP} channel axis was evident in both normotensive and hypertensive pregnant rats (Souza-Paula et al. 2020). Similarly, in a model of hypertension in pregnant rats, the distinct roles of PVAT and endothelium in the vasodilatory responses induced by endogenous H_2S triggered by L-cysteine were evaluated. In aortic rings where PVAT was preserved, the vasorelaxation response to L-cysteine was unaffected by pregnancy (Preg)-induced hypertension. In contrast, aortic rings with intact endothelium but lacking PVAT exhibited impaired responses. To verify the role of H_2S derived from PVAT, a cumulative concentration curves of L-cysteine in aortic rings pre-incubated with CSE inhibitor were performed. The results showed no relaxation in any of the preparations from the Non-Preg, Hypertensive Non-Preg, and Preg groups, thereby supporting the findings reported by Fang et al. (2009). However, in the aortic rings of hypertensive pregnant rats, there was a slight relaxation when both PVAT and endothelium were intact. In contrast, the aortic rings lacking PVAT but retaining the endothelium did not exhibit any relaxation. These observations suggest that other H_2S-producing enzymes may be activated, leading to the modest vasorelaxant responses induced by L-cysteine, as previously demonstrated. This data highlights the significance of PVAT-derived H_2S in regulating vascular tone during pregnancy hypertension. Thus, it indicates that L-cysteine may induce a slight vasorelaxation independently of the H_2S produced by CSE in both PVAT and the vascular smooth muscle layers of aortas from hypertensive pregnant rats. Instead, the endothelium showed no response to L-cysteine in the presence of PAG, CSE inhibitor, indicating that H_2S produced by CSE is the primary enzyme in the endothelium. In contrast, H_2S released by PVAT may be generated by other enzymes, such as CBS. Therefore, the stimulation of endogenous PVAT-derived H_2S with L-cysteine could serve as a potential clinical strategy to alleviate hypertensive disorders of pregnancy linked to endothelial dysfunction (Oliveira et al. 2023). In summary, these findings highlight the role of PVAT-derived H_2S in regulating vascular tone, exhibiting both anti-contractile and pro-contractile actions depending on the vascular state, and influencing the development of hypertension. Thus, under hypertensive conditions, the activity of PVAT and the vasoactive effects of H_2S may be modified.

In this scenario, a differing regulatory role of PVAT was also found in diabetic rats (Emilova et al. 2015). Specifically, this study reveals the regulatory role of CSE-derived H_2S in the PVAT of diabetic rats in the *gracilis* artery. In normoglycemic rats, the contraction response to serotonin in the *gracilis* artery in the presence of PVAT (denuded of endothelium) was reduced and such difference disappeared in diabetic rats.

In detail, two different experimental models were employed: streptozocin injection and the incubation of healthy vessels in high glucose concentrations. A distinct profile for serotonin-induced contraction was observed; specifically, in hyperglycaemic conditions, the serotonin-induced contraction response was reduced in the presence of PVAT compared to its absence, and this effect was reversed by CSE inhibition. This finding suggests an anticontractile role of CSE-derived H_2S in PVAT, indicating a potential protective effect in diabetes. In *gracilis* arteries harvested from diabetic rats (induced by streptozocin), the contraction response to serotonin was similar whether PVAT was present or absent. However, at higher concentrations, the presence of PVAT enhanced the contraction, leading to the hypothesis that PVAT releases contractile agents with detrimental effects in diabetes. Notably, when CSE was inhibited by PAG, an increase in serotonin-induced contraction was observed, indicating an anticontractile effect of H_2S; this effect shifted to a reduction in contraction at higher serotonin concentrations, suggesting the release of a pro-contractile influence related to H_2S (likely produced only in its presence). Another possible explanation for this phenomenon is that blocking H_2S production generates another dilatory factor from PVAT.

Nevertheless, the nature of this signalling molecule, as well as the mechanism of its paracrine regulation and tissue specificity, requires further investigation. In our view, the data obtained from vessels harvested from diabetic rats exhibit greater similarities to diabetes in humans, thus making them more credible. However, a limitation of this study is that the authors only used one contracting agent, i.e. serotonin; this raises the question of whether similar effects could be observed with different contracting agents. Additionally, the high concentrations of serotonin used are far from physiological levels.

Recent studies have also reported the expression of CBS and CSE genes in PVAT surrounding internal arteries in humans for the first time. Tissue samples were obtained from patients with renal cell carcinoma, renal pelvic tumours, or ureteral tumours undergoing radical nephrectomy and divided into normotensive and hypertensive groups. The authors observed significantly higher CBS gene expression than CSE in PVAT in the arterial wall of both groups, while CSE gene expression was only detectable in PVAT, being below the limit of detection in the arterial wall of these patients (Cacanyiova et al. 2020).

4 Therapeutic Potential

Given the significant role of H_2S in PVAT and vascular health, restoring or enhancing H_2S signalling within PVAT represents a potential therapeutic strategy for preventing or treating cardiovascular diseases. Various approaches are being explored to increase endogenous H_2S production, including the use of H_2S donors, enzyme modulators, or gene therapy targeting H_2S-producing enzymes. Furthermore, lifestyle interventions such as exercise and dietary modifications have been shown to improve H_2S production in PVAT, suggesting that non-pharmacological strategies could complement pharmacological treatments.

5 Conclusion

H_2S-derived PVAT plays an essential role in maintaining vascular homeostasis and protecting against cardiovascular diseases. The production of H_2S in PVAT helps to regulate vascular tone, to reduce inflammation, and to mitigate oxidative stress, all of which are critical in preventing cardiovascular diseases. Dysfunction of PVAT, characterized by reduced H_2S production, contributes to the pathogenesis of conditions like atherosclerosis, hypertension, diabetes, and heart failure. Therefore, strategies to restore or enhance H_2S signalling in PVAT may offer promising therapeutic avenues for the prevention and treatment of cardiovascular diseases.

References

Aalbaek F, Bonde L, Kim S, Boedtkjer E (2015) The Journal of Physiology Perivascular tissue inhibits rho-kinase-dependent smooth muscle Ca_{2+} sensitivity and endothelium-dependent H_2S signalling in rat coronary arteries. J Physiol 593:4747–4764

Abramavicius S, Petersen AG, Renaltan NS, Prat-Duran J, Torregrossa R, Stankevicius E, Whiteman M, Simonsen U (2021) GYY4137 and sodium hydrogen sulfide relaxations are inhibited by L-cysteine and KV7 channel blockers in rat small mesenteric arteries. Front Pharmacol 12:613989

Ahmed A, Bibi A, Valoti M, Fusi F (2023) Perivascular adipose tissue and vascular smooth muscle tone: friends or foes? Cells 12:1196

Almabrouk TA, Ewart MA, Salt IP, Kennedy S (2014) Perivascular fat, AMP-activated protein kinase and vascular diseases. Br J Pharmacol 171:595–617

Antoniades C, Tousoulis D, Vavlukis M, Fleming I, Duncker DJ, Eringa E, Manfrini O, Antonopoulos AS, Oikonomou E, Padró T, Trifunovic-Zamaklar D, De Luca G, Guzik T, Cenko E, Djordjevic-Dikic A, Crea F (2023) Perivascular adipose tissue as a source of therapeutic targets and clinical biomarkers. Eur Heart J 12(44):3827–3844

Arita Y, Kihara S, Ouchi N, Takahashi M, Maeda K, Miyagawa J, Hotta K, Shimomura I, Nakamura T, Miyaoka K, Kuriyama H, Nishida M, Yamashita S, Okubo K, Matsubara K, Muraguchi M, Ohmoto Y, Funahashi T, Matsuzawa Y (1999) Paradoxical decrease of an adipose-specific protein, adiponectin, in obesity. Biochem Biophys Res Commun 257:79–83

Asimakopoulou A, Panopoulos P, Chasapis CT, Coletta C, Zhou Z, Cirino G, Giannis A, Szabo C, Spyroulias GA, Papapetropoulos A (2013) Selectivity of commonly used pharmacological inhibitors for cystathionine β synthase (CBS) and cystathionine γ lyase (CSE). Br J Pharmacol 169:922–932

Auguet T, Aragones G, Guiu-Jurado E, Berlanga A, Curriu M, Martinez S, Alibalic A, Aguilar C, Camara M-L, Hernàndez E, Ruyra X, Martin-Paredero V, Richart C (2016) Adipo/cytokines in atherosclerotic secretomes: increased visfatin levels in unstable carotid plaque. BMC Cardiovasc Disord 16:149

Bełtowski J (2013) Endogenous hydrogen sulfide in perivascular adipose tissue: role in the regulation of vascular tone in physiology and pathology. Can J Physiol Pharmacol 91:889–898

Beltowski J (2015) High fat diet-induced obesity augments H_2S in perivascular adipose tissue by cannabinoid-mediated decrease in mitochondrial biogenesis; Proceedings of the 83rd Congress of European Atherosclerosis Society. Atherosclerosis 241(1):e28

Bełtowski J, Jamroz-Wiśniewska A (2012) Modulation of h(2)s metabolism by statins: a new aspect of cardiovascular pharmacology. Antioxid Redox Signal 17:81–94

Bełtowski J, Jamroz-Wiśniewska A (2016) Hydrogen sulfide in the adipose tissue-physiology, pathology and a target for pharmacotherapy. Molecules 22:63

Bełtowski J, Guranowski A, Jamroz-Wiśniewska A, Wolski A, Hałas K (2015) Hydrogen-sulfide-mediated vasodilatory effect of nucleoside 5′-monophosphorothioates in perivascular adipose tissue. Can J Physiol Pharmacol 93:585–595

Burnett MS, Lee CW, Kinnaird TD, Stabile E, Durrani S, Dullum MK, Devaney JM, Fishman C, Stamou S, Canos D, Zbinden S, Clavijo LC, Jang GJ, Andrews JA, Zhu J, Epstein SE (2005) The potential role of resistin in atherogenesis. Atherosclerosis 182:241–248

Cacanyiova S, Majzunova M, Golas S, Berenyiova A (2019) The role of perivascular adipose tissue and endogenous hydrogen sulfide in vasoactive responses of isolated mesenteric arteries in normotensive and spontaneously hypertensive rats. J Physiol Pharmacol 70:295–306

Cacanyiova S, Krskova K, Zorad S, Frimmel K, Drobna M, Valaskova Z, Misak A, Golas S, Breza J, Breza J Jr, Berenyiova A (2020) Arterial hypertension and plasma glucose modulate the vasoactive effects of nitroso-sulfide coupled signaling in human intrarenal arteries. Molecules 25:2886

Chang J, Li Y, Huang Y, Lam KS, Hoo RL, Wong WT, Cheng KK, Wang Y, Vanhoutte PM, Xu A (2010) Adiponectin prevents diabetic premature senescence of endothelial progenitor cells and promotes endothelial repair by suppressing the p38 MAP kinase/p16INK4A signaling pathway. Diabetes 59:2949–2959

Chang L, Garcia-Barrio MT, Chen YE (2020) Perivascular adipose tissue regulates vascular function by targeting vascular smooth muscle cells. Arterioscler Thromb Vasc Biol 40:1094–1109

Chen C, Jiang J, Lü J-M, Chai H, Wang X, Lin PH, Yao Q (2010) Resistin decreases expression of endothelial nitric oxide synthase through oxidative stress in human coronary artery endothelial cells. Am J Physiol Heart Circ Physiol 299:H193–H201

Cheng KK, Lam KS, Wang Y, Huang Y, Carling D, Wu D, Wong C, Xu A (2007) Adiponectin-induced endothelial nitric oxide synthase activation and nitric oxide production are mediated by APPL1 in endothelial cells. Diabete 56:1387–1394

Chu S, Ding W, Li K, Pang Y, Tang C (2008) Plasma resistin associated with myocardium injury in patients with acute coronary syndrome. Circ J 72:1249–1253

Donovan J, Wong PS, Roberts R, Garle MJ, Alexander SPH, Dunn WR, Ralevic V (2017) A critical role for cystathionine-β-synthase in hydrogen sulfide-mediated hypoxic relaxation of the coronary artery. Vascular Pharmacol 93-95:20–32

Donovan J, Wong PS, Garle MJ, Alexander SPH, Dunn WR, Ralevic V (2018) Coronary artery hypoxic vasorelaxation is augmented by perivascular adipose tissue through a mechanism involving hydrogen sulphide and cystathionine-β-synthase. Acta Physiol 224:13126

Emilova R, Dimitrova D, Mladenov M, Daneva T, Schubert R, Gagov H (2015) Cystathionine gamma-lyase of perivascular adipose tissue with reversed regulatory effect in diabetic rat artery. Biotechnol Biotechnol Equip 29:147–151

Fang L, Zhao J, Chen Y, Ma T, Xu G, Tang C, Liu X, Geng B (2009) Hydrogen sulfide derived from periadventitial adipose tissue is a vasodilator. J Hypertens 27:2174–2185

Filková M, Haluzík M, Gay S, Senolt L (2009) The role of resistin as a regulator of inflammation: Implications for various human pathologies. Clin Immunol 133:157–170

Galvez B, de Castro J, Herold D, Dubrovska G, Arribas S, Gonzàlez MC, Aranguez I, Luft FC, Ramos MP, Gollasch M, Alfonso MSF (2006) Perivascular adipose tissue and mesenteric vascular function in spontaneously hypertensive rats. Arterioscler Thromb Vasc Biol 26: 1297–1302

Galvez-Prieto B, Bolbrinker J, Stucchi P, de Las Heras AI, Merino B, Arribas S, Ruiz-Gayo M, Huber M, Wehland M, Kreutz M, Fernandez-Alfonso MS (2008) Comparative expression analysis of the renin-angiotensin system components between white and brown perivascular adipose tissue. J Endocrinol 197:55–64

Gao YJ, Lu C, Su LY, Sharma AM, Lee RM (2007) Modulation of vascular function by perivascular adipose tissue: the role of endothelium and hydrogen peroxide. Br J Pharmacol 151:323–331

Gil-Ortega M, Somoza B, Huang Y, Gollasch M, Fernandez-Alfonso MS (2015) Regional differences in perivascular adipose tissue impacting vascular homeostasis. Trends Endocrinol Metab 26:367–375

Golas S, Berenyiova A, Majzunova M, Drobna M, Tuorkey MJ, Cacanyiova S (2022) The vasoactive effect of perivascular adipose tissue and hydrogen sulfide in thoracic aortas of normotensive and spontaneously hypertensive rats. Biomolecules 16(12):457

Gollasch M (2012) Vasodilator signals from perivascular adipose tissue. Br J Pharmacol 165:633–642

Greenstein AS, Khavandi K, Withers SB, Sonoyama K, Clancy O, Jeziorska M, Laing I, Yates AP, Pemberton PW, Malik RA et al (2009) Local inflammation and hypoxia abolish the protective anticontractile properties of perivascular fat in obese patients. Circulation 119:1661–1670

Guo W, Kan JT, Cheng ZY, Chen JF, Shen YQ, Xu J, Wu D, Zhu YZ (2012) Hydrogen sulfide as an endogenous modulator in mitochondria and mitochondria dysfunction. Oxidative Med Cell Longev 2012:878052

Guzik T, Mangalat D, Korbut R (2007) Adipocytokines—novel link between inflammation and vascular function? J Physiol Pharmacol 54:505–528

Hagberg CE, Spalding KL (2024) White adipocyte dysfunction and obesity-associated pathologies in humans. Nat Rev Mol Cell Biol 25:270–289. Erratum in: Nat Rev Mol Cell Biol. 2024;25: 333

Harms M, Seale P (2013) Brown and beige fat: development, function and therapeutic potential. Nat Med 19:1252–1263

Hedegaard ER, Gouliaev A, Winther AK, Arcanjo DD, Aalling M, Renaltan NS, Wood ME, Whiteman M, Skovgaard N, Simonsen U (2016) Involvement of potassium channels and calcium-independent mechanisms in hydrogen sulfide-induced relaxation of rat mesenteric small arteries. J Pharmacol Exp Ther 356:53–63

Herz CT, Kiefer FW (2019) Adipose tissue browning in mice and humans. J Endocrinol 241:R97–R109

Hotta K, Funahashi T, Arita Y, Takahashi M, Matsuda M, Okamoto Y, Iwahashi H, Kuriyama H, Ouchi N, Maeda K, Nishida M, Kihara S, Sakai N, Nakajima T, Hasegawa K, Muraguchi M, Ohmoto Y, Nakamura T, Yamashita S, Hanafusa T, Matsuzawa Y (2000) Plasma concentrations of a novel, adipose-specific protein, adiponectin, in type 2 diabetic patients. Arterioscler Thromb Vasc Biol 20:1595–1599

Huang F, Lezama MAR, Ontiveros JAP, Bravo G, Villafana S, Estela del-Rio-Navarro B, Hong E (2010) Effect of losartan on vascular function in fructose-fed rats: the role of perivascular adipose tissue. Clin Exp Hypertens 32:98–104

Huang D, Jing G, Zhu S (2023) Regulation of mitochondrial respiration by hydrogen sulfide. Antioxidants (Basel). 12:1644

Hulsmans M, Van Dooren E, Mathieu C, Holvoet P (2012) Decrease of miR-146b-5p in monocytes during obesity is associated with loss of the anti-inflammatory but not insulin signaling action of adiponectin. PLoS One 7:32794

Kawahito H, Yamada H, Irie D, Kato T, Akakabe Y, Kishida S, Takata H, Wakana N, Ogata T, Ikeda K, Ueyama T, Matoba S, Mori Y, Matsubara H (2013) Periaortic adipose tissue-specific activation of the renin-angiotensin system contributes to atherosclerosis development in uninephrectomized apoE−/− mice. Am J Physiol Heart Circ Physiol 305:667–675

Kawanami D, Maemura K, Takeda N, Harada T, Nojiri T, Imai Y, Manabe I, Utsunomiya K, Nagai R (2004) Direct reciprocal effects of resistin and adiponectin on vascular endothelial cells: a new insight into adipocytokine-endothelial cell interactions. Biochem Biophys Res Commun 314: 415–419

Kazama K, Okada M, Hara Y, Yamawaki H (2013) A novel adipocytokine, omentin, inhibits agonists-induced increases of blood pressure in rats. J Vet Med Sci 75:1029–1034

Kazama K, Okada M, Yamawaki H (2014a) A novel adipocytokine, omentin, inhibits platelet-derived growth factor-BB-induced vascular smooth muscle cell migration through antioxidative mechanism. Am J Physiol Heart Circ Physiol 306:H1714

Kazama K, Okada M, Yamawaki H (2014b) A novel adipocytokine, omentin, inhibits monocrotaline-induced pulmonary arterial hypertension in rats. Biochem Biophys Res Commun 452:142–146

Ketonen J, Shi J, Martonen E, Mervaala E (2010) Periadventitial adipose tissue promotes endothelial dysfunction via oxidative stress in diet-induced obese C57Bl/6 mice. Circ J 74:1479–1487

Kim S, Iwao H (2000) Molecular and cellular mechanisms of angiotensin II-mediated cardiovascular and renal diseases. Pharmacol Rev 52:11–34

Kim HW, Shi H, Winkler MA, Lee R, Weintraub NL (2020) Perivascular adipose tissue and vascular perturbation/atherosclerosis. Arterioscler Thromb Vasc Biol 40:2569–2576

Kohn C, Schleifenbaum J, Szijarto IA, Marko L, Dubrovska G, Huang Y, Gollasch M (2012) Differential effects of cystathionine-gamma-lyase dependent vasodilatory H_2S in periadventitial vasoregulation of rat and mouse aortas. PLoS One 7:41951

Lee RM, Lu C, Su LY, Gao YJ (2009) Endothelium-dependent relaxation factor released by perivascular adipose tissue. J Hypertens 27:782–790

Lee YC, Chang HH, Chiang CL, Liu CH, Yeh JI, Chen MF, Chen PY, Kuo JS, Lee TJ (2011) Role of perivascular adipose tissue-derived methyl palmitate in vascular tone regulation and pathogenesis of hypertension. Circulation 124:1160–1171

Lee RH, Couto E Silva A, Possoit HE, Lerner FM, Chen PY, Azizbayeva R, Citadin CT, Wu CY, Neumann JT, Lin HW (2019) Palmitic acid methyl ester is a novel neuroprotective agent against cardiac arrest. Prostaglandins Leukot Essent Fatty Acids 147:6–14

Li R, Andersen I, Aleke J, Golubinskaya V, Gustafsson H, Nilsson H (2013) Reduced anti-contractile effect of perivascular adipose tissue on mesenteric small arteries from spontaneously hypertensive rats: role of Kv7 channels. Eur J Pharmacol 698:310–315

Li H, Wang YP, Zhang LN, Tian G (2014) Perivascular adipose tissue-derived leptin promotes vascular smooth muscle cell phenotypic switching via p38 mitogen-activated protein kinase in metabolic syndrome rats. Exp Biol Med (Maywood) 239:954–965

Li X, Ma Z, Zhu YZ (2021) Regional heterogeneity of perivascular adipose tissue: morphology, origin, and secretome. Front Pharmacol 12:697720

Liu R, Wang X, Bu P (2011) Omentin-1 is associated with carotid atherosclerosis in patients with metabolic syndrome. Diabetes Res Clin Pract 93:21–25

Liu CH, Hsu HJ, Tseng TL, Lin T, Weng WH, Chen MF, Lee TJF (2020) COMT-catalyzed palmitic acid methyl ester biosynthesis in perivascular adipose tissue and its potential role against hypertension. J Pharmacol Exp Ther 373:175–183

Löhn M, Dubrovska G, Lauterbach B, Luft FC, Gollasch M, Sharma AM (2002) Periadventitial fat releases a vascular relaxing factor. FASEB J 16:1057–1063

Lu C, Su L-Y, Lee RMKW, Gao Y-J (2010) Mechanisms for perivascular adipose tissue-mediated potentiation of vascular contraction to perivascular neuronal stimulation: the role of adipocyte-derived angiotensin II. Eur J Pharmacol 634:107–112

Lu C, Su LY, Lee RM, Gao YJ (2011) Alterations in perivascular adipose tissue structure and function in hypertension. Eur J Pharmacol 656:68–73

Luo J, Huang L, Wang A, Liu Y, Cai R, Li W, Zhou MS (2018) Resistin-induced endoplasmic reticulum stress contributes to the impairment of insulin signaling in endothelium. Front Pharmacol 26(9):1226. Erratum in: Front Pharmacol. 2018; 9:1446

Lynch FM, Withers SB, Yao Z, Werner ME, Edwards G, Weston AH, Heagerty AM (2013) Perivascular adipose tissue-derived adiponectin activates BK(Ca) channels to induce anticontractile responses. Am J Physiol Heart Circ Physiol 304:786–795

Ma L, Ma S, He H, Yang D, Chen X, Luo Z, Liu D, Zhu Z (2010) Perivascular fat-mediated vascular dysfunction and remodeling through the AMPK/mTOR pathway in high-fat diet-induced obese rats. Hypertens Res 33:446–453

Malinowski M, Deja MA, Golba KS, Roleder T, Biernat J, Wos S (2008) Perivascular tissue of internal thoracic artery releases potent nitric oxide and prostacyclin-independent anticontractile factor. Eur J Cardiothorac Surg 33:225–231

Margaritis M, Antonopoulos AS, Digby J, Lee R, Reilly S, Coutinho P, Shirodaria C, Sayeed R, Petrou M, De Silva R, Jalilzadeh S, Demosthenous M, Bakogiannis C, Tousoulis D, Stefanadis C, Choudhury RP, Casadei B, Channon KM, Antoniades C (2013) Interactions between vascular wall and perivascular adipose tissue reveal novel roles for adiponectin in the regulation of endothelial nitric oxide synthase function in human vessels. Circulation 127: 2209–2221

Miao CY, Li ZY (2012) The role of perivascular adipose tissue in vascular smooth muscle cell growth. Br J Pharmacol 165:43–658

Mitidieri E et al (2022) The role of perivascular adipose tissue-derived hydrogen sulfide in the control of vascular homeostasis. Antioxid Redox Signal 37:84–97

Mu H, Ohashi R, Yan S, Chai H, Yang H, Lin P, Yao Q, Chen C (2006) Adipokine resistin promotes in vitro angiogenesis of human endothelial cells. Cardiovasc Res 70:146–157

Nóbrega N, Araújo NF, Reis D, Facine LM, Miranda CAS, Mota GC, Aires RD, Capettini LDSA, Cruz JDS, Bonaventura D (2019) Hydrogen peroxide and nitric oxide induce anticontractile effect of perivascular adipose tissue via renin angiotensin system activation. Nitric Oxide 84: 50–59

Nosalski R, Guzik TJ (2017) Perivascular adipose tissue inflammation in vascular disease. Br J Pharmacol 174:3496–3513

Oliveira PB, Zochio GP, Caetano ESP, da Silva MLS, Dias-Junior CA (2023) Vasodilator responses of perivascular adipose tissue-derived hydrogen sulfide stimulated with L-cysteine in pregnancy hypertension-induced endothelial dysfunction in rats. Antioxidants (Basel) 26(12): 1919

Owen MK, Witzmann FA, McKenney ML, Lai X, Berwick ZC, Moberly SP, Alloosh M, Sturek M, Tune JD (2013) Perivascular adipose tissue potentiates contraction of coronary vascular smooth muscle: Influence of obesity. Circulation 128:9–18

Peri-Okonny PA, Ayers C, Maalouf N, Das SR, de Lemos JA, Berry JD, Turer AT, Neeland IJ, Scherer PE, Vongpatanasin W (2016) Adiponectin protects against incident hypertension independent of body fat distribution: observations from the Dallas Heart Study. Diabetes Metab Res Rev 33:10.1002

Ramirez JG, O'Malley EJ, Ho WSV (2017) Pro-contractile effects of perivascular fat in health and disease. Br J Pharmacol 174:3482–3495

Reilly MP, Lehrke M, Wolfe ML, Rohatgi A, Lazar MA, Rader DJ (2005) Resistin is an inflammatory marker of atherosclerosis in humans. Circulation 111:932–939

Sarvottam K, Magan D, Yadav RK, Mehta N, Mahapatra SC (2013) Adiponectin, interleukin-6, and cardiovascular disease risk factors are modified by a short-term yoga-based lifestyle intervention in overweight and obese men. J Altern Complement Med 19:397–402

Schleifenbaum J, Köhn C, Voblova N, Dubrovska G, Zavarirskaya O, Gloe T, Crean CS, Luft FC, Huang Y, Schubert R, Gollasch M (2010) Systemic peripheral artery relaxation by KCNQ channel openers and hydrogen sulfide. J Hypertens 28:1875–1882

Shibata R, Ouchi N, Kikuchi R, Takahashi R, Takeshita K, Kataoka Y, Ohashi K, Ikeda N, Kihara S, Murohara T (2011) Circulating omentin is associated with coronary artery disease in men. Atherosclerosis 219:811–814

Shuldiner AR, Fried SK, McLenithan JC, Gong DW (2006) Identification of omentin as a novel depot-specific adipokine in human adipose tissue: possible role in modulating insulin action. Am J Physiol Endocrinol Metab 290:E1253–E1261

Small HY, McNeilly S, Mary S, Sheikh AM, Delles C (2019) Resistin mediates sex-dependent effects of perivascular adipose tissue on vascular function in the shrsp. Sci Rep 9:6897

Soltis EE, Cassis LA (1991) Influence of perivascular adipose tissue on rat aortic smooth muscle responsiveness. Clin Exp Hypertens A 13:277–296

Song A, Dai W, Jang MJ, Medrano L, Li Z, Zhao H, Shao M, Tan J, Li A, Ning T, Miller MM, Armstrong B, Huss JM, Zhu Y, Liu Y, Gradinaru V, Wu X, Jiang L, Scherer PE, Wang QA (2020) Low- and high-thermogenic brown adipocyte subpopulations coexist in murine adipose tissue. J Clin Invest 130:247–257

Sousa AS, Sponton ACS, Delbin MA (2021) Perivascular adipose tissue and microvascular endothelial dysfunction in obese mice: Beneficial effects of aerobic exercise in adiponectin receptor (AdipoR1) and peNOS(Ser1177). Clin Exp Pharmacol Physiol 48:1430–1440

Souza-Paula E, Polonio LCC, Zochio GP, Pinho da Silva K, Kushima H, Dias-Junior CA (2020) Anticontractile effect of perivascular adipose tissue but not of endothelium is enhanced by hydrogen sulfide stimulation in hypertensive pregnant rat aortae. J Cardiovasc Pharmacol 76: 715–729

Steppan CM, Bailey ST, Bhat S, Brown EJ, Banerjee RR, Wright CM, Patel HR, Ahima RS, Lazar MA (2001) The hormone resistin links obesity to diabetes. Nature 409:307–312

Stern JH, Rutkowski JM, Scherer PE (2016) Adiponectin, leptin, and fatty acids in the maintenance of metabolic homeostasis through adipose tissue crosstalk. Cell Metab 23:770–784

Szabo C, Ransy C, Módis K, Andriamihaja M, Murghes B, Coletta C, Olah G, Yanagi K, Bouillaud F (2014) Regulation of mitochondrial bioenergetic function by hydrogen sulfide. Part I. Biochemical and physiological mechanisms. Br J Pharmacol 171:2099–2122

Szasz T, Webb RC (2012) Perivascular adipose tissue: more than just structural support. Clin Sci (Lond) 122:1–12

Szasz T, Bomfim GF, Webb RC (2013) The influence of perivascular adipose tissue on vascular homeostasis. Vasc Health Risk Manag 9:105–116

Tahergorabi Z, Khazaei M (2015) Leptin and its cardiovascular effects: focus on angiogenesis. Adv Biomed Res 4:79

Tano JY, Schleifenbaum J, Gollasch M (2014) Perivascular adipose tissue, potassium channels, and vascular dysfunction. Arterioscler Thromb Vasc Biol 34:1827–1830

Tedesco L, Valerio A, Dossena M, Cardile A, Ragni M, Pagano C, Pagotto U, Carruba MO, Vettor R, Nisoli E (2010) Cannabinoid receptor stimulation impairs mitochondrial biogenesis in mouse white adipose tissue, muscle, and liver: the role of eNOS, p38 MAPK, and AMPK pathways. Diabetes 59:2826–2836

Testai L, Montanaro R, Flori L, Pagnotta E, Vellecco V, Gorica E, Ugolini L, Righetti L, Brancaleone V, Bucci M, Piragine E, Martelli A, Di Cesare Mannelli L, Ghelardini C, Calderone V (2023) Persulfidation of mitoKv7.4 channels contributes to the cardioprotective effects of the H2S-donor Erucin against ischemia/reperfusion injury. Biochem Pharmacol 215: 115728

Trayhurn P (2013) Hypoxia and adipose tissue function and dysfunction in obesity. Physiol Rev 93: 1–21

Uemura Y, Shibata R, Kanemura N, Ohashi K, Kambara T, Hiramatsu-Ito M, Enomoto T, Yuasa D, Joki Y, Matsuo K, Ito M, Hayakawa S, Ogawa H, Murohara T, Ouchi N (2015) Adipose-derived protein omentin prevents neointimal formation after arterial injury. FASEB J 29:141–151

Van de Voorde J, Boydens C, Pauwels B, Decaluwe K (2014) Perivascular adipose tissue, inflammation and vascular dysfunction in obesity. Curr Vasc Pharmacol 12:403–411

Verlohren S, Dubrovska G, Tsang SY, Essin K, Luft FC, Huang Y, Gollasch M (2004) Visceral periadventitial adipose tissue regulates arterial tone of mesenteric arteries. Hypertension 44: 271–276

Wang P, Xu TY, Guan YF, Su DF, Fan GR, Miao CY (2009) Perivascular adipose tissue-derived visfatin is a vascular smooth muscle cell growth factor: role of nicotinamide mononucleotide. Cardiovasc Res 81:370–380

Weston AH, Egner I, Dong Y, Porter EL, Heagerty AM, Edwards G (2013) Stimulated release of a hyperpolarizing factor (ADHF) from mesenteric artery perivascular adipose tissue: involvement of myocyte BKCa channels and adiponectin. Br J Pharmacol 169:1500–1509

Weyer C, Funahashi T, Tanaka S, Hotta K, Matsuzawa Y, Pratley RE, Tataranni PA (2001) Hypoadiponectinemia in obesity and type 2 diabetes: close association with insulin resistance and hyperinsulinemia. J Clin Endocrinol Metab 86:1930–1935

Withers SB, Bussey CE, Saxton SN, Melrose HM, Watkins AE, Heagerty AM (2014) Mechanisms of adiponectin-associated perivascular function in vascular disease. Arterioscler Thromb Vasc Biol 34:1637–1642

Wolk R, Somers VK (2006) Leptin and vascular function: friend or foe? Eur Heart J 27:2263–2265

Xu A, Vanhoutte PM (2012) Adiponectin and adipocyte fatty acid binding protein in the pathogenesis of cardiovascular disease. Am J Physiol Heart Circ Physiol 302:1231–1240

Yamauchi T, Iwabu M, Okada-Iwabu M, Kadowaki T (2014) Adiponectin receptors: a review of their structure, function and how they work. Best Pract Res Clin Endocrinol Metab 28:15–23

Yamawaki H, Tsubaki N, Mukohda M, Okada M, Hara Y (2010) Omentin, a novel adipokine, induces vasodilation in rat isolated blood vessels. Biochem Biophys Res Commun 393:668–672

Yaturu S, Daberry RP, Rains J, Jain S (2006) Resistin and adiponectin levels in subjects with coronary artery disease and type 2 diabetes. Cytokine 34:219–223

Yu PL, Wang C, Li W, Zhang FX (2019) Visfatin level and the risk of hypertension and cerebrovascular accident: a systematic review and meta-analysis. Horm Metab Res 51:220–229

Zavaritskaya O, Zhuravleva N, Schleifenbaum J, Gloe T, Devermann L, Kluge R, Mladenov M, Frey M, Gagov H, Fésus G, Gallosch M, Schubert R (2013) Role of KCNQ channels in skeletal muscle arteries and periadventitial vascular dysfunction. Hypertension 61:151–159

Zhang P, Wang Y, Fan Y, Tang Z, Wang N (2009) Overexpression of adiponectin receptors potentiates the antiinflammatory action of subeffective dose of globular adiponectin in vascular endothelial cells. Arterioscler Thromb Vasc Biol 29:67–74

Naturally Occurring H_2S-Donors in Vascular Diseases

Lorenzo Flori, Jacopo Spezzini, Eugenia Piragine, Valentina Citi, Vincenzo Calderone, and Alma Martelli

Contents

Lorenzo Flori and Jacopo Spezzini contributed equally with all other contributors.

L. Flori · J. Spezzini
Department of Pharmacy, University of Pisa, Pisa, Italy

E. Piragine
Department of Pharmacy, University of Pisa, Pisa, Italy

Interdepartmental Research Centre "Nutraceuticals and Food for Health (NUTRAFOOD)", University of Pisa, Pisa, Italy

V. Citi · V. Calderone · A. Martelli (✉)
Department of Pharmacy, University of Pisa, Pisa, Italy

Interdepartmental Research Centre "Nutraceuticals and Food for Health (NUTRAFOOD)", University of Pisa, Pisa, Italy

Interdepartmental Research Centre of "Ageing Biology and Pathology," University of Pisa, Pisa, Italy
e-mail: alma.martelli@unipi.it

A. Papapetropoulos, C. Szabo (eds.), *Hydrogen Sulfide: A Toxicant with Healing Potential*, Handbook of Experimental Pharmacology 292,
https://doi.org/10.1007/164_2025_753

Abstract

Hydrogen sulfide (H_2S) is an endogenous gasotransmitter able to exert a pivotal role in different organs and systems, strongly influencing cardiovascular health. It is an endowed antioxidant, a vasorelaxant, and has cardioprotective properties, thanks to the activation of different classes of potassium channels and the interaction with several pathways including those involving sirtuins, nuclear factor erythroid 2-related factor 2 (Nrf2), nuclear factor kappa B (NF-κB), and nitric oxide (NO) production. Alterations in the homeostasis of H_2S represent an etiopathogenetic factor in the onset and progression of cardiovascular diseases, such as hypertension, atherosclerosis, and vascular diabetic complications, highlighting the therapeutic potential of exogeneous H_2S-donors in H_2S-poor diseases. Several natural H_2S-donor compounds, or their precursors, derive from plants belonging to the *Alliaceae* (e.g., garlic and onion), *Brassicaceae* (e.g., broccoli and rocket salad), or *Moringaceae* (e.g., moringa) families. Preclinical studies demonstrated the antioxidant, vasoprotective, and anti-hypertensive properties of both plant extracts and isolated polysulfides or isothiocyanates (ITCs). In fact, polysulfides and ITCs are able to exert vascular effects superimposable to those induced by H_2S. Preclinical and clinical studies successfully demonstrated that garlic extracts decrease systolic and diastolic blood pressure and contrast endothelial dysfunction and atherosclerosis. Similarly, preclinical studies highlighted the anti-inflammatory, anti-hypertensive, and vasoprotective properties of ITCs. However, clinical studies only demonstrated the improvement of the lipid profile in healthy patients, with no effects on systolic or diastolic blood pressure. Taken together, these findings suggest that natural H_2S-donors could restore H_2S homeostasis, therefore preventing and/or contrasting cardiovascular diseases.

Keywords

Alliaceae · *Brassicaceae* · Hydrogen sulfide · Isothiocyanates · Natural H_2S-donors · Polysulfides · Vascular disease

1 Introduction

1.1 Hydrogen Sulfide

The identification of hydrogen sulfide (H_2S) in mammalian tissues led to consider this gaseous mediator not exclusively as a toxic gas but also an endogenous gasotransmitter, representing a turning point in the cardiovascular research (Abe and Kimura 1996). In the wake of the most known and characterized endogenous gasotransmitter nitric oxide (NO), the importance of H_2S in the physiological regulation of several systems has been described over the past three decades (Kimura 2021).

The extreme biological potential of H_2S is linked to its lipophilic structure which allows it to easily cross biological membranes (Li and Moore 2008). Considering that H_2S is a weak acid, at physiological pH more than 80% is found in the dissociated form of hydrosulfide anion (HS^-). Therefore, the effects of H_2S are due to a balance between the undissociated (H_2S) and the first dissociation anionic form (HS^-) (Li and Lancaster 2013).

Enzymatic and non-enzymatic pathways contribute to the production of endogenous H_2S. The enzymatic one is the most relevant from a quantitative point of view. Basal levels of H_2S in tissues and plasma are regulated by different enzymes which are cystathionine-β-synthase (CBS), cystathionine-γ-lyase (CSE), and 3-mercaptopyruvate sulfotransferase (3-MST), together with cysteine aminotransferase (CAT) (Bełtowski 2015; Li et al. 2011). The non-enzymatic pathway provides a less significant H_2S production and uses reducing elements obtained from glucose oxidation to produce H_2S through the reduction of elemental sulfur (S^0) (Searcy and Lee 1998).

H_2S has antioxidant activity mainly, but not exclusively, as a reducing agent. Many mechanisms underlying the pleiotropic effects of H_2S are not yet fully understood. In this context, S-persulfidation, a post-translational modification of proteins, is one of the proposed explanations underlying the beneficial effects of H_2S (Cirino et al. 2023).

The aim of this chapter is to describe the H_2S-derived vasoprotective properties of natural H_2S-donor compounds, or their precursor, derived from plants belonging to the *Alliaceae*, *Brassicaceae,* or *Moringaceae* family against hypertension and in the maintenance of vascular wall integrity.

1.2 H_2S in Hypertension

Over the last few years, experimental evidence on the importance of H_2S in regulating vascular function has grown exponentially (Flori et al. 2024a; Holwerda et al. 2015). H_2S controls vascular tone and maintains the integrity of the endothelial layer, ensuring the proper functioning of vascular homeostatic processes, also through a close crosstalk with NO. H_2S owes its vasorelaxant properties to the modulation of several targets, including the inhibition of 5-phosphodiesterase (5-PDE) enzyme, the opening of potassium channels (e.g., Kv7 and KATP) and transient receptor potential vanilloid 4 (TRPV4) channels, and the activation of the vascular endothelial growth factor receptor 2 (VEGFR2) (Dongó et al. 2018).

Potassium channels are the main modulators of cardiovascular function, and their opening represents one of the first described anti-hypertensive mechanisms of action of H_2S. This effect was first reported by Zhao in 2001 and subsequently confirmed by several preclinical studies by the administration of H_2S-donors in the presence or absence of selective KATP channel blockers (Zhao et al. 2001). Experimental evidence suggests that the interaction of H_2S both with the SUR1 and SUR2B subunits of KATP channels determines S-persulfidation of extracellular cysteine residues (Jiang et al. 2010; Mustafa et al. 2011). In both cases, H_2S alters the Kir6.1 subunit by forming channel opening pores (Kang et al. 2015).

Acting through post-translational modifications of target proteins, a nonspecific vasorelaxant mechanism is quite likely. Indeed, experimental evidence indicates that H_2S may act by opening other potassium channels. H_2S activates endothelial calcium channels (i.e., SKCa and IKCa) working as an endothelium-derived hyperpolarizing factor (EDHF). Furthermore, in vivo studies have supported the hypothesis that H_2S also acts as an adipose-derived relaxing factor (ADRF), leading to the discovery that H_2S activates voltage-gated potassium channels (Kv) in the arteries (Hedegaard et al. 2016). In particular, the detailed analysis of the effects mediated by the release of H_2S by NaHS, thioureas, and isothiocyanates (ITCs) allowed to identify the opening of Kv7 channels, and specifically the Kv7.4 isoform, as one of the main mechanisms of the H_2S-mediated vasodilation, as demonstrated in human aortic smooth muscle cells (HASMCs) and rat aortic rings (Citi et al. 2020; Martelli et al. 2013a, 2014). The antihypertensive and vasorelaxant properties of H_2S are also linked to the activation of the eNOS/cGMP/PKG pathway. H_2S deficiency was associated with a reduction in endothelial nitric oxide synthase (eNOS) activity (King et al. 2014). H_2S increases cyclic guanosine monophosphate (cGMP) levels both by stimulating eNOS to produce NO and by inhibiting 5-PDE (Bucci et al. 2010). The accumulation of cGMP leads to the activation of protein kinase G (PKG), promoting angiogenesis and vasodilation. It has also been proposed that H_2S directly acts on both cGMP and PKG. Experimental data suggest close cooperation between H_2S and NO. As mentioned above, crosstalk between H_2S and NO plays an important role in the mechanisms of vasodilation. H_2S stimulates eNOS activity via direct S-persulfidation, mobilizing Ca^{2+} and phosphorylating protein kinase B (Akt) (Jin et al. 2017). Furthermore, several reactive species derived from H_2S and NO react with each other to form intermediates. Among these, the molecule HSSNO, resistant to reducing agents such as cysteine, reduced glutathione (GSH), and dithiolethione (DTT), has its own vasorelaxant effect (Kimura 2016). Other reaction products such as nitrosopersulfide ($SSNO^-$) and polysulfides have been described for their vasorelaxant effect in the rat thoracic aorta and mesenteric artery rings (Berenyiova et al. 2015).

The HNO-TRPA1-CGRP is another pathway that confirms the crosstalk between H_2S and NO. Eberhardt and colleagues demonstrated in 2014 how nitroxyl (HNO), an intermediate of the reaction between H_2S and NO, activates the transient receptor potential ankyrin 1 (TRPA1) channel. The activation through amino-terminal disulfide bonds determines an influx of Ca^{2+} and the consequent release of calcitonin gene-related peptide (CGRP) that induces local and systemic vasodilation (Eberhardt et al. 2014; Pozsgai et al. 2019).

Furthermore, H_2S can promote NO-induced vasodilation via soluble guanylate cyclase (sGC). Specific analyses have ruled out a direct effect of H_2S-donors on the increase of sGC levels (Coletta et al. 2012), and a crosstalk between NO and H_2S on its redox state has also been demonstrated. Zhou and colleagues in 2016 pointed out that H_2S acts as a reducing agent able to promote the ferrous state of sGC (Fe^{2+}), increasing the reactivity of sGC toward NO and the consequent vasorelaxant effects (Zhou et al. 2016).

The complexity of the endogenous H_2S system may provide vasodilation even in endothelial dysfunction. Endothelial impairment, typical of vascular diseases such as hypertension, gradually reduces NO production. Under such conditions, H_2S production from smooth muscle cells may trigger a series of processes described above and still provide vasodilation.

More recent studies describe additional mechanisms of action regarding the vasorelaxant effect of H_2S. VEGFR2 activation is one of them. H_2S-donors, such as NaHS and GYY4137, have demonstrated an effect on VEGFR2 in rat cerebral basilar arteries and spontaneously hypertensive rats (SHRs), respectively. In particular, the endothelial protection was improved, and the antihypertensive effect was enhanced. The action of H_2S through the involvement of VEGFR2 was confirmed by the use of selective blockers and VEGFR2 knockdown, resulting in a reduction of the H_2S-induced vasorelaxant effects both in in vitro and in vivo studies (Chen et al. 2022; Zhu et al. 2021).

H_2S also appears to act on TRPV4 channels resulting in the modulation of Ca^{2+} currents and activation of large-conductance Ca^{2+}-activated potassium channels expressed on endothelial cells (eBK$_{Ca}$s), triggering endothelial hyperpolarization and subsequent vasodilation (Naik et al. 2016).

1.3 H_2S and Vascular Wall Protection

Pathophysiology of cardiovascular diseases (CVD), including hypertension, atherosclerosis, and vascular diabetic complications, is characterized by excessive oxidative stress in the vascular wall determining chronic and subclinical inflammation (Siti et al. 2015). Furthermore, endothelial and smooth muscle dysfunction in the vascular tree exacerbates the degenerative processes in the vascular wall, leading to impaired vascular tone, platelet aggregation, and macrophage infiltration. The loss of vascular wall integrity ultimately contributes to increased inflammation also in other districts (Citi et al. 2021). In this field, H_2S and H_2S-donor compounds exert antioxidant effects directly by scavenging reactive oxygen species (ROS), increasing the amount of GSH and promoting the expression of antioxidant enzymes (Xie et al. 2016b). In addition, they contrast inflammation by reducing cytokine levels like interleukin (IL)-1β, IL-6, and tumor necrosis factor-α (TNF-α) (Wang 2012). Indeed, H_2S activates the Nrf2/ARE (nuclear erythroid 2-related factor 2/antioxidant response element) pathway by the S-persulfidation of Cys151 on the Nrf2 suppressor Keap1, leading to dissociation of Keap1 from Nrf2 and subsequent nuclear translocation and antioxidant effects (Xie et al. 2016a). Moreover, H_2S is able to directly activate sirtuin 1 (SIRT1), an enzyme with deacetylase activity deeply involved in cell metabolism, further stimulating the activation of the Nrf2/ARE pathway and eNOS (El-Sheakh et al. 2016). SIRT1, in turn, regulates mechanistic target of rapamycin (mTOR) pathway, thus repressing oxidative stress and promoting autophagy (Meng et al. 2020). H_2S also contributes to the homeostasis of vascular wall by increasing eNOS phosphorylation through the activation of peroxisome proliferator-activated receptor δ/ protein kinase B/AMP-activated protein kinase (PPAR-δ/Akt/AMPK) signaling pathway, thus enhancing eNOS activity and increasing the production of NO, inducing vasorelaxation, increasing protein nitrosylation, and inhibiting the proliferation and migration of vascular smooth muscle cells (VSMCs) (Lin et al. 2016). Furthermore, H_2S inhibits the expression of intercellular adhesion molecule-1 (ICAM-1) through the reduced degradation of IκB-α and thus inhibits the activation of the pro-inflammatory transcription factor NF-κB (Ganster et al. 2010). ICAM-1 inhibition represents a pivotal step in reducing the recruitment of immune cells like monocytes and T cells, limiting cell migration and contrasting inflammation (Wang et al. 2009). As a result, the lack of endogenous H_2S represents a risk factor for vascular wall dysfunction. Of note, many CVDs are associated with a reduction in CSE levels and consequently with a reduced endogenous biosynthesis of H_2S (Calderone et al. 2016). H_2S deficiency determines vascular remodeling with thickened aortic walls, impaired contraction, collagen deposition, and increased phosphorylation of the signal transducer and activator of transcription 3 (STAT3), further determining a reduction in the aortic production of both PPARδ and suppressor of cytokine signaling 3 (SOCS3). In particular, it was demonstrated that SOCS3 preserves endothelial function by inhibiting tyrosine–protein kinase JAK1–STAT signaling (Tian et al. 2021) (Fig. 1).

2 Naturally Occurring Organosulfur Compounds: H_2S as the "Hidden" Vasoactive Mediator

In recent decades, H_2S has gained great interest as a small endogenous gasotransmitter able to modulate several pathophysiological pathways (Kolluru et al. 2013). Consistently, the reduction in plasma H_2S levels has been associated with the onset of several age-related diseases, including hypertension, atherosclerosis, and diabetes (Piragine et al. 2023). For this reason, the use of molecules capable of donating H_2S in a physiological manner may represent a pharmacological strategy for treating H_2S-poor diseases (Citi et al. 2022). In this scenario, several synthetic compounds capable of donating H_2S have been developed (for example, thioamides, synthetic ITCs, iminothioethers, dithiolthiones), demonstrating their effectiveness in lowering systolic blood pressure, reducing vascular wall inflammation, and restoring endothelial function (Barresi et al. 2017; Martelli et al. 2013b; Testai et al. 2016; Tomasova et al. 2015; Zaorska et al. 2019). Looking at natural compounds, traditional medicine teaches that a diet rich in garlic (*Allium sativum* L.), which is an edible plant belonging to the *Alliaceae* family, helps in maintaining a healthy cardiovascular system, mainly promoting anti-hypertensive, antioxidant, and anti-inflammatory effects. More recently, edible plants belonging to the *Brassicaceae* family, such as broccoli, rocket salad, and cauliflower, have been reported to exhibit

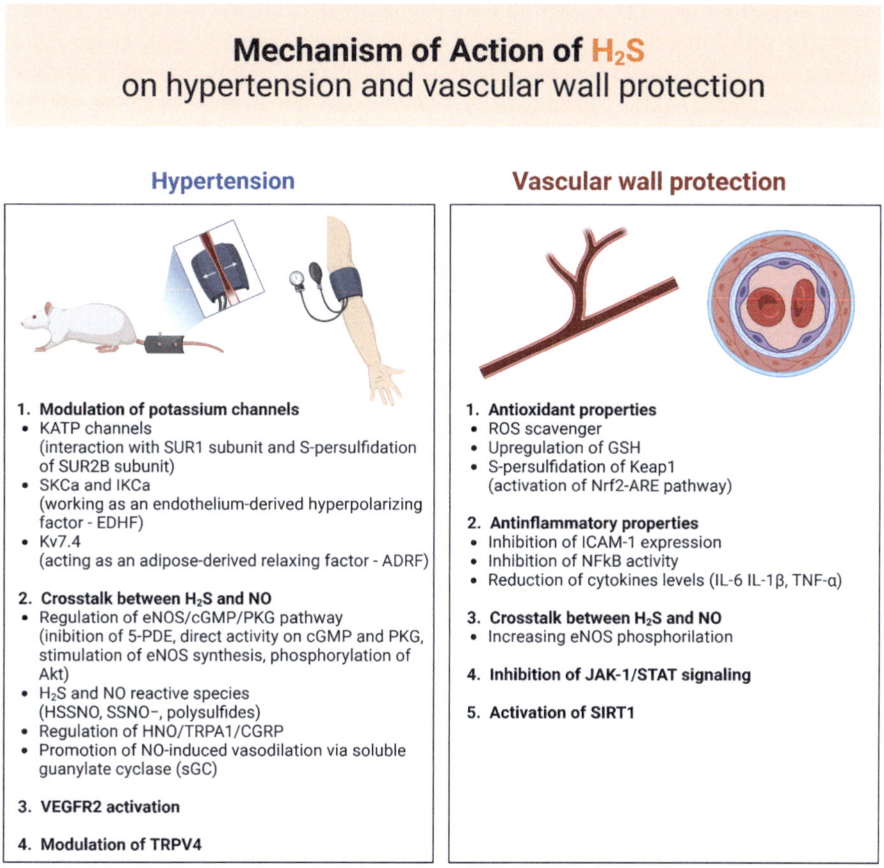

Fig. 1 Mechanisms of action of H_2S on hypertension and vascular wall protection

similar beneficial effects in the cardiovascular system (Connolly et al. 2024; Ried 2020).

The characteristic that *Alliaceae* and *Brassicaceae* share is the high content of organosulfur compounds: *Alliaceae* are characterized by the presence of allicin, which is rapidly converted due to the activity of alliinase into the unstable derivative alliin that, in turn, decomposes into polysulfide compounds such as diallyl disulfide (DADS) and diallyl trisulfide (DATS), while *Brassicaceae* contains glucosinolates (GLSs). GLSs are metabolized into ITCs when the plants are cut or chopped, allowing the enzyme myrosinase, or thioglucosidase, to be activated, cleaving off the glucose group from the glucosinolate. After a consequent spontaneous reaction, called Lossen-like rearrangement, ITCs are generated (Borlinghaus et al. 2014; Prieto et al. 2019).

Notably, polysulfides and ITCs, despite their completely different chemical structure, exhibit vascular pharmacological effects similar to those exerted by

H_2S. Polysulfides and ITCs showed KATP channel-dependent vasodilation in phenylephrine precontracted rat aortic rings (Ashraf et al. 2004; Wilson et al. 2002); sulforaphane (SFN, the ITC derived from glucoraphanin, highly present in *Brassica oleracea* L.) and DADS (Elosta et al. 2017) inhibited the increase of the receptor of advanced glycated end products (RAGE), monocyte chemoattractant protein-1 (MCP-1), ICAM-1, and vascular cell adhesion molecule-1 (VCAM-1) (crucial molecules that contribute to the development of vascular inflammation) gene expression, in endothelial cells treated with advanced glycated end products (AGEs), preventing the adhesion of monocytes (THP-1) to human umbilical vein endothelial cells (HUVECs) (Yamagishi and Matsui 2016). Another overlap between polysulfides and ITCs is related to the vascular antioxidant and anti-inflammatory effects, since they are able to activate Nrf2, promoting the activation of ARE genes, which are responsible for the expression of several endogenous antioxidant molecules, including GSH (Barrera et al. 2012; Padiya et al. 2014; Recinella et al. 2023; Rodrigues and Percival 2019; Zhang et al. 2022). Furthermore, also the protective effects against oxidized low-density lipoprotein (ox-LDL) injury have been reported, since polysulfides and ITCs inhibited ox-LDL-induced ROS production and NF-κB translocation in the nucleus, reducing the expression of VCAM-1 and ICAM-1, resulting in the suppression of inflammatory mediators production, including prostaglandin E2 (PGE2), TNF-α, and IL-1β (Lau 2006; Lei et al. 2010; Liu et al. 2023). Another study demonstrated that aged black garlic significantly reduced the expression of VCAM-1 in TNF-α-stimulated HUVECs, exerting its anti-inflammatory effects primarily through the inhibition of VCAM-1 expression and by counteracting NF-κB activation in endothelial cells (Lu et al. 2024). Worthy of mention is also the hormetic behavior shared by ITCs, polysulfides, and H_2S: low concentrations show beneficial effects and are endowed with antioxidant properties and vasoprotective effects, while increasing concentrations promote cytotoxic effects, widely explored for the potential anticancer effect (Citi et al. 2019; Yi and Su 2013).

The clear overlap about the effects exerted by ITCs, polysulfides, and H_2S, provided by experimental evidence, led the researchers to investigate whether H_2S could be the "hidden" player behind the beneficial effects attributed to the consumption of *Alliaceae* and *Brassicaceae*.

Benavides and coworkers were the first to hypothesize a possible role of H_2S in the vasoactive effects of garlic. They demonstrated that the vasorelaxant effects promoted by DADS and DATS in isolated rat aortic rings were due to the release of H_2S. Indeed, DADS and DATS induced vasodilation in phenylephrine-precontracted rat aortic rings in a Tyrode solution added with GSH, simultaneously with the H_2S formation, suggesting a connection between garlic-derived polysulfides and H_2S. Beside this pharmacological activity, the authors reported for the first time that GSH reacts with polysulfides, leading to a nucleophilic substitution at the α carbon yielding the formation of S-allyl-GSH and allyl perthiol derivatives. Following another nucleophilic substitution at the S-atom, finally allyl-GSH disulfide and H_2S are formed (Fig. 2) (Benavides et al. 2007).

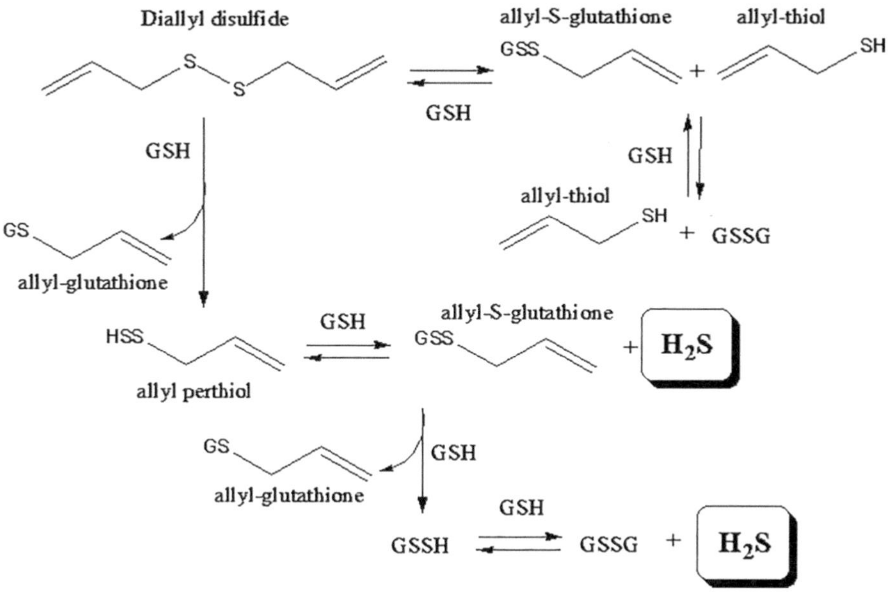

Fig. 2 Reaction between DADS and GSH, responsible for the formation of H_2S

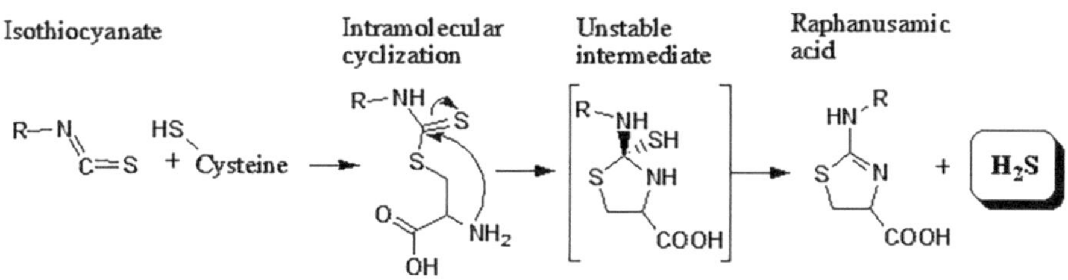

Fig. 3 Organic ITCs react with L-cysteine yielding the formation of H_2S

As reported for DADS and DATS, ITCs also exhibit effects similar to those reported for H_2S (Martelli et al. 2020a). This observation led Citi and co-workers to demonstrate, by using an amperometric method, that several natural ITCs, i.e., allyl ITC (from *Brassica nigra* L.), 4-hydroxybenzyl ITC (from *Sinapis alba* L.), benzyl ITC (from *Lepidium sativum* L.), and erucin (from *Eruca sativa* Mill.), release H_2S in a thiol-dependent manner (Citi et al. 2014). More recently, Lucarini et al. also reported the H_2S-releasing properties of SFN, which is the ITC derived from glucoraphanin, abundantly present in *Brassica oleracea* L. (Lucarini et al. 2018). Besides the observation that ITCs release H_2S when they are incubated with organic thiols, in 2021, the exact mechanism leading to the formation of the gaseous compound has been described by Lin and coworkers: ITCs quickly react with cysteine forming adducts which spontaneously cyclize. This process yields the release of organic amine and raphanusamic acid as major products, and H_2S and

2-carbylamino-4,5-dihydrothiazole-4-carboxylic acids as minor products (Fig. 3) (Lin et al. 2019).

The evidence about the H_2S-donating properties of polysulfides and ITCs, respectively derived from *Alliaceae* and *Brassicaceae* plants, and the comprehension of the exact chemical reaction between organic thiols and the sulfur moieties responsible for the generation of significant amount of H_2S, established the basis for specific pharmacological investigations, directly correlating the H_2S-releasing properties to the vasorelaxant effect. In particular, a very recent study evaluated the vascular effects of erucin, an ITC derived from *Eruca sativa* Mill., in association with its H_2S-releasing properties. In this work, the authors demonstrated that erucin released H_2S inside human HASMCs, evoking hyperpolarizing effect in a concentration-dependent manner. Furthermore, erucin promoted vasorelaxation in endothelium-intact and -denuded vessels (aortic rings), inhibited the vasoconstriction induced by noradrenaline, and reduced systolic pressure in SHRs (Martelli et al. 2020b). More recently, erucin has been also demonstrated to protect vascular cells against lipopolysaccharide (LPS)-induced inflammation, significantly preserving cell viability and preventing the increase in ROS and TNF-α levels (Ciccone et al. 2022). Moreover, erucin reduced endothelial hyperpermeability, counteracted the loss of vascular endothelial (VE)-cadherin levels, and decreased the expression of VCAM-1, cyclooxygenase-2 (COX-2), and microsomal prostaglandin E-synthase 1 (mPGES-1) induced by high glucose concentrations (Martelli et al. 2021). Finally, erucin reverted the angiogenic impairment induced by high glucose in HUVECs (Genah et al. 2024).

3 Polysulfides in Hypertension and in Vascular Wall Protection

3.1 Anti-hypertensive and Vasoprotective Properties of *Alliaceae* Extracts and Isolated Polysulfides: Results of Preclinical Studies In Vivo

The growing interest in natural H_2S-donors is demonstrated by the large number of studies focusing on the pharmacological effects of *Alliaceae* extracts and their bioactive compounds in vivo (Ansary et al. 2020; Imaizumi et al. 2023; Kianian et al. 2021; Piragine et al. 2022a, b; Spezzini et al. 2024; Varade et al. 2024). In the cardiovascular system, an important "trigger" for the pharmacological characterization of garlic, onion, and isolated polysulfides was the discovery of reduced H_2S biosynthesis in animal models of hypertension (Liang et al. 2015; Szlęzak et al. 2022) and in hypertensive patients (Wang 2023). In this regard, Liang and coworkers showed that a 3-week treatment with the H_2S-donor DATS (40 mg/kg/day, orally) significantly lowered blood pressure levels in rats with metabolic syndrome, in part by restoring serum levels of H_2S (Liang et al. 2015). Further evidence for the role of H_2S in the pharmacology of polysulfides from *Alliaceae* was provided by Cui and coworkers, who demonstrated that the blood pressure-lowering effects promoted by the precursor of polysulfides allicin (7–14 mg/kg/day for 4 weeks) in hypertensive rats were significantly reduced in the presence of the H_2S biosynthesis inhibitor DL-propargylglycine (PAG) (Cui et al. 2020). The anti-hypertensive properties of allicin were also confirmed in hypertensive rats with chronic kidney disease (40 mg/kg/day of allicin for 6 weeks, orally) (García-Trejo et al. 2016), as wells as in rat models of high-fructose diet-induced hypertension (8 mg/kg/day of allicin for 2 weeks, orally) (Elkayam et al. 2001) and dexamethasone-induced hypertension (8 mg/kg/day of allicin for 8 weeks) (Dubey et al. 2017). Finally, in rats with metabolic syndrome, the natural H_2S-donor DATS reduced the risk of developing hyperhomocysteinemia (Jeremic et al. 2020), a metabolic disorder that progressively leads to arterial stiffness and hypertension and that may result from an imbalance in H_2S metabolism (Hermann and Sitdikova 2021). Therefore, restoring physiological levels of H_2S with exogenous H_2S-donors could rebalance metabolic pathways that impact blood vessel function, including those involved in homocysteine metabolism.

The study of isolated bioactive compounds is essential for identifying the mechanism of action of nutraceuticals or, for natural H_2S-donors, "sulfanutraceuticals" (Martelli et al. 2023). However, to provide a complete overview of the clinical potential of organosulfur-containing natural products, the biological effects of the phytocomplex must also be described. Many preclinical studies have demonstrated the potential vasorelaxant and anti-hypertensive effects of garlic extracts. Among "physiological" and reliable animal models of hypertension, SHR is the most representative. In this regard, Harauma and coworkers described the blood pressure-lowering effects of a 10-week treatment of SHRs with an aged garlic extract (Harauma and Moriguchi 2006), while Han and coworkers found that processed garlic (30–50 mg/kg/day, orally) prevented the development of hypertension in SHRs after 8 weeks of treatment (Han et al. 2011). Besides SHRs, many other animal models of hypertension have been used to study the anti-hypertensive properties of garlic. For instance, both acute (5–20 mg/kg, intravenously) (Nwokocha et al. 2011) and sub-chronic (50 mg/kg/day for 1 month, orally) (Sharifi et al. 2003) treatment with garlic extract significantly lowered blood pressure levels in two-kidney-one-clip (2K1C) hypertensive rats. In another work, a 7-week treatment with black garlic extract (50 and 100 mg/kg/day, orally), standardized in DADS and DATS, promoted anti-hypertensive properties in a rat model of deoxycorticosterone acetate salt-induced hypertension (Chen et al. 2021). Finally, garlic oil supplementation (100 mg/kg/day, orally) in female rats during pregnancy and lactation significantly prevented high-fat diet-induced hypertension in male rat offspring, in part through modulation of the endogenous "H_2S system" (Hsu et al. 2021). Similar results were shown for onion extracts, further suggesting the existence of a possible "hidden" player in the pharmacology of organosulfur-containing plants. For instance, dried onion (5% w/w for 4 weeks) reduced systolic blood pressure in NG-nitro-L-arginine methyl ester (L-NAME)-induced hypertensive rats and stroke-prone spontaneously hypertensive rats by enhancing eNOS activity in the rat aorta (Sakai et al. 2003). In another study, onion peel hydroalcoholic extract (200–800 mg/kg/day for 3 weeks, orally) reduced systolic blood pressure in a rat model of diet-induced hypertension and prevented KCl- and phenylephrine-induced vasoconstriction in the aorta of hypertensive rats in a concentration-dependent manner (Naseri et al. 2008).

One of the prodromal conditions for hypertension is endothelial dysfunction, which can be described as the loss of physiological function of the endothelium that results in a reduced release of vasoactive mediators, including H_2S (Wang et al. 2015). This sub-clinical and silent feature unites hypertension with other pathological conditions, such as atherosclerosis and type 2 diabetes. Therefore, identifying compounds capable of preventing endothelial dysfunction could be an effective strategy to "fight" many cardiovascular risk factors simultaneously. About 20 years ago, Sun and coworkers demonstrated that dietary supplementation with freeze-dried garlic extract (1% w/w for 3 weeks, orally) significantly prevented coronary endothelial dysfunction and improved vasoactivity of isolated coronaries in rats with monocrotaline-induced pulmonary hypertension. On the contrary, the "allicin-free" boiled garlic or aged garlic were ineffective, thus highlighting the crucial role of organosulfur compounds in the bioactive properties of garlic (Sun and Ku 2006). Furthermore, aqueous extracts of garlic and onion (150 and 400 mg/kg/day for 8 weeks, orally) attenuated vascular inflammation in mesenteric arteries of fructose-fed rats, in part by reducing the expression of VCAM-1 (Vazquez-Prieto et al. 2011). The potential contribution of natural H_2S-donors in the vascular benefits of *Alliaceae* extracts was confirmed a few years later by Liu and coworkers, who found that DATS (5 mg/kg/day for 7 days, intravenously) protected vascular endothelium and significantly improved endothelial-dependent vasorelaxation of the aorta of obese diabetic rats (Liu et al. 2014). In addition, daily treatment with allicin (10 mg/kg/day for 4 weeks, orally) promoted antioxidant and anti-inflammatory effects in the aorta of diet-induced hypercholesterolemic rabbits by reducing the expression of the pro-inflammatory transcription factor NF-κB (El-Sheakh et al. 2016). In the same experimental model of dyslipidemia, allicin (10–14 mg/kg/day for 4 weeks, orally) improved vascular remodeling (i.e., reduced aortic thickness and smooth muscle cell proliferation) and endothelial dysfunction (Liu et al. 2022), while dried garlic powder standardized by allicin content (300 mg/day for 35 days, orally) prevented the neointima formation and lipid accumulation in the rabbit aorta, thus showing anti-atherogenic activity (Sobenin et al. 2016).

Taken together, preclinical evidence suggests that *Alliaceae* extracts and isolated polysulfides could be useful for the prevention and treatment of hypertension and endothelial dysfunction, probably through the release of H_2S (Fig. 4).

3.2 Potential Effects of Garlic on Vascular Health: Clinical Evidence

About 10 years ago, an observational study demonstrated a negative association between regular intake of vegetables from the *Alliaceae* botanical family and the incidence of cardiovascular outcomes and hypertension (Bahadoran et al. 2017). This "real-world" finding has reinforced the results of numerous interventional

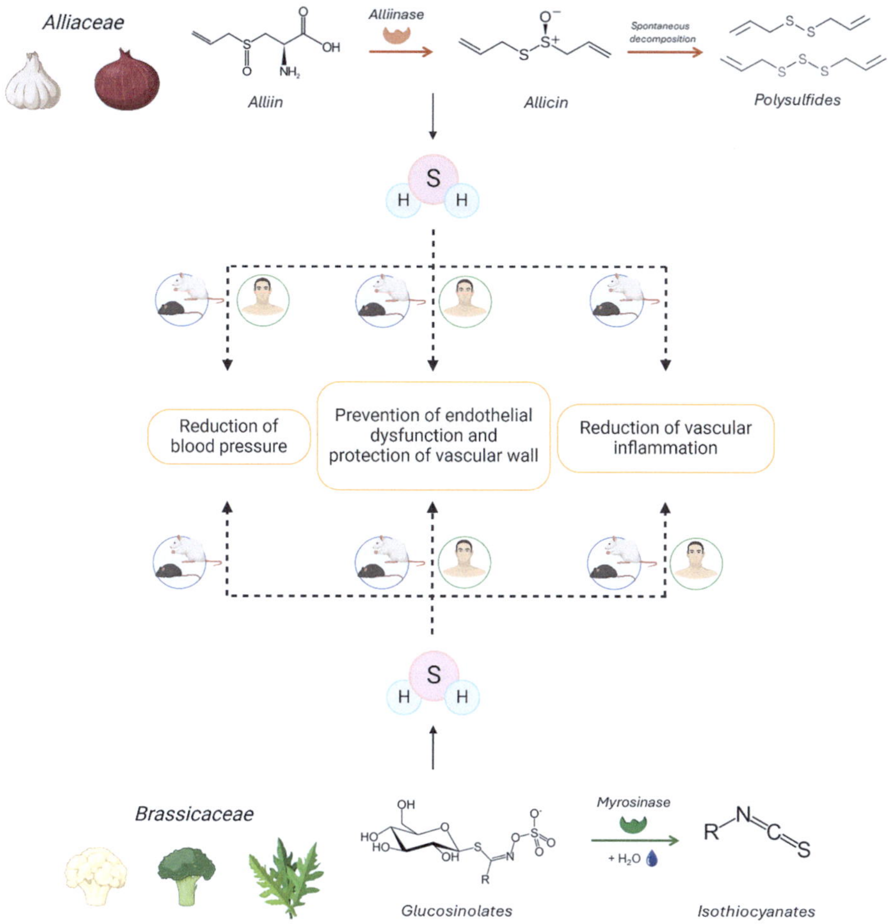

Fig. 4 The image represents which molecules deriving from plants belonging to the *Alliaceae* (alliin, allicin, and polysulfides) and *Brassicaceae* (Glucosinolates and Isothiocyanates) can release H_2S. Furthermore, the positive effects exerted by the H_2S released in the vessels are pointed out by emphasizing the presence of preclinical and/or clinical studies

studies. In this regard, a meta-analysis of twelve randomized controlled trials published in 2020 reported a mean reduction of 8.3 and 5.5 mmHg in systolic and diastolic blood pressure, respectively, in hypertensive patients receiving garlic extracts (Ried 2020). More recently, Kravchuk and coworkers showed that garlic extract (400 mg/day for 4 weeks) significantly reduced systolic and diastolic blood pressure levels in hypertensive men (mean change from baseline: −16.5 and −12.5 mmHg, respectively), in part by restoring physiological H_2S levels, which were almost halved in patients with hypertension (Kravchuk et al. 2021). Furthermore, processed garlic (1 g/day for 2 months, ~ 0.8 mg/day S-allilcysteine, SAC) lowered systolic blood pressure (mean change from baseline: −8 mmHg) in hypertensive patients (Han et al. 2011), while garlic tablets (300–1,500 mg/day for 6 months) reduced systolic and diastolic blood pressure levels in patients with essential hypertension in a dose-dependent manner (about 8 mmHg at the highest dose tested) (Ashraf et al. 2013). Finally, an aged garlic extract (0.25 mg/day for 3 months) promoted a further reduction in blood pressure levels (about 1.8 mmHg in systolic blood pressure and 1.5 mmHg in diastolic blood pressure) in patients with grade 1 hypertension treated with antihypertensive drugs (Serrano et al. 2023). It should be noted that in a very recent randomized, double-blind clinical trial, capsules of freeze-dried garlic extract (twice daily for 8 weeks) significantly reduced blood pressure levels in subjects at risk of hypertension, suggesting that garlic could be a valuable ally not only in the treatment but also in the prevention of this common cardiovascular risk factor (Rahmatinia et al. 2024).

The anti-hypertensive effects of garlic could result from both vascular remodeling and improved endothelial function. A recent systematic review concluded that garlic treatment can restore carotid intima-media thickness and microcirculation (Imaizumi et al. 2023), while the GarGIC trial showed that supplementation with an aged garlic extract (1.2 g/day, ~1.2 mg SAC for 12 weeks) prevented arterial stiffness and reversed vessel aging in patients with hypertension (Ried et al. 2018). The clinical evidence of the beneficial effects of garlic on vascular endothelium, rather than vascular remodeling, is more consistent. About 20 years ago, William and coworkers demonstrated that an aged garlic extract (2.4 g/day for 2 weeks) improved endothelial function, measured as brachial artery flow mediated endothelium-dependent dilation (FMD), in men with coronary artery disease (Williams et al. 2005). More recently, a randomized double-blind clinical trial showed that garlic extract (400 mg/day for 12 weeks) preserved endothelial integrity in obese patients (Szulińska et al. 2018). Finally, coenzyme Q10 plus aged garlic extract (1.2 g/day for 12 months) significantly improved vascular elasticity and endothelial function in people with high work stress (Larijani et al. 2013), encapsulated purple garlic oil (4–8 mg/day for 5 weeks) promoted microvascular function in subjects with cardiometabolic alterations (Bara-Ledesma et al. 2024), and aged garlic extract (4 mg/day for 6 weeks) prevented microvascular endothelial dysfunction induced by hyperhomocysteinemia in healthy people, as indicated by decreased FMD (Weiss et al. 2006).

Endothelial dysfunction is not only a prerogative of hypertension but is also a harmful feature of atherosclerosis, which is a common consequence of dyslipidemia and metabolic syndrome (Aboonabi et al. 2019; Du and Qin 2023). In this regard, a recent systematic review demonstrated that garlic prevents cardiovascular risk by reducing total cholesterol, triglycerides, body mass index, and waist circumference (Imaizumi et al. 2023). These findings were confirmed by a meta-analysis of controlled studies, which showed that treatment with garlic extracts significantly reduced total cholesterol and LDL levels in patients with hypercholesterolemia (Piragine et al. 2024). In addition, two randomized, double-blind clinical trials independently demonstrated that daily treatment with an aged garlic extract (2.4 g/day for 12 months) significantly slowed coronary artery calcification in subjects with an intermediate-high risk of cardiovascular events (Wlosinska et al. 2020) and reduced low-attenuation plaque in coronary arteries of patients with type 2 diabetes (Shaikh et al. 2020).

In conclusion, garlic could be an effective option for maintaining vascular health, but the experimental conditions used in current trials are too heterogeneous to provide high-quality evidence. Furthermore, there are no clinical studies on onion extracts, thus limiting the generalizability of the results to all organosulfur-containing natural products. Further studies are needed to confirm the clinical potential of *Alliaceae* extracts in the management of vascular diseases.

4 Isothiocyanates in Hypertension and Vascular Wall Protection

In recent years, many in vivo studies and clinical trials focused on the evaluation of the potential beneficial properties of organic ITCs on the vascular wall and on their ability to reduce vascular inflammation, with SFN being the main ITC investigated. Daily supplementation of SFN reduced TNF-α-induced vascular inflammation and VCAM-1 expression in the mouse aorta, and preserved the vascular endothelium integrity (Nallasamy et al. 2014). It was also able to lower blood pressure in hypertensive rats and reduce the arterial wall/lm ratio (Senanayake et al. 2012). In rats injected with AGEs, the daily administration of SFN reduced the expression of ICAM-1, VCAM-1, and aortic RAGE (Matsui et al. 2016). In two in vivo mouse models of pulmonary arterial hypertension, the administration of SFN contrasted the hypoxia-induced pulmonary arterioles thickening and reduced the expression of inflammatory markers, fibrosis, and macrophage recruitment around pulmonary arteries. Furthermore, SFN upregulated Nrf2 expression and reduced oxidative stress (Kang et al. 2020; Pan et al. 2023). Lastly, daily administration of SFN significantly ameliorated the vascular profile of rats fed with high fructose diet, enhancing the responsiveness to acetylcholine-induced vasorelaxation while ameliorating hepatic oxidative stress levels and reducing serum levels of C-reactive protein and lactate dehydrogenase (Shawky et al. 2019). Another ITC found in a widespread plant is erucin, derived from the GLS glucoerucin, present in *Eruca sativa* Mill. This ITC exerted hyperpolarizing, vasorelaxant, anti-hypertensive properties in vivo in SHRs (Martelli et al. 2020a) and exhibited cardioprotective effects in an in vivo model of myocardial infarction (Martelli et al. 2020b; Testai et al. 2023). Interestingly, both KATP and Kv7.4 channels contributed to the cardioprotective effects of erucin (Flori et al. 2023; Testai et al. 2023). Moreover, also a defatted seed meal (DSM) extract of *Eruca sativa* Mill. was able to release H_2S in a cell-free assay. This extract induced a significative reduction of the blood pressure in phenylephrine-induced hypertensive rats and in SHRs together with a dose-dependent cardioprotection in an in vivo model of acute myocardial infarction (Testai et al. 2022). Another source of natural ITCs is represented by *Moringa oleifera* Lam. (*Moringaceae*), a subtropical tree native to Asia and Africa. The daily administration for 3 weeks of its aqueous leaf extract was able to significantly reduce blood pressure in L-NAME-infused hypertensive rats (Aekthammarat et al. 2019). A further study from the same author demonstrated that the anti-hypertensive activity was mediated by the increased expression of eNOS, thus increasing NO production (Aekthammarat et al. 2020). In rats fed with a high fat-high carbohydrate diet, supplementation with wasabi (*Eutrema japonicum* (Miq.) Koidz., belonging to the *Brassicaceae* family) rhizome dry powder (5% w/w), standardized in 6-(methylsulfinyl)hexyl ITC, normalized systolic blood pressure, decreased inflammatory cytokine levels, and significantly reduced the body weight gain, together with an amelioration of the lipid profile (Thomaz et al. 2022). In humans, two randomized clinical trials in 2010 and 2013 enrolled respectively 40 hypertensive patients and 48 volunteers with a 10-year CVD risk profile of 10% to 20%, estimated by using the Joint British Societies 2 (JBS2) CVD risk assessor to evaluate the impact of a diet rich in high-glucosinolate broccoli or dried broccoli sprouts, both standardized in glucoraphanin, on CVDs. However, both studies did not show a significant reduction in systolic or diastolic blood pressure (Armah et al. 2013; Christiansen et al. 2010). Conversely, in a phase 1 study conducted in 2004, a significant reduction in both total cholesterol, LDL levels, and DNA damage was noted in 12 healthy subjects who consumed 100 g/day of fresh broccoli sprouts for 1 week (Murashima et al. 2004). Taken together, despite the clear vasoprotective properties of the ITCs, future studies are needed to standardize the formulation of most supplements or extracts used in current preclinical studies and clinical trials (Fig. 4).

5 Conclusions

Experimental evidence that has grown exponentially in recent decades has officially elected H_2S as the third member of the family of endogenous gasotransmitters. Thanks to its ability to modulate different targets and molecular pathways, it has a high potential in the control of different physio-pathological processes (Flori et al. 2024b; Testai et al. 2021). H_2S deficiency in plasma and vascular district has been associated with reduced functionality of vascular smooth muscle and reduced efficiency of the endothelial layer with consequent risk of developing hypertension and atherosclerosis. H_2S exerts antioxidant properties directly by scavenging ROS or increasing the expression of antioxidant enzymes and exerts antihypertensive and vasorelaxant effects by modulating potassium channels and through a crosstalk with NO. For this reason, the use of compounds capable of releasing H_2S in a physiological manner can represent an effective pharmacological strategy for treating vascular impairments. In this regard, several synthetic H_2S-donors have demonstrated their efficacy in lowering systolic blood pressure, reducing vascular inflammation and restoring endothelial function (Barresi et al. 2017; Citi et al. 2020; Martelli et al. 2013b, 2014; Severino et al. 2018).

As described in detail in this chapter, edible plants belonging to the *Alliaceae* and *Brassicaceae* families and natural molecules or plant extracts derived therefrom have shown beneficial effects on vascular function by releasing H_2S. Isolated polysulfides, such as DADS and DATS, or their precursor alliin, GLs such as glucoraphanin and glucoerucin, ITCs like SFN and erucin, *Alliaceae* or

Brassicaceae extracts are described in depth for the anti-hypertensive and vasoprotective properties by several preclinical investigations.

The several preclinical evidence on isolated polysulfide and *Alliaceae* extracts clearly indicated antihypertensive effects resulting from both vascular remodeling and improved endothelial function revealing the key role of H_2S as the "hidden" vasoactive mediator. These results were confirmed by clinical evidence. Real-world data confirmed a lowering of blood pressure in both hypertensive and at-risk patients due to the intake of garlic extracts. Furthermore, garlic extracts promoted endothelial function in patients with coronary artery disease and in obese patients and vascular elasticity in people with high work stress and subjects with cardiometabolic alterations. Finally, garlic extracts prevented CVD risk by improving the lipid and glycemic profile and significantly slowing down coronary artery calcification.

Similarly, several preclinical evidence confirmed potential beneficial properties of organic ITCs on the vascular wall and the ability to reduce vascular inflammation, especially for SFN and erucin. Furthermore, as for *Alliaceae*, also the *Bassicaceae* and *Moringaceae* extracts, have shown antihypertensive effects, an improvement of the vascular inflammation, lipid profile, and a significant reduction in body weight gain in obesogenic conditions. Unfortunately, unlike the clinical evidence reported for garlic extracts, the limited clinical evidence currently available and collected in this chapter regarding plants belonging to the *Brassicaceae* family does not seem to confirm a significant effect in the regulation of systolic and diastolic blood pressure in hypertensive patients but highlighted a significant effect in the reduction of total cholesterol and LDL levels in patients with CVD risk profile.

In conclusion, the increasing experimental evidence about natural compounds derived from *Alliaceae* and *Brassicaceae* edible plants, capable of releasing H_2S, in the regulation of vascular physio-pathological processes confirms the antihypertensive and vasorelaxant effects, the anti-inflammatory activity, and the enhancement of antioxidant defense pathways, deserving the nickname of "sulfanutraceuticals." Further investigations through clinical studies are necessary to confirm the effects of both plant extracts and H_2S-donor derivatives and to better understand the ability of H_2S to positively modulate blood pressure and vascular function in hypertensive and high CVD risk profile patients.

References

Abe K, Kimura H (1996) The possible role of hydrogen sulfide as an endogenous neuromodulator. J Neurosci 16:1066–1071. https://doi.org/10.1523/jneurosci.16-03-01066.1996

Aboonabi A, Meyer RR, Singh I (2019) The association between metabolic syndrome components and the development of atherosclerosis. J Hum Hypertens 33:844–855

Aekthammarat D, Pannangpetch P, Tangsucharit P (2019) Moringa oleifera leaf extract lowers high blood pressure by alleviating vascular dysfunction and decreasing oxidative stress in L-NAME hypertensive rats. Phytomedicine 54:9–16

Aekthammarat D, Tangsucharit P, Pannangpetch P, Sriwantana T, Sibmooh N (2020) Moringa oleifera leaf extract enhances endothelial nitric oxide production leading to relaxation of resistance artery and lowering of arterial blood pressure. Biomed Pharmacother 130:110605

Ansary J, Forbes-Hernández TY, Gil E, Cianciosi D, Zhang J, Elexpuru-Zabaleta M, Simal-Gandara J, Giampieri F, Battino M (2020) Potential health benefit of garlic based on human intervention studies: a brief overview. Antioxidants 9:619

Armah CN, Traka MH, Dainty JR, Defernez M, Janssens A, Leung W, Doleman JF, Potter JF, Mithen RF (2013) A diet rich in high-glucoraphanin broccoli interacts with genotype to reduce discordance in plasma metabolite profiles by modulating mitochondrial function. Am J Clin Nutr 98:712–722. https://doi.org/10.3945/ajcn.113.065235

Ashraf MZ, Hussain ME, Fahim M (2004) Endothelium mediated vasorelaxant response of garlic in isolated rat aorta: role of nitric oxide. J Ethnopharmacol 90:5–9. https://doi.org/10.1016/j.jep.2003.06.001

Ashraf R, Khan RA, Ashraf I, Qureshi AA (2013) Effects of Allium sativum (garlic) on systolic and diastolic blood pressure in patients with essential hypertension. Pak J Pharm Sci 26:859–863

Bahadoran Z, Mirmiran P, Momenan AA, Azizi F (2017) Allium vegetable intakes and the incidence of cardiovascular disease, hypertension, chronic kidney disease, and type 2 diabetes in adults: a longitudinal follow-up study. J Hypertens 35:1909–1916

Bara-Ledesma N, Jimenez-Esteban J, Fabregate M, Fabregate-Fuente R, Cymberknop LJ, Castillo-Martinez P, Navarro-Fayos MT, Gomez del Olmo V, Saban-Ruiz J (2024) Effect of encapsulated purple garlic oil on microvascular function and the components of metabolic syndrome: a randomized placebo-controlled study—the ENDOTALLIUM study. Nutrients 16:1755

Barrera LN, Cassidy A, Johnson IT, Bao Y, Belshaw NJ (2012) Epigenetic and antioxidant effects of dietary isothiocyanates and selenium: potential implications for cancer chemoprevention. Proc Nutr Soc 71:237–245. https://doi.org/10.1017/s002966511200016x

Barresi E, Nesi G, Citi V, Piragine E, Piano I, Taliani S, Da Settimo F, Rapposelli S, Testai L, Breschi MC (2017) Iminothioethers as hydrogen sulfide donors: from the gasotransmitter release to the vascular effects. J Med Chem 60:7512–7523

Bełtowski J (2015) Hydrogen sulfide in pharmacology and medicine--An update. Pharmacol Rep 67:647–658. https://doi.org/10.1016/j.pharep.2015.01.005

Benavides GA, Squadrito GL, Mills RW, Patel HD, Isbell TS, Patel RP, Darley-Usmar VM, Doeller JE, Kraus DW (2007) Hydrogen sulfide mediates the vasoactivity of garlic. Proc Natl Acad Sci USA 104:17977–17982. https://doi.org/10.1073/pnas.0705710104

Berenyiova A, Grman M, Mijuskovic A, Stasko A, Misak A, Nagy P, Ondriasova E, Cacanyiova S, Brezova V, Feelisch M, Ondrias K (2015) The reaction products of sulfide and S-nitrosoglutathione are potent vasorelaxants. Nitric Oxide 46:123–130. https://doi.org/10.1016/j.niox.2014.12.008

Borlinghaus J, Albrecht F, Gruhlke MC, Nwachukwu ID, Slusarenko AJ (2014) Allicin: chemistry and biological properties. Molecules 19:12591–12618

Bucci M, Papapetropoulos A, Vellecco V, Zhou Z, Pyriochou A, Roussos C, Roviezzo F, Brancaleone V, Cirino G (2010) Hydrogen sulfide is an endogenous inhibitor of phosphodiesterase activity. Arterioscler Thromb Vasc Biol 30:1998–2004. https://doi.org/10.1161/atvbaha.110.209783

Calderone V, Martelli A, Testai L, Citi V, Breschi MC (2016) Using hydrogen sulfide to design and develop drugs. Expert Opin Drug Discov 11:163–175. https://doi.org/10.1517/17460441.2016.1122590

Chen C-Y, Tsai T-Y, Chen B-H (2021) Effects of black garlic extract and nanoemulsion on the deoxy corticosterone acetate-salt induced hypertension and its associated mild cognitive impairment in rats. Antioxidants 10:1611

Chen J, Ding X, Chen W, Chen S, Guan Q, Wen J, Chen Z (2022) VEGFR2 in vascular smooth muscle cells mediates H2S-induced dilation of the rat cerebral basilar artery. Microvasc Res 141:104309

Christiansen B, Bellostas Muguerza N, Petersen AM, Kveiborg B, Madsen CR, Thomas H, Ihlemann N, Sørensen JC, Køber L, Sørensen H, Torp-Pedersen C, Domínguez H (2010)

Ingestion of broccoli sprouts does not improve endothelial function in humans with hypertension. PLoS One 5:e12461. https://doi.org/10.1371/journal.pone.0012461
Ciccone V, Piragine E, Gorica E, Citi V, Testai L, Pagnotta E, Matteo R, Pecchioni N, Montanaro R, Di Cesare ML (2022) Anti-inflammatory effect of the natural H2S-donor erucin in vascular endothelium. Int J Mol Sci 23:15593
Cirino G, Szabo C, Papapetropoulos A (2023) Physiological roles of hydrogen sulfide in mammalian cells, tissues, and organs. Physiol Rev 103:31–276
Citi V, Martelli A, Testai L, Marino A, Breschi MC, Calderone V (2014) Hydrogen sulfide releasing capacity of natural isothiocyanates: is it a reliable explanation for the multiple biological effects of Brassicaceae? Planta Med 80:610–613. https://doi.org/10.1055/s-0034-1368591
Citi V, Piragine E, Pagnotta E, Ugolini L, Di Cesare ML, Testai L, Ghelardini C, Lazzeri L, Calderone V, Martelli A (2019) Anticancer properties of erucin, an H2S-releasing isothiocyanate, on human pancreatic adenocarcinoma cells (AsPC-1). Phytother Res 33:845–855
Citi V, Martelli A, Bucci M, Piragine E, Testai L, Vellecco V, Cirino G, Calderone V (2020) Searching for novel hydrogen sulfide donors: the vascular effects of two thiourea derivatives. Pharmacol Res 159:105039
Citi V, Martelli A, Gorica E, Brogi S, Testai L, Calderone V (2021) Role of hydrogen sulfide in endothelial dysfunction: pathophysiology and therapeutic approaches. J Adv Res 27:99–113
Citi V, Piragine E, Calderone V, Martelli A (2022) Hydrogen sulfide: the hidden player of isothiocyanates pharmacology. In: Hydrogen sulfide: chemical biology basics, detection methods, therapeutic applications, and case studies, pp 261–292
Coletta C, Papapetropoulos A, Erdelyi K, Olah G, Módis K, Panopoulos P, Asimakopoulou A, Gerö D, Sharina I, Martin E, Szabo C (2012) Hydrogen sulfide and nitric oxide are mutually dependent in the regulation of angiogenesis and endothelium-dependent vasorelaxation. Proc Natl Acad Sci USA 109:9161–9166. https://doi.org/10.1073/pnas.1202916109
Connolly EL, Liu AH, Radavelli-Bagatini S, Shafaei A, Boyce MC, Wood LG, McCahon L, Koch H, Sim M, Hill CR (2024) Cruciferous vegetables lower blood pressure in adults with mildly elevated blood pressure in a randomized, controlled, crossover trial: the VEgetableS for vaScular hEaLth (VESSEL) study. BMC Med 22:353
Cui T, Liu W, Chen S, Yu C, Li Y, Zhang J-Y (2020) Antihypertensive effects of allicin on spontaneously hypertensive rats via vasorelaxation and hydrogen sulfide mechanisms. Biomed Pharmacother 128:110240
Dongó E, Beliczai-Marosi G, Dybvig AS, Kiss L (2018) The mechanism of action and role of hydrogen sulfide in the control of vascular tone. Nitric Oxide 81:75–87
Du Z, Qin Y (2023) Dyslipidemia and cardiovascular disease: current knowledge, existing challenges, and new opportunities for management strategies. J Clin Med 12:363
Dubey H, Singh A, Patole AM, Tenpe CR (2017) Antihypertensive effect of allicin in dexamethasone-induced hypertensive rats. Integr Med Res 6:60–65
Eberhardt M, Dux M, Namer B, Miljkovic J, Cordasic N, Will C, Kichko TI, de la Roche J, Fischer M, Suárez SA, Bikiel D, Dorsch K, Leffler A, Babes A, Lampert A, Lennerz JK, Jacobi J, Martí MA, Doctorovich F, Högestätt ED, Zygmunt PM, Ivanovic-Burmazovic I, Messlinger K, Reeh P, Filipovic MR (2014) H2S and NO cooperatively regulate vascular tone by activating a neuroendocrine HNO-TRPA1-CGRP signalling pathway. Nat Commun 5:4381. https://doi.org/10.1038/ncomms5381
Elkayam A, Mirelman D, Peleg E, Wilchek M, Miron T, Rabinkov A, Sadetzki S, Rosenthal T (2001) The effects of allicin and enalapril in fructose-induced hyperinsulinemic hyperlipidemic hypertensive rats. Am J Hypertens 14:377–381. https://doi.org/10.1016/s0895-7061(00)01298-x
Elosta A, Slevin M, Rahman K, Ahmed N (2017) Aged garlic has more potent antiglycation and antioxidant properties compared to fresh garlic extract in vitro. Sci Rep 7:39613

El-Sheakh AR, Ghoneim HA, Suddek GM, Ammar ESM (2016) Attenuation of oxidative stress, inflammation, and endothelial dysfunction in hypercholesterolemic rabbits by allicin. Can J Physiol Pharmacol 94:216–224. https://doi.org/10.1139/cjpp-2015-0267
Flori L, Montanaro R, Pagnotta E, Ugolini L, Righetti L, Martelli A, Di Cesare ML, Ghelardini C, Brancaleone V, Testai L (2023) Erucin exerts cardioprotective effects on ischemia/reperfusion injury through the modulation of mitoKATP channels. Biomedicine 11:3281
Flori L, Benedetti G, Calderone V, Testai L (2024a) Hydrogen sulfide and irisin, potential allies in ensuring cardiovascular health. Antioxidants 13:543. https://doi.org/10.3390/antiox13050543
Flori L, Piragine E, Calderone V, Testai L (2024b) Role of hydrogen sulfide in the regulation of lipid metabolism: implications on cardiovascular health. Life Sci 341:122491. https://doi.org/10.1016/j.lfs.2024.122491
Ganster F, Burban M, de La Bourdonnaye M, Fizanne L, Douay O, Loufrani L, Mercat A, Calès P, Radermacher P, Henrion D (2010) Effects of hydrogen sulfide on hemodynamics, inflammatory response and oxidative stress during resuscitated hemorrhagic shock in rats. Crit Care 14:1–11
García-Trejo EM, Arellano-Buendía AS, Argüello-García R, Loredo-Mendoza ML, García-Arroyo FE, Arellano-Mendoza MG, Castillo-Hernández MC, Guevara-Balcázar G, Tapia E, Sánchez-Lozada LG (2016) Effects of allicin on hypertension and cardiac function in chronic kidney disease. Oxidative Med Cell Longev 2016:3850402
Genah S, Ciccone V, Filippelli A, Simonis V, Martelli A, Piragine E, Pagnotta E, Pecchioni N, Calderone V, Morbidelli L (2024) Erucin, a natural isothiocyanate, exerts pro-angiogenic properties in cultured endothelial cells and reverts angiogenic impairment induced by high glucose. Phytother Res 38:2641–2655
Han C-H, Liu J-C, Chen K-H, Lin Y-S, Chen C-T, Fan C-T, Lee H-L, Liu D-Z, Hou W-C (2011) Antihypertensive activities of processed garlic on spontaneously hypertensive rats and hypertensive humans. Bot Stud 52
Harauma A, Moriguchi T (2006) Aged garlic extract improves blood pressure in spontaneously hypertensive rats more safely than raw garlic. J Nutr 136:769s–773s. https://doi.org/10.1093/jn/136.3.769S
Hedegaard ER, Gouliaev A, Winther AK, Arcanjo DD, Aalling M, Renaltan NS, Wood ME, Whiteman M, Skovgaard N, Simonsen U (2016) Involvement of potassium channels and calcium-independent mechanisms in hydrogen sulfide-induced relaxation of rat mesenteric small arteries. J Pharmacol Exp Ther 356:53–63. https://doi.org/10.1124/jpet.115.227017
Hermann A, Sitdikova G (2021) Homocysteine: biochemistry, molecular biology and role in disease, vol 11. MDPI, p 737
Holwerda KM, Karumanchi SA, Lely AT (2015) Hydrogen sulfide: role in vascular physiology and pathology. Curr Opin Nephrol Hypertens 24:170–176. https://doi.org/10.1097/mnh.0000000000000096
Hsu CN, Hou CY, Chang-Chien GP, Lin S, Tain YL (2021) Maternal garlic oil supplementation prevents high-fat diet-induced hypertension in adult rat offspring: implications of H2S-generating pathway in the gut and kidneys. Mol Nutr Food Res 65:2001116
Imaizumi VM, Laurindo LF, Manzan B, Guiguer EL, Oshiiwa M, Otoboni AMMB, Araujo AC, Tofano RJ, Barbalho SM (2023) Garlic: a systematic review of the effects on cardiovascular diseases. Crit Rev Food Sci Nutr 63:6797–6819
Jeremic JN, Jakovljevic VL, Zivkovic VI, Srejovic IM, Bradic JV, Milosavljevic IM, Mitrovic SL, Jovicic NU, Bolevich SB, Svistunov AA (2020) Garlic derived diallyl trisulfide in experimental metabolic syndrome: metabolic effects and cardioprotective role. Int J Mol Sci 21:9100
Jiang B, Tang G, Cao K, Wu L, Wang R (2010) Molecular mechanism for H(2)S-induced activation of K(ATP) channels. Antioxid Redox Signal 12:1167–1178. https://doi.org/10.1089/ars.2009.2894
Jin S, Teng X, Xiao L, Xue H, Guo Q, Duan X, Chen Y, Wu Y (2017) Hydrogen sulfide ameliorated L-NAME-induced hypertensive heart disease by the Akt/eNOS/NO pathway. Exp Biol Med 242:1831–1841

Kang M, Hashimoto A, Gade A, Akbarali HI (2015) Interaction between hydrogen sulfide-induced sulfhydration and tyrosine nitration in the KATP channel complex. Am J Physiol Gastrointest Liver Physiol 308:G532–G539. https://doi.org/10.1152/ajpgi.00281.2014

Kang Y, Zhang G, Huang EC, Huang J, Cai J, Cai L, Wang S, Keller BB (2020) Sulforaphane prevents right ventricular injury and reduces pulmonary vascular remodeling in pulmonary arterial hypertension. Am J Phys Heart Circ Phys 318:H853–H866

Kianian F, Marefati N, Boskabady M, Ghasemi SZ, Boskabady MH (2021) Pharmacological properties of Allium cepa, preclinical and clinical evidences; a review. Iran J Pharm Res 20:107

Kimura H (2016) Hydrogen polysulfide (H 2 S n) signaling along with hydrogen sulfide (H 2 S) and nitric oxide (NO). J Neural Transm 123:1235–1245

Kimura H (2021) Hydrogen sulfide (H2S) and polysulfide (H2Sn) signaling: the first 25 years. Biomolecules 11:896

King AL, Polhemus DJ, Bhushan S, Otsuka H, Kondo K, Nicholson CK, Bradley JM, Islam KN, Calvert JW, Tao YX, Dugas TR, Kelley EE, Elrod JW, Huang PL, Wang R, Lefer DJ (2014) Hydrogen sulfide cytoprotective signaling is endothelial nitric oxide synthase-nitric oxide dependent. Proc Natl Acad Sci USA 111:3182–3187. https://doi.org/10.1073/pnas.1321871111

Kolluru GK, Shen X, Bir SC, Kevil CG (2013) Hydrogen sulfide chemical biology: pathophysiological roles and detection. Nitric Oxide 35:5–20. https://doi.org/10.1016/j.niox.2013.07.002

Kravchuk OM, Goshovska YV, Korkach YP, Sagach VF, Nagre SW, Bhosle KN (2021) Garlic supplement lowers blood pressure in 40-60 years old hypertensive individuals, regulates oxidative stress, plasma cholesterol and protrombin index. J Cardiovasc Med Cardiol 8:041–047

Larijani VN, Ahmadi N, Zeb I, Khan F, Flores F, Budoff M (2013) Beneficial effects of aged garlic extract and coenzyme Q10 on vascular elasticity and endothelial function: the FAITH randomized clinical trial. Nutrition 29:71–75. https://doi.org/10.1016/j.nut.2012.03.016

Lau BH (2006) Suppression of LDL oxidation by garlic compounds is a possible mechanism of cardiovascular health benefit. J Nutr 136:765s–768s. https://doi.org/10.1093/jn/136.3.765S

Lei YP, Liu CT, Sheen LY, Chen HW, Lii CK (2010) Diallyl disulfide and diallyl trisulfide protect endothelial nitric oxide synthase against damage by oxidized low-density lipoprotein. Mol Nutr Food Res 54(Suppl 1):S42–S52. https://doi.org/10.1002/mnfr.200900278

Li Q, Lancaster JR Jr (2013) Chemical foundations of hydrogen sulfide biology. Nitric Oxide 35: 21–34. https://doi.org/10.1016/j.niox.2013.07.001

Li L, Moore PK (2008) Putative biological roles of hydrogen sulfide in health and disease: a breath of not so fresh air? Trends Pharmacol Sci 29:84–90. https://doi.org/10.1016/j.tips.2007.11.003

Li L, Rose P, Moore PK (2011) Hydrogen sulfide and cell signaling. Annu Rev Pharmacol Toxicol 51:169–187. https://doi.org/10.1146/annurev-pharmtox-010510-100505

Liang D, Wu H, Wong MW, Huang D (2015) Diallyl Trisulfide is a fast H2S donor, but Diallyl Disulfide is a slow one: the reaction pathways and intermediates of glutathione with polysulfides. Org Lett 17:4196–4199. https://doi.org/10.1021/acs.orglett.5b01962

Lin Y, Chen Y, Zhu N, Zhao S, Fan J, Liu E (2016) Hydrogen sulfide inhibits development of atherosclerosis through up-regulating protein S-nitrosylation. Biomed Pharmacother 83:466–476. https://doi.org/10.1016/j.biopha.2016.07.003

Lin Y, Yang X, Lu Y, Liang D, Huang D (2019) Isothiocyanates as H2S donors triggered by cysteine: reaction mechanism and structure and activity relationship. Org Lett 21:5977–5980

Liu LL, Yan L, Chen YH, Zeng GH, Zhou Y, Chen HP, Peng WJ, He M, Huang QR (2014) A role for diallyl trisulfide in mitochondrial antioxidative stress contributes to its protective effects against vascular endothelial impairment. Eur J Pharmacol 725:23–31. https://doi.org/10.1016/j.ejphar.2014.01.010

Liu W, Xu S, Liang S, Duan C, Xu Z, Zhao L, Wen F, Li Q, Li Y, Zhang J (2022) Hypertensive vascular and cardiac remodeling protection by allicin in spontaneous hypertension rats via CaMK II/NF-κB pathway. Biomed Pharmacother 155:113802

Liu S, Zhang Y, Zheng X, Wang Z, Wang P, Zhang M, Shen M, Bao Y, Li D (2023) Sulforaphane inhibits foam cell formation and atherosclerosis via mechanisms involving the modulation of macrophage cholesterol transport and the related phenotype. Nutrients 15:2117

Lu M, Pan J, Hu Y, Ding L, Li Y, Cui X, Zhang M, Zhang Z, Li C (2024) Advances in the study of vascular related protective effect of garlic (Allium sativum) extract and compounds. J Nutr Biochem 124:109531

Lucarini E, Micheli L, Trallori E, Citi V, Martelli A, Testai L, De Nicola GR, Iori R, Calderone V, Ghelardini C (2018) Effect of glucoraphanin and sulforaphane against chemotherapy-induced neuropathic pain: Kv7 potassium channels modulation by H2S release in vivo. Phytother Res 32:2226–2234

Martelli A, Testai L, Breschi MC, Lawson K, McKay NG, Miceli F, Taglialatela M, Calderone V (2013a) Vasorelaxation by hydrogen sulphide involves activation of Kv7 potassium channels. Pharmacol Res 70:27–34. https://doi.org/10.1016/j.phrs.2012.12.005

Martelli A, Testai L, Citi V, Marino A, Pugliesi I, Barresi E, Nesi G, Rapposelli S, Taliani S, Da Settimo F, Breschi MC, Calderone V (2013b) Arylthioamides as H2S donors: l-cysteine-activated releasing properties and vascular effects in vitro and in vivo. ACS Med Chem Lett 4:904–908. https://doi.org/10.1021/ml400239a

Martelli A, Testai L, Citi V, Marino A, Bellagambi FG, Ghimenti S, Breschi MC, Calderone V (2014) Pharmacological characterization of the vascular effects of aryl isothiocyanates: is hydrogen sulfide the real player? Vasc Pharmacol 60:32–41

Martelli A, Citi V, Testai L, Brogi S, Calderone V (2020a) Organic isothiocyanates as hydrogen sulfide donors. Antioxid Redox Signal 32:110–144

Martelli A, Piragine E, Citi V, Testai L, Pagnotta E, Ugolini L, Lazzeri L, Di Cesare Mannelli L, Manzo OL, Bucci M (2020b) Erucin exhibits vasorelaxing effects and antihypertensive activity by H2S-releasing properties. Br J Pharmacol 177:824–835

Martelli A, Piragine E, Gorica E, Citi V, Testai L, Pagnotta E, Lazzeri L, Pecchioni N, Ciccone V, Montanaro R (2021) The H2S-donor erucin exhibits protective effects against vascular inflammation in human endothelial and smooth muscle cells. Antioxidants 10:961

Martelli A, di Villa Bianca RE, Cirino G, Sorrentino R, Calderone V, Bucci M (2023) Hydrogen sulfide and sulfaceutic or sulfanutraceutic agents: classification, differences and relevance in preclinical and clinical studies. Pharmacol Res 196:106947

Matsui T, Nakamura N, Ojima A, Nishino Y, Yamagishi SI (2016) Sulforaphane reduces advanced glycation end products (AGEs)-induced inflammation in endothelial cells and rat aorta. Nutr Metab Cardiovasc Dis 26:797–807. https://doi.org/10.1016/j.numecd.2016.04.008

Meng T, Qin W, Liu B (2020) SIRT1 antagonizes oxidative stress in diabetic vascular complication. Front Endocrinol 11:568861

Murashima M, Watanabe S, Zhuo XG, Uehara M, Kurashige A (2004) Phase 1 study of multiple biomarkers for metabolism and oxidative stress after one-week intake of broccoli sprouts. Biofactors 22:271–275. https://doi.org/10.1002/biof.5520220154

Mustafa AK, Sikka G, Gazi SK, Steppan J, Jung SM, Bhunia AK, Barodka VM, Gazi FK, Barrow RK, Wang R, Amzel LM, Berkowitz DE, Snyder SH (2011) Hydrogen sulfide as endothelium-derived hyperpolarizing factor sulfhydrates potassium channels. Circ Res 109:1259–1268. https://doi.org/10.1161/circresaha.111.240242

Naik JS, Osmond JM, Walker BR, Kanagy NL (2016) Hydrogen sulfide-induced vasodilation mediated by endothelial TRPV4 channels. Am J Phys Heart Circ Phys 311:H1437–H1444

Nallasamy P, Si H, Babu PV, Pan D, Fu Y, Brooke EA, Shah H, Zhen W, Zhu H, Liu D, Li Y, Jia Z (2014) Sulforaphane reduces vascular inflammation in mice and prevents TNF-α-induced monocyte adhesion to primary endothelial cells through interfering with the NF-κB pathway. J Nutr Biochem 25:824–833. https://doi.org/10.1016/j.jnutbio.2014.03.011

Naseri MK, Arabian M, Badavi M, Ahangarpour A (2008) Vasorelaxant and hypotensive effects of Allium cepa peel hydroalcoholic extract in rat. Pak J Biol Sci 11:1569–1575. https://doi.org/10.3923/pjbs.2008.1569.1575

Nwokocha CR, Ozolua RI, Owu DU, Nwokocha MI, Ugwu AC (2011) Antihypertensive properties of Allium sativum (garlic) on normotensive and two kidney one clip hypertensive rats. Niger J Physiol Sci 26:213–218

Padiya R, Chowdhury D, Borkar R, Srinivas R, Pal Bhadra M, Banerjee SK (2014) Garlic attenuates cardiac oxidative stress via activation of PI3K/AKT/Nrf2-Keap1 pathway in fructose-fed diabetic rat. PLoS One 9:e94228. https://doi.org/10.1371/journal.pone.0094228

Pan J, Wang R, Pei Y, Wang D, Wu N, Ji Y, Tang Q, Liu L, Cheng K, Liu Q (2023) Sulforaphane alleviated vascular remodeling in hypoxic pulmonary hypertension via inhibiting inflammation and oxidative stress. J Nutr Biochem 111:109182

Piragine E, Citi V, Lawson K, Calderone V, Martelli A (2022a) Potential effects of natural H2S-donors in hypertension management. Biomolecules 12:581

Piragine E, Petri D, Giometto S, Martelli A, Lucenteforte E, Calderone V (2022b) Potential effects of Alliaceae and Brassicaceae edible plants on blood glucose levels in patients with type 2 diabetes: a systematic review and meta-analysis of clinical trials. Pharmacol Res 185:106519

Piragine E, Malanima MA, Lucenteforte E, Martelli A, Calderone V (2023) Circulating levels of hydrogen sulfide (H2S) in patients with age-related diseases: a systematic review and meta-analysis. Biomolecules 13:1023

Piragine E, Malanima MA, Ceccanti C, Guidi L, Martelli A, Lucenteforte E, Calderone V (2024) Alliaceae versus Brassicaceae for Dyslipidemia: state of the art and future perspectives. Systematic review and meta-analysis of clinical studies. Phytother Res

Pozsgai G, Bátai IZ, Pintér E (2019) Effects of sulfide and polysulfides transmitted by direct or signal transduction-mediated activation of TRPA1 channels. Br J Pharmacol 176:628–645

Prieto M, López CJ, Simal-Gandara J (2019) Glucosinolates: molecular structure, breakdown, genetic, bioavailability, properties and healthy and adverse effects. Adv Food Nutr Res 90: 305–350

Rahmatinia E, Amidi B, Naderi N, Ahmadipour S, Ahmadvand H, Pahlevan-Fallahy M-T, Ghorbanzadeh V, Nazari A (2024) Randomized, double-blind clinical trial evaluating the impact of freeze-dried garlic extract capsules on blood pressure, lipid profile, and nitric oxide levels in individuals at risk for hypertension. Horm Mol Biol Clin Invest 45(4):139–147

Recinella L, Libero ML, Citi V, Chiavaroli A, Martelli A, Foligni R, Mannozzi C, Acquaviva A, Di Simone S, Calderone V (2023) Anti-inflammatory and vasorelaxant effects induced by an aqueous aged black garlic extract supplemented with vitamins D, C, and B12 on cardiovascular system. Foods 12:1558

Ried K (2020) Garlic lowers blood pressure in hypertensive subjects, improves arterial stiffness and gut microbiota: a review and meta-analysis. Exp Ther Med 19:1472–1478

Ried K, Travica N, Sali A (2018) The effect of kyolic aged garlic extract on gut microbiota, inflammation, and cardiovascular markers in hypertensives: the GarGIC trial. Front Nutr 5:122

Rodrigues C, Percival SS (2019) Immunomodulatory effects of glutathione, garlic derivatives, and hydrogen sulfide. Nutrients 11:295

Sakai Y, Murakami T, Yamamoto Y (2003) Antihypertensive effects of onion on NO synthase inhibitor-induced hypertensive rats and spontaneously hypertensive rats. Biosci Biotechnol Biochem 67:1305–1311. https://doi.org/10.1271/bbb.67.1305

Searcy DG, Lee SH (1998) Sulfur reduction by human erythrocytes. J Exp Zool 282:310–322. https://doi.org/10.1002/(sici)1097-010x(19981015)282:3<310::aid-jez4>3.0.co;2-p

Senanayake GV, Banigesh A, Wu L, Lee P, Juurlink BH (2012) The dietary phase 2 protein inducer sulforaphane can normalize the kidney epigenome and improve blood pressure in hypertensive rats. Am J Hypertens 25:229–235. https://doi.org/10.1038/ajh.2011.200

Serrano JC, Castro-Boqué E, García-Carrasco A, Morán-Valero MI, González-Hedström D, Bermúdez-López M, Valdivielso JM, Espinel AE, Portero-Otín M (2023) Antihypertensive effects of an optimized aged garlic extract in subjects with Grade I hypertension and antihypertensive drug therapy: a randomized, triple-blind controlled trial. Nutrients 15:3691

Severino B, Corvino A, Fiorino F, Luciano P, Frecentese F, Magli E, Saccone I, Di Vaio P, Citi V, Calderone V (2018) 1, 2, 4-Thiadiazolidin-3, 5-diones as novel hydrogen sulfide donors. Eur J Med Chem 143:1677–1686

Shaikh K, Kinninger A, Cherukuri L, Birudaraju D, Nakanishi R, Almeida S, Jayawardena E, Shekar C, Flores F, Hamal S (2020) Aged garlic extract reduces low attenuation plaque in

coronary arteries of patients with diabetes: a randomized, double-blind, placebo-controlled study. Exp Ther Med 19:1457–1461
Sharifi AM, Darabi R, Akbarloo N (2003) Investigation of antihypertensive mechanism of garlic in 2K1C hypertensive rat. J Ethnopharmacol 86:219–224. https://doi.org/10.1016/s0378-8741(03)00080-1
Shawky NM, Shehatou GS, Suddek GM, Gameil NM (2019) Comparison of the effects of sulforaphane and pioglitazone on insulin resistance and associated dyslipidemia, hepatosteatosis, and endothelial dysfunction in fructose-fed rats. Environ Toxicol Pharmacol 66:43–54
Siti HN, Kamisah Y, Kamsiah J (2015) The role of oxidative stress, antioxidants and vascular inflammation in cardiovascular disease (a review). Vasc Pharmacol 71:40–56. https://doi.org/10.1016/j.vph.2015.03.005
Sobenin IA, Andrianova IV, Lakunin KY, Karagodin VP, Bobryshev YV, Orekhov AN (2016) Anti-atherosclerotic effects of garlic preparation in freeze injury model of atherosclerosis in cholesterol-fed rabbits. Phytomedicine 23:1235–1239. https://doi.org/10.1016/j.phymed.2015.10.014
Spezzini J, Piragine E, Flori L, Calderone V, Martelli A (2024) Natural H2S-donors: a new pharmacological opportunity for the management of overweight and obesity. Phytother Res 38:2388–2405
Sun X, Ku DD (2006) Allicin in garlic protects against coronary endothelial dysfunction and right heart hypertrophy in pulmonary hypertensive rats. Am J Physiol Heart Circ Physiol 291:H2431–H2438. https://doi.org/10.1152/ajpheart.00384.2006
Szlęzak D, Hutsch T, Ufnal M, Wróbel M (2022) Heart and kidney H2S production is reduced in hypertensive and older rats. Biochimie 199:130–138
Szulińska M, Kręgielska-Narożna M, Świątek J, Styś P, Kuźnar-Kamińska B, Jakubowski H, Walkowiak J, Bogdański P (2018) Garlic extract favorably modifies markers of endothelial function in obese patients–randomized double blind placebo-controlled nutritional intervention. Biomed Pharmacother 102:792–797
Testai L, Marino A, Piano I, Brancaleone V, Tomita K, Mannelli LDC, Martelli A, Citi V, Breschi MC, Levi R (2016) The novel H2S-donor 4-carboxyphenyl isothiocyanate promotes cardioprotective effects against ischemia/reperfusion injury through activation of mitoKATP channels and reduction of oxidative stress. Pharmacol Res 113:290–299
Testai L, Brancaleone V, Flori L, Montanaro R, Calderone V (2021) Modulation of EndMT by hydrogen sulfide in the prevention of cardiovascular fibrosis. Antioxidants 10:910
Testai L, Pagnotta E, Piragine E, Flori L, Citi V, Martelli A, Mannelli LDC, Ghelardini C, Matteo R, Suriano S (2022) Cardiovascular benefits of Eruca sativa mill. Defatted seed meal extract: potential role of hydrogen sulfide. Phytother Res 36:2616–2627
Testai L, Montanaro R, Flori L, Pagnotta E, Vellecco V, Gorica E, Ugolini L, Righetti L, Brancaleone V, Bucci M (2023) Persulfidation of mitoKv7. 4 channels contributes to the cardioprotective effects of the H2S-donor Erucin against ischemia/reperfusion injury. Biochem Pharmacol 215:115728
Thomaz FS, Tan YP, Williams CM, Ward LC, Worrall S, Panchal SK (2022) Wasabi (eutrema japonicum) reduces obesity and blood pressure in diet-induced metabolic syndrome in rats. Foods 11:3435
Tian D, Teng X, Jin S, Chen Y, Xue H, Xiao L, Wu Y (2021) Endogenous hydrogen sulfide improves vascular remodeling through PPARδ/SOCS3 signaling. J Adv Res 27:115–125
Tomasova L, Pavlovicova M, Malekova L, Misak A, Kristek F, Grman M, Cacanyiova S, Tomasek M, Tomaskova Z, Perry A (2015) Effects of AP39, a novel triphenylphosphonium derivatised anethole dithiolethione hydrogen sulfide donor, on rat haemodynamic parameters and chloride and calcium Cav3 and RyR2 channels. Nitric Oxide 46:131–144
Varade S, Nadella M, Hirake A, Bhausaheb Mungase S, Ali A, Adela R (2024) Effect of garlic on the components of metabolic syndrome: a systematic review and meta-analysis of randomized controlled trials. J Ethnopharmacol 318:116960

Vazquez-Prieto MA, Rodriguez Lanzi C, Lembo C, Galmarini CR, Miatello RM (2011) Garlic and onion attenuates vascular inflammation and oxidative stress in fructose-fed rats. J Nutr Metab 2011:475216. https://doi.org/10.1155/2011/475216

Wang R (2012) Physiological implications of hydrogen sulfide: a whiff exploration that blossomed. Physiol Rev 92:791–896. https://doi.org/10.1152/physrev.00017.2011

Wang R (2023) Roles of hydrogen sulfide in hypertension development and its complications: what, so what, now what. Hypertension 80:936–944

Wang Y, Zhao X, Jin H, Wei H, Li W, Bu D, Tang X, Ren Y, Tang C, Du J (2009) Role of hydrogen sulfide in the development of atherosclerotic lesions in apolipoprotein E knockout mice. Arterioscler Thromb Vasc Biol 29:173–179. https://doi.org/10.1161/atvbaha.108.179333

Wang R, Szabo C, Ichinose F, Ahmed A, Whiteman M, Papapetropoulos A (2015) The role of H2S bioavailability in endothelial dysfunction. Trends Pharmacol Sci 36:568–578

Weiss N, Ide N, Abahji T, Nill L, Keller C, Hoffmann U (2006) Aged garlic extract improves homocysteine-induced endothelial dysfunction in macro- and microcirculation. J Nutr 136: 750s–754s. https://doi.org/10.1093/jn/136.3.750S

Williams MJ, Sutherland WH, McCormick MP, Yeoman DJ, de Jong SA (2005) Aged garlic extract improves endothelial function in men with coronary artery disease. Phytother Res 19:314–319. https://doi.org/10.1002/ptr.1663

Wilson RK, Kwan TK, Kwan CY, Sorger GJ (2002) Effects of papaya seed extract and benzyl isothiocyanate on vascular contraction. Life Sci 71:497–507. https://doi.org/10.1016/s0024-3205(02)01708-3

Wlosinska M, Nilsson A-C, Hlebowicz J, Hauggaard A, Kjellin M, Fakhro M, Lindstedt S (2020) The effect of aged garlic extract on the atherosclerotic process–a randomized double-blind placebo-controlled trial. BMC Complement Med Ther 20:1–10

Xie L, Gu Y, Wen M, Zhao S, Wang W, Ma Y, Meng G, Han Y, Wang Y, Liu G, Moore PK, Wang X, Wang H, Zhang Z, Yu Y, Ferro A, Huang Z, Ji Y (2016a) Hydrogen sulfide induces Keap1 S-sulfhydration and suppresses diabetes-accelerated atherosclerosis via Nrf2 activation. Diabetes 65:3171–3184. https://doi.org/10.2337/db16-0020

Xie ZZ, Liu Y, Bian JS (2016b) Hydrogen sulfide and cellular redox homeostasis. Oxidative Med Cell Longev 2016:6043038. https://doi.org/10.1155/2016/6043038

Yamagishi S, Matsui T (2016) Protective role of sulphoraphane against vascular complications in diabetes. Pharm Biol 54:2329–2339. https://doi.org/10.3109/13880209.2016.1138314

Yi L, Su Q (2013) Molecular mechanisms for the anti-cancer effects of diallyl disulfide. Food Chem Toxicol 57:362–370. https://doi.org/10.1016/j.fct.2013.04.001

Zaorska E, Hutsch T, Gawryś-Kopczyńska M, Ostaszewski R, Ufnal M, Koszelewski D (2019) Evaluation of thioamides, thiolactams and thioureas as hydrogen sulfide (H2S) donors for lowering blood pressure. Bioorg Chem 88:102941

Zhang Y, Wu Q, Liu J, Zhang Z, Ma X, Zhang Y, Zhu J, Thring RW, Wu M, Gao Y (2022) Sulforaphane alleviates high fat diet-induced insulin resistance via AMPK/Nrf2/GPx4 axis. Biomed Pharmacother 152:113273

Zhao W, Zhang J, Lu Y, Wang R (2001) The vasorelaxant effect of H(2)S as a novel endogenous gaseous K(ATP) channel opener. EMBO J 20:6008–6016. https://doi.org/10.1093/emboj/20.21.6008

Zhou Z, Martin E, Sharina I, Esposito I, Szabo C, Bucci M, Cirino G, Papapetropoulos A (2016) Regulation of soluble guanylyl cyclase redox state by hydrogen sulfide. Pharmacol Res 111: 556–562. https://doi.org/10.1016/j.phrs.2016.06.029

Zhu M-L, Zhao F-R, Zhu T-T, Wang Q-Q, Wu Z-Q, Song P, Xu J, Wan G-R, Yin Y-L, Li P (2021) The antihypertension effect of hydrogen sulfide (H2S) is induced by activating VEGFR2 signaling pathway. Life Sci 267:118831

Crosstalk Between H_2S and Hormones: The Bilateral Relationship and Molecular Mechanisms

Guangdong Yang

Contents

Abstract

Hydrogen sulfide (H_2S) is increasingly recognized as a key gasotransmitter with diverse physiological roles across a range of organisms. Accumulating evidence has clearly shown that disruptions in H_2S homeostasis are linked to various endocrine disorders. Hormones are crucial in regulating H_2S metabolism and signaling, while H_2S influences hormones by modulating their biosynthesis, secretion, transport, and signaling pathways. H_2S and hormones function

G. Yang (✉)
School of Natural Sciences, Laurentian University, Sudbury, Ontario, Canada

Cardiovascular and Metabolic Research Unit, Laurentian University, Sudbury, Ontario, Canada
e-mail: gyang2@laurentian.ca

A. Papapetropoulos, C. Szabo (eds.), *Hydrogen Sulfide: A Toxicant with Healing Potential*, Handbook of Experimental Pharmacology 292,
https://doi.org/10.1007/164_2025_756

cooperatively to regulate cell survival, metabolism, reproduction, development, and stress responses, etc. The interaction between H_2S and hormones exhibits a certain degree of complexity due to the fact that they interact dynamically and through multiple signaling pathways. This book chapter summarizes the current state of knowledge on the bilateral relationship of H_2S and several important hormones, detailing the molecular mechanisms of H_2S-mediated hormone release and signal pathways, hormone-modulated H_2S synthesis and metabolism, and the related pathological implications. Insights into the reciprocal interactions between H_2S and various hormones can facilitate the development of therapeutic strategies for hormone-related disorders.

Keywords

Endocrine disorders · Estrogen · H_2S · Hormones · Insulin

Abbreviations

CBS	cystathionine β-synthase
CSE	cystathionine gamma-lyase
ECs	endothelial cells
ER	estrogen receptor
GI	gastrointestinal
GLP-1	glucagon-like peptide-1
H_2S	hydrogen sulfide
IGF-1	insulin-like growth factor-1
KO	knockout
PPG	propargylglycine
SMCs	smooth muscle cells
UA	uterine artery
WT	wild-type
3-MST	3-mercaptopyruvate sulfurtransferase

1 Introduction

Hydrogen sulfide (H_2S), historically known for its foul odor reminiscent of rotten eggs, has been recognized as a significant gasotransmitter with essential roles in various physiological processes (Arif et al. 2022; Ju et al. 2013; Wang 2012). While H_2S at physiological levels is crucial for numerous biological functions, elevated concentrations can be detrimental (Kabil and Banerjee 2014; Olson 2012). Consequently, organisms have evolved complex mechanisms to maintain H_2S homeostasis (Giuffrè and Vicente 2018; Zhu et al. 2021). In mammalians, the mechanisms underlying the modulations of H_2S metabolism and signals are complex and diverse.

Hormones, as vital chemical messengers, are precisely regulated to facilitate communication between cells and their target organs (Emanuele and Emanuele 1997; Hiller-Sturmhöfel and Bartke 1998). Hormonal signaling is essential for various physiological processes, including signal transduction, growth, metabolism, reproduction, development, stress responses, and mood regulation (Hoffmann et al. 2023; Kraemer et al. 2020). Disruptions in hormonal regulation can lead to a range of pathological conditions, such as metabolic syndromes, immune dysfunctions, vascular diseases, inflammation, and age-related disorders (Rahman et al. 2021; Sabatino et al. 2014).

Extensive research using cell and animal models has explored the interactions between H_2S and hormones (Chen et al. 2021; Pichette and Gagnon 2016; Song et al. 2024). These interactions are crucial for regulating numerous physiological and pathological processes. H_2S and hormones function cooperatively to regulate a wide variety of physiological and pathological processes. H_2S can affect hormones by regulating their biosynthesis, catabolism, transport, and signal pathways, etc. Conversely, hormones act as important factors in regulating H_2S metabolism and signaling. The interaction between H_2S and hormones exhibits a certain degree of complexity due to the fact that they interact dynamically and through multiple signaling pathways. In this review, the importance of H_2S as a novel gasotransmitter, its bilateral relationship with hormones, the molecular mechanisms of H_2S-mediated hormone release and signal pathway, and hormone-modulated H_2S metabolism and functions, as well as their pathological implications, will be discussed. Additionally, some future perspectives toward a more complete understanding of the mechanisms implicated in H_2S-hormone in both health and diseases are also highlighted. Understanding the interplays between H_2S and various hormones can definitely help develop therapeutic agents for treatment of hormone-related disorders.

2 Hormones: The Types and Functions

Hormones are non-nutritive chemical messengers secreted by endocrine organs that affect virtually every organ system in our body. Hormones are released directly into the bloodstream and act on target organs to regulate physiological functions in response to specific stimuli (Hiller-Sturmhöfel and Bartke 1998). Upon reaching their target tissues, hormones bind to specific receptors either on the cell surface or within the target cells, initiating biochemical reactions that modify cellular activity or function (Bassett et al. 2003; Font-Díaz et al. 2021).

Hormones can be categorized into three main types based on their chemical structures: lipid-derived, peptide-derived, and amino acid-derived hormones (Costa-e-Sousa and Hollenberg 2012; Xu et al. 2023). Lipid-derived hormones are largely made up of steroid hormones that are produced by the gonads (testes and ovaries) and the adrenal cortex and have a similar chemical structure to that of cholesterol (Dušková 2023). Estrogen, testosterone, and glucocorticoids are some examples of steroid hormones. Lipid-derived hormones are insoluble in water; therefore, must be transported by the carrier proteins in the blood. Lipid-derived hormones diffuse across the plasma membrane to bind to receptors either in the cytoplasm or the nucleus. Once bound to the receptor, the activated receptor-hormone complex can initiate a series of downstream actions and can also move to the nucleus to bind directly to the DNA, where it can act as a transcription factor for gene expression (Schwartz et al. 2016). Peptide-derived hormones are structurally composed of one or multiple polypeptide chains of various lengths (Bettio et al. 2020). They can also be found in the form of pro-hormones, which require cleaving to produce one or more active hormones (Rahman et al. 2021). These hormones are more prominently found in the pancreas and gastrointestinal (GI) system, including insulin and glucagon-like peptide-1 (GLP-1), etc. Peptide-derived hormones are water-soluble and therefore unable to enter cells, and require receptors on the surface of target cells to communicate (Costa-e-Sousa and Hollenberg 2012). Once the receptor-hormone complex is formed, this activates secondary messengers inside the cell, which allows the signal to be amplified and transduced to initiate cellular activity (Emanuele and Emanuele 1997). Amino acid-derived hormones are derivatives of the amino acids tryptophan and tyrosine (Bassett et al. 2003). They are mostly found in the thyroid gland and adrenal medulla gland (Braun and Schweizer 2018). Amino acid-derived hormones include thyroid, epinephrine, nor-epinephrine, melatonin, etc. Similarly to steroid hormones (lipid-derived), amino acid-derived hormones can cross the plasma membrane, diffuse to nucleus, and bind to nucleoprotein receptors for initiating de novo synthesis of proteins (Bassett et al. 2003; Braun and Schweizer 2018).

The actions of hormones must be precisely regulated to maintain tissue homeostasis (Costa-e-Sousa and Hollenberg 2012; Kraemer et al. 2020). Dysregulation of hormone signals can lead to endocrinal disorders, which may result from excessive or insufficient hormone secretion, abnormal hormone transport or catabolism, or impaired signal transduction (Dušková 2023). Endocrine disorders are classified as primary, originating from dysfunctions within the peripheral endocrine glands, or secondary, stemming from abnormalities in the pituitary gland (Emanuele and Emanuele 1997; Xu et al. 2023). Understanding the regulatory mechanisms of hormone signaling is crucial for accurate laboratory testing and effective treatment of endocrine disorders.

3 H_2S as a Novel Gasotransmitter

Extensive research has been conducted for gaining the importance of H_2S within the mammalian body (Gadalla and Snyder 2010; Wang 2012). As a gaseous messenger molecule, H_2S exhibits higher lipid solubility, allowing it to diffuse readily across cellular membranes. This property facilitates its rapid intracellular and intercellular transport and enables H_2S to interact with a diverse range of targets [Olson KR, 2012; Wang 2002]. H_2S plays a critical role in regulating various physiological functions, including gene expression, protein modification, energy metabolism, redox homeostasis, cell survival, inflammation, angiogenesis, vasodilation, and neuromodulation (Li et al. 2011; Szabo et al. 2013).

The body maintains H_2S levels within a precise range, as deviations can be harmful (Ju et al. 2013; Olson 2012; Zhu et al. 2021). Disruptions in H_2S regulation

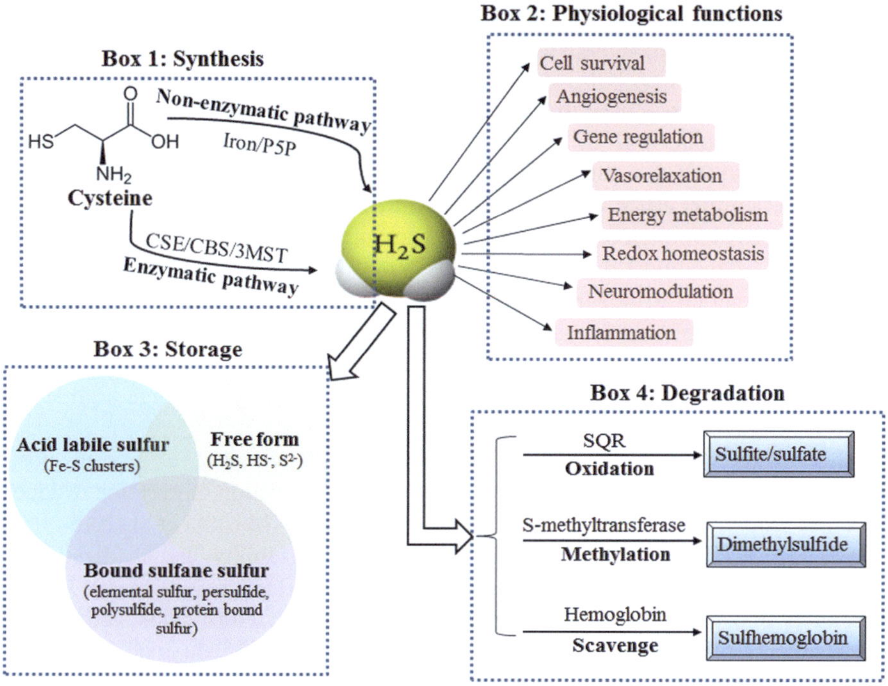

Fig. 1 The synthesis, physiological functions, storage, and degradation of H_2S in mammalian cells. H_2S can be endogenously produced from cysteine via enzymatic and non-enzymatic pathways (Box 1). H_2S exerts multiple physiological functions (Box 2) via various forms, including free form of H_2S gas, acid labile sulfur, and bound sulfane sulfur (Box 3). H_2S can be degraded through oxidation, methylation, and scavenging (Box 4). With the aid of SQR, H_2S can be oxidized in mitochondria for forming sulfite/sulfate; H_2S can also be converted to dimethylsulfide via *S*-methyltransferase-mediated methylation; by interaction with hemoglobin, H_2S can be scavenged for forming sulfhemoglobin (Box 4). SQR, sulfur:quinone oxidoreductase

and metabolism are associated with a range of pathological conditions, including cardiovascular diseases, neurodegenerative disorders, diabetes, and cancer, etc. (Mani et al. 2013; Panagaki et al. 2019; Yang et al. 2008). Several mechanisms contribute to H_2S homeostasis, including its synthesis, storage, and degradation (Fig. 1).

H_2S is endogenously produced from cysteine, with pyridoxal-5′-phosphate serving as a cofactor, through both enzymatic and non-enzymatic pathways (Bełtowski 2019; Kabil and Banerjee 2014; Wang 2012). Enzymatic production involves three key enzymes: cystathionine gamma-lyase (CSE), cystathionine beta-synthase (CBS), and 3-mercaptopyruvate sulfurtransferase (3-MST) (Giuffrè and Vicente 2018; Panagaki et al. 2019; Shibuya et al. 2013). These enzymes are tissue-specific. CSE is predominantly found in the vasculature and peripheral tissues such as blood vessels, kidneys, pancreas, liver, intestines, and uterus (Li et al. 2011; Mani et al. 2013), while CBS is mainly expressed in the brain (Kimura 2019). All three enzymes are present in the liver and kidneys (Szabo et al. 2013; Tomita et al. 2016).

Non-enzymatic production occurs when cysteine and pyridoxal-5′-phosphate interact with iron under physiological conditions to generate H_2S (Yang et al. 2019).

Similar to hormones, H_2S can be produced in specific organs and transported in the blood, where it is stored in the form of bound sulfane sulfur (Kasamatsu et al. 2016). Within cells, H_2S exists in various forms, including free H_2S gas, HS^-, persulfides, polysulfides, and elemental sulfur, which can interconvert under specific conditions (Kimura 2015; Olson 2012). To prevent excessive accumulation, H_2S is rapidly degraded through oxidation or methylation. In mitochondria, sulfide-quinone oxidoreductase facilitates the oxidation of H_2S to sulfite (SO_3^{2-}) and sulfate (SO_4^{2-}) (Kabil and Banerjee 2014; Murphy et al. 2019). Additionally, H_2S can be methylated by thiol-S-methyltransferase to form methanethiol (CH_3SH) and dimethylsulfide (CH_3SCH_3) (Maldonato et al. 2021; Pietri et al. 2011). In some cells, H_2S binds to hemoglobin, forming sulfhemoglobin, which helps to rapidly scavenge high levels of free H_2S gas (Zhu et al. 2021).

H_2S exerts its effects primarily through *S*-sulfhydration, a protein post-translational modification that involves a covalent modification of cysteine residues in target proteins for forming persulfides or polysulfides (Mustafa et al. 2009; Paul and Snyder 2015). This modification alters the activity, stability, and localization of proteins, thereby mediating H_2S signaling (Filipovic 2015; Yang et al. 2013). The specificity of *S*-sulfhydration is influenced by the location, redox status, and cellular environment of the targeted cysteine residues (Iciek et al. 2015; Ju et al. 2017). The dynamics of protein *S*-sulfhydration can vary with intracellular stress responses and different pathological conditions (Peng et al. 2017; Zivanovic et al. 2019). Additionally, H_2S possesses antioxidant properties, interacting with reactive oxygen and nitrogen species to protect cells from excessive oxidative stress (Hancock and Whiteman 2014; Kabil et al. 2014).

4 Crosstalk Between H_2S and Hormones: General Mechanisms and Implications

The crosstalk between H_2S and hormones has been recently studied in more detail. These two signaling molecules interact by modulating respective homeostatic levels and bioactivity through a variety of mechanisms (Table 1). Accumulated evidence indicates that diseases associated with abnormal hormone secretion and signaling often involve the dysregulation of H_2S (Fig. 2). This section focuses on the interplay between H_2S and several key hormones, including insulin, thyroid hormones, estrogen, testosterone, glucocorticoids, and GI hormones.

4.1 H_2S and Insulin

Insulin is a polypeptide hormone predominantly secreted by pancreatic beta cells to regulate glucose metabolism. Dysregulation of insulin secretion and sensitivity are hallmark features of diabetes, a global health concern. Research has extensively

Table 1 Hormonal regulation of H_2S signals

Type of hormones (dose and duration of treatment	Changes of H_2S signals	Cell and/or animal models	Targeted signal pathways	Path-physiological outcomes	References
Insulin (8 U/kg/day for 5 days, s.c.)	Lower mRNA levels of both CSE and CBS in pancreases and liver tissues	Streptozotocin-induced diabetic rats	N/A	Positive correlation between increased H_2S generation and diabetic development	Yusuf et al. (2005)
Insulin (1–10 μg/ml for 6 h)	Upregulation of CSE and CBS protein levels in myoblasts	Isolated chicken myoblasts	mTOR/p70S6 kinase	Enhanced H_2S generation by insulin contributes to protein synthesis and muscle development	Wang et al. (2019)
Thyroid (25 μg/100 g body weight/day for 4 days)	Reduced levels of CSE and CBS in liver tissues	C57BL6 mice	Down regulation of IGF-1	Reduced H_2S by thyroid negatively correlates with stress resistance and health.	Hine et al. (2017)
Estrogen (1 μg/kg body weight, i.v.)	Increased CBS expression and H_2S generation in uterine artery endothelium and smooth muscle	Ovariectomized nonpregnant ewes	N/A	High level of H_2S by estrogen improves vasodilation and postmenopausal health	Lechuga et al. (2015)
Estrogen (10 nM for 24 h)	Upregulation of CBS mRNA/protein and H2S production	Pregnant human uterine artery	Recruitment of ERα and ERβ to CBS gene promoter for inducing its transcription	High level of H_2S by estrogen induces uterine artery dilation	Bai et al. (2023)
Estrogen (10 nM for 24 h)	Increased CBS protein and H_2S production	Human endometrial stromal cells	Estrogen receptor-mediated CBS transcription	High level of H_2S by estrogen increases cell migration and endometrial angiogenesis	Qi et al. (2020)
Estrogen (1–10 nM for 5–60 min)	Increased CSE phosphorylation and H_2S production	Human umbilical vein endothelial cells	pGC-A/cGMP/PKG signaling	High level of H_2S by estrogen stimulates endothelial cell-related angiogenic properties	Xu et al. (2019)

(continued)

Table 1 (continued)

Type of hormones (dose and duration of treatment	Changes of H_2S signals	Cell and/or animal models	Targeted signal pathways	Path-physiological outcomes	References
Estrogen (200 nM for 24 h)	Induced CBS and CSE expressions and H_2S production	Human umbilical vein endothelial cells	N/A	High level of H_2S by estrogen alleviates homocysteine-induced cell injury	Zhang et al. (2021)
Estrogen (0.25 mg over 60-day release, implanted subcutaneously in mice; 10 nM for 48 h in cells)	Enhanced blood H_2S level and endothelial CSE gene expression	Ovariectomized apoE knockout female mice fed with an atherogenic diet and human umbilical vein endothelial cells	ERα-Sp1 interaction to binding to CSE gene promoter	High level of H_2S by estrogen attenuates atherosclerosis development	Fu et al. (2013)
Estrogen (10 nM for 24 h)	Higher CSE transcription	Human primary osteoblasts	ERα stimulation of Sp1 binding to CSE promoter	High level of H_2S by estrogen provides bone protection and healing	Lambertini et al. (2017)
Testosterone (10 nM-30 μM for 15 min)	Enhanced activities of aortic CSE and CBS and higher blood H_2S level	Isolated rat aortic rings	Positive modulation of CSE/CBS activity through post-translational modification and also possibly through androgen receptor mediated-CSE transcription	High level of H_2S by testosterone induces vasodilation	Brancaleone et al. (2015), Bucci et al. (2009)
Dexamethasone (5 mg/kg body weight/day for 4 weeks)	Reduced CBS and CSE expressions in the bone marrow and lower level of serum H_2S	A rat model of secondary osteoporosis	Activated Wnt signaling	Low level of H_2S by glucocorticoids contributes to bone loss	Ma et al. (2019)
Dexamethasone (1 mg/kg body weight, i.p.)	Reversal of lipopolysaccharide-evoked expression of CSE in liver tissues	A rat model of lipopolysaccharide-induced endotoxic shock	Reduced NF-κB-activation	Low level of H_2S by glucocorticoids provides anti-inflammatory effect	Li et al. (2009)

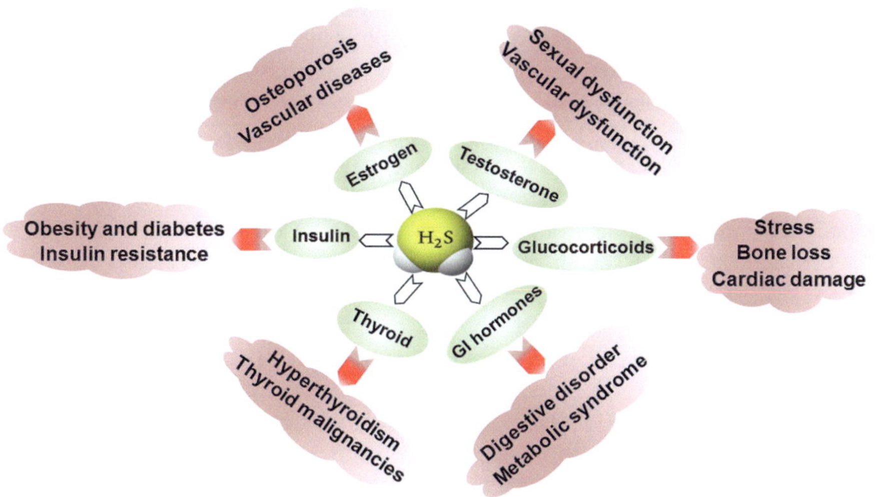

Fig. 2 Involvement of H_2S and hormones in various diseases. H_2S mutually influence hormone's biosynthesis, catabolism, transport, and signal pathways and contribute to a wide range of pathophysiological processes

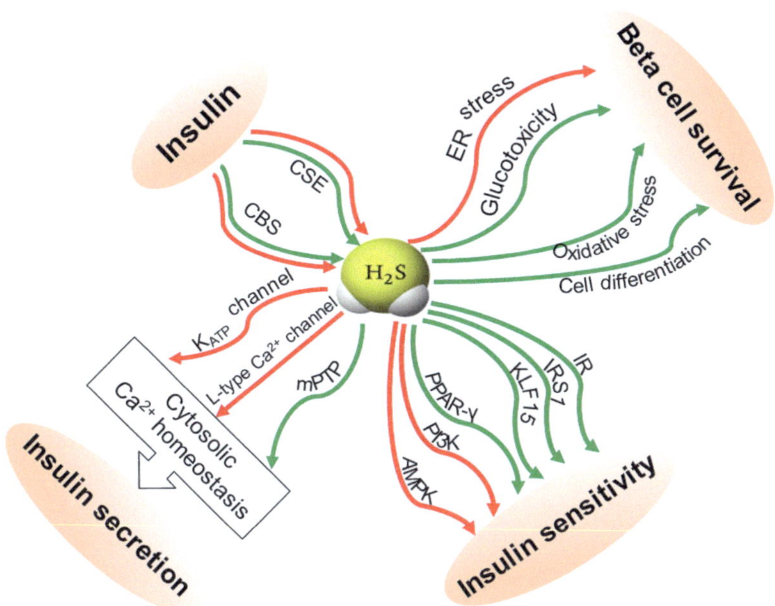

Fig. 3 The reciprocal interactions between H_2S and insulin. The interactions between H_2S and insulin are complex and multifaceted. H_2S biosynthesis can be regulated by insulin. H_2S can have both protective and toxic effects on beta cells, and H_2S can either inhibit or stimulate insulin secretion depending on various factors. The impact of H_2S on insulin sensitivity also varies, with evidence suggesting both positive and negative effects

investigated the role of H_2S in beta cell survival, insulin secretion, and insulin sensitivity, revealing reciprocal effects of H_2S on insulin metabolism (Fig. 3).

H_2S is essential for determining the cell fate of insulin-secreting beta cells. The application of exogenous H_2S or overexpression of CSE has been shown to promote apoptotic cell death in INS-1 cells through increased endoplasmic reticulum stress (Yang et al. 2007). Elevated H_2S levels in pancreatic tissues have been implicated in beta cell death in models of streptozotocin-induced diabetic mice and Zucker diabetic fatty rats (Wu et al. 2009; Yusuf et al. 2005). Conversely, CSE deficiency protected against beta cell death induced by streptozotocin (Yang et al. 2011). On the other side, H_2S has been reported to protect beta cells from oxidative damage and cell death. Beta cell apoptosis and DNA fragmentation were enhanced in high-fat diet-fed CSE knockout (CSE-KO) mice (Okamoto et al. 2013). Supplementation of H_2S could protect beta cells from glucotoxicity and prevent type 2 diabetes development (Kaneko et al. 2008; Okamoto et al. 2013; Okamoto et al. 2015). Furthermore, H_2S has been shown to induce beta cell differentiation from human dental pulp stem cells, enhancing insulin release and reducing glucotoxicity (Matei et al. 2017). The varying reports of H_2S as either protective or toxic in beta cells are likely influenced by factors such as dosage, type of H_2S donors, exposure duration, and the specific cell or animal models used.

Generally, H_2S appears to inhibit insulin secretion. Exogenous H_2S administration significantly reduced insulin secretion in various pancreatic beta cell lines, including rat INS-1E, mouse MIN6, and hamster HIT-T15 cells, as well as cultured mouse primary islets (Ali et al. 2007; Kaneko et al. 2006; Yang et al. 2005). Long-term H_2S exposure to obese db/db mice and type 2 diabetic rats also resulted in decreased serum insulin levels (Gheibi et al. 2019; Zheng et al. 2018). Enhanced endogenous H_2S production from CSE or CBS overexpression further decreased insulin secretion, whereas inhibition of these enzymes promoted insulin secretion (Ali et al. 2007). Polysulfide salts, such as Na_2S_2, Na_2S_3, and Na_2S_4, have also been shown to suppress glucose-stimulated insulin secretion (Shoji et al. 2019). The inhibition of insulin secretion by H_2S is believed to involve the activation of plasma membrane K_{ATP} channels and the inhibition of L-type voltage-dependent calcium channels (Ali et al. 2007; Kaneko et al. 2006; Yang et al. 2005). However, Lu et al. recently suggested that H_2S may stimulate insulin secretion by inducing Ca^{2+} release from mitochondria through the mitochondrial permeability transition pore (Lu et al. 2019). Thus, the effect of H_2S on insulin release is closely related to cytosolic Ca^{2+} homeostasis. Okamoto et al. observed that CSE deficiency reduced endogenous H_2S and inhibited insulin release in high-fat diet-fed mice, both before and after oral glucose administration (Okamoto et al. 2013). Additionally, Takahashi et al. found that reduced cysteine hydropersulfide levels due to decreased H_2S synthesis led to higher oxidative stress, impairing insulin secretion and causing glucose intolerance (Takahashi et al. 2017). Cysteine hydropersulfide, a highly reactive sulfur species, may stimulate insulin secretion in beta cells and mouse models by promoting tRNA methylthiolation. These findings highlight the complex role of H_2S in regulating insulin secretion. Notably, there is no direct evidence of H_2S affecting insulin gene expression or protein quality control.

H_2S also influences insulin resistance, particularly in adipose tissue, an insulin-sensitive organ involved in glucose uptake and fatty acid synthesis. H_2S has been reported to improve insulin sensitivity in adipocytes by increasing insulin receptor substrate 1 protein levels, reversing insulin resistance in high-fat diet-fed mice (Xue et al. 2013). H_2S facilitated the conversion of glucose to triglycerides in fat tissues through *S*-sulfhydration of PPAR-γ, alleviating insulin resistance (Cai et al. 2016). Recent studies suggest that H_2S promoted the browning of white adipocytes through the KLF15 signaling pathway (Lee et al. 2019). Moreover, 3-MST-derived H_2S also contributed to insulin sensitivity in adipocytes, with 3-MST knockout leading to cellular lipid accumulation and obesity in mice (Casili et al. 2022). H_2S supplementation in Goto-Kakizaki rats, a model of type 2 diabetes, improved insulin sensitivity and reduced muscle atrophy (Bitar et al. 2018). Several H_2S donors, including GYY4137, SG1002, Na_2S_3, and diallyl trisulfide, have been shown to enhance insulin signaling and reduce obesity in mice (Geng et al. 2013; Katsouda et al. 2022; Wang et al. 2022). Metformin, an insulin sensitizer, has been demonstrated to increase H_2S levels through enhanced CSE expression (Ma et al. 2020; Wiliński et al. 2013). Chen et al. found that Desulfovibrio-mediated H_2S release from chondroitin sulfate in the gut activated the AKT signaling pathway and improved insulin sensitivity in nonalcoholic fatty liver disease mice (Chen et al. 2022). Padiya et al. reported that garlic enhanced insulin sensitivity and mitigated metabolic syndromes in fructose-fed rats by stimulating H_2S production (Padiya et al. 2011). More importantly, H_2S may also improve insulin sensitivity by *S*-sulfhydrating PTP1B, a major negative regulator for insulin receptor signaling (Lin et al. 2023). Decreased blood H_2S levels were associated with increased risks of insulin resistance and vascular disease in type 2 diabetic patients (Jain et al. 2012). H_2S donor propargyl-Cysteine alleviated diabetic cardiomyopathy in db/db mice through activation of cardiac IR/Akt/GSK-3β signaling (Li et al. 2021b). Additionally, H_2S supplementation protected against AKT2 knockout-induced insulin resistance and cardiac complications by mitigating PTEN-Akt-GSK3β-mediated mitochondrial injury and apoptosis (Hu et al. 2014). Chronic H_2S treatment has been shown to enhance insulin receptor kinase activity and improve kidney lesions in Goto-Kakizaki diabetic rats (Xue et al. 2013). However, H_2S does not appear to influence insulin resistance development in Wistar rats (Patel and Shah 2010). Conversely, some studies report that H_2S may induce insulin resistance. H_2S treatment or CSE overexpression could cause insulin resistance and impair glucose utilization in hepatocytes via the AMPK pathway (Zhang et al. 2013). Donoso-Barraza et al. observed that exogenous H_2S reduced insulin-induced glucose uptake in L6 skeletal muscle cells (Donoso-Barraza et al. 2022). At high doses (up to 1 mM), NaHS could decrease insulin-stimulated glucose consumption and uptake in rat adipocytes and adipose tissue explants by modulating the PI3K pathway (Feng et al. 2009). Similarly, Huang et al. reported that increased H_2S generation contributed to TNF-α-stimulated insulin resistance in 3 T3-L1 cells (Huang et al. 2013). Clinically, a significant positive correlation between insulin resistance parameters and thiol levels in blood from rosacea patients has been observed (Pektas et al. 2022).

In summary, the interactions between H_2S and insulin reveal complex and sometimes contradictory effects on beta cell survival, insulin secretion, and insulin sensitivity. Variations in H_2S donor types, concentrations, the specific H_2S-synthesizing enzymes and their inhibitors, cell lines, tissue types, and animal models contribute to these discrepancies. The regulatory roles of H_2S in insulin signaling and diabetes pathophysiology are intricate and warrant further investigation. The mechanisms through which H_2S regulates beta cell survival, insulin release, and insulin signaling need to be elucidated, and the therapeutic potential of H_2S-releasing agents for diabetes treatment requires clinical exploration.

It also has been reported that insulin affects H_2S metabolism in a tissue-dependent manner (Fig. 3). In pancreatic and liver tissues from streptozotocin-induced type 1 diabetic rats, H_2S synthesis was elevated, likely due to reduced insulin levels (Yusuf et al. 2005). Further studies have shown that insulin treatment reduced H_2S generation in these tissues by inhibiting the expression of CBS and CSE (Yusuf et al. 2005). In contrast, blood and skeletal muscle H_2S levels were lower in Goto-Kakizaki rats, a model of type 2 diabetes. Insulin administration increased H_2S synthesis by upregulating CBS and CSE expression in myoblasts through the mTOR/p70S6 kinase (p70S6K) pathway, thus promoting protein synthesis (Wang et al. 2019). Correspondingly, blood H_2S levels were significantly lower in diabetic mice and patients compared to age-matched healthy controls (Lin et al. 2023). Abnormal H_2S synthesis may contribute to diabetes risk. Further research is needed to fully elucidate the mechanisms by which insulin regulates H_2S metabolism across different tissues and pathophysiological conditions.

4.2 H_2S and Thyroid

Thyroid hormones, primarily synthesized by the thyroid gland, regulate numerous metabolic processes involved in cell growth, differentiation, and the physiological functions of nearly all tissues (Mullur et al. 2014). Dysregulation of thyroid hormones can lead to conditions such as hyperthyroidism or hypothyroidism, which are associated with various diseases, including diabetes, heart failure, and hypertension (Chaker et al. 2017; Hughes and Eastman 2021). Emerging evidence highlights a significant interplay between H_2S and thyroid function, suggesting that H_2S may influence thyroid hormone levels and vice versa.

All three major H_2S-producing enzymes, including CBS, CSE, and 3-MST, were expressed in thyrocytes, indicating a potential role of H_2S in thyroid regulation (Zhao et al. 2022). Studies have demonstrated that administering H_2S donors can increase thyroid hormone levels in hypothyroid rats. In human primary thyrocytes, H_2S has been shown to enhance thyroid hormone synthesis and secretion, potentially through activation of SIRT1, a regulator of various cellular processes (Du et al. 2019). SIRT1 may be activated directly by H_2S via *S*-sulfhydration. Additionally, overexpression of CBS in the paraventricular nucleus of the hypothalamus has been linked to elevated serum thyroid hormone levels by stimulating thyrotropin-releasing hormone (Zheng et al. 2018). However, there are contrasting findings;

for instance, garlic, a natural H_2S donor, has been reported to decrease serum thyroid hormone levels in hyperthyroid rats (Tahiliani and Kar 2003). The discrepancies in these results might be related to variations in the types of H_2S donors used, their concentrations, or experimental conditions.

Thyroid hormones also impact H_2S biosynthesis. Direct thyroid hormone administration could reduce the expression of CSE and CBS, thereby decreasing H_2S production in liver tissues of mice (Hine et al. 2017). Studies have observed lower blood H_2S levels in hypothyroid patients and rats compared to normal controls (Zhao et al. 2022). Similarly, CSE expression and H_2S production were reduced in the aortas of fetal hypothyroidism rats (Nourabadi et al. 2021). Nechiporuk et al. also reported decreased activities of CBS and CSE in the liver and kidneys of hypothyroid rats (Nechiporuk et al. 2017). Conversely, Jeddi et al. found elevated H_2S levels in the liver of hyperthyroid rats, but lower levels in the serum, aorta, heart, and muscles, indicating differential regulation of H_2S-generating enzymes across tissues (Jeddi et al. 2019). These observations suggest that thyroid hormones can influence the expression and activity of H_2S-producing enzymes in a tissue-specific manner. Thyroid hormones may alter CSE transcription through thyroid receptors, as evidenced by the presence of thyroid receptor binding sites in the CSE promoter region (Hine et al. 2017). Disruption of H_2S production in mice exposed to thyroid receptor antagonists, such as Bisphenol A, further supports the role of thyroid receptors in modulating H_2S metabolism (Yang et al. 2021).

H_2S also plays a protective role in thyroid dysfunction-related cardiovascular conditions. Liu et al. found that H_2S ameliorated thyroxine-induced myocardial fibrosis in rats by upregulating the PI3K/AKT signaling pathway and downregulating microRNAs such as miR-21, miR-34a, and miR-214 (Liu et al. 2018). Similarly, Song et al. demonstrated that H_2S improved hypothyroidism-induced myocardial fibrosis through inhibition of the TGF-β1/SMAD2 signaling pathway (Song et al. 2023). These findings suggest that targeting H_2S signaling could offer therapeutic benefits for preventing and treating cardiovascular issues associated with both hyperthyroidism and hypothyroidism.

4.3 H_2S and Estrogen

Estrogen, a key female sex hormone primarily produced by the ovaries, regulates a variety of physiological processes, including reproductive functions, cardiovascular health, cognitive function, and bone health (Kiyama and Wada-Kiyama 2015). The biological effects of estrogen are mediated through estrogen receptors (ERs), including ERα, ERβ, and G protein-coupled ER (Bai et al. 2020). Recent research has uncovered a complex interplay between H_2S and estrogen, influencing cellular processes such as proliferation, migration, vasodilation, oxidative stress, and inflammation.

Studies have shown variable effects of H_2S on estrogen levels. For instance, adult female CSE-KO mice exhibited lower blood estrogen levels compared to wild-type (WT) littermates, which exacerbates early atherosclerotic lesions in the CSE-KO mice fed an atherogenic diet (Li et al. 2017). Despite this, NaHS, an H_2S donor, did not alter estrogen levels in either WT or CSE-KO female mice, suggesting that H_2S does not directly induce estrogen synthesis. Conversely, Zhang et al. reported that H_2S supplementation could increase serum estrogen levels and stimulate mammary gland development in prepubescent female mice through the PI3K/Akt-mTOR signaling pathway (Zhang et al. 2020). These findings imply that H_2S might stimulate estrogen production in younger mice, indicating age-dependent effects.

Estrogen is known to regulate H_2S biosynthesis. Both CSE and CBS, key enzymes in H_2S production, were sensitive to estrogen and play roles in regulating vascular tone and redox status (Lechuga et al. 2015; Sheibani et al. 2017). Elevated estrogen levels during pregnancy enhance maternal-fetal interface vasodilation and angiogenesis by stimulating H_2S production from the uterine artery (UA) through selective upregulation of CBS (Bai et al. 2023). During ovine and human pregnancies, endogenous estrogen increased H_2S biosynthesis in UA endothelial cells (UAECs) and smooth muscle cells (UASMCs) (Lechuga et al. 2015; Sheibani et al. 2017). Exogenous estrogen further boosted H_2S levels in these cells (Sheibani et al. 2017). Similarly, Qi et al. found that estrogen stimulated CBS-derived H_2S biosynthesis and promoted angiogenesis in human endometrial microvascular endothelial cells (Qi et al. 2020). In contrast, blockade of H_2S biosynthesis partially attenuated estrogen-induced and pregnancy-associated uterine vasodilation. These studies provide a novel mechanism for the mediation of H_2S in estrogen-induced uterine vasodilation and hemodynamics in pregnancy.

Estrogen's effects on H_2S are also evident in vascular health. For example, estrogen treatment increased H_2S levels and mitigated atherosclerotic lesions in WT mice fed an atherosclerotic diet, but this effect was not observed in CSE-KO mice (Li et al. 2017). Estrogen protected against atherogenesis in apolipoprotein E-KO mice by upregulating CSE and H_2S production in endothelial cells (Fu et al. 2013; Zhou et al. 2013). Blocking H_2S production with CSE inhibitors reversed estrogen's protective effects on atherosclerotic lesions in ovariectomized apolipoprotein E-KO mice (Fu et al. 2013; Teoh et al. 2020). Moreover, Bok et al. demonstrated that CSE was crucial for estrogen-induced blood flow stimulation, as CSE knockout abolishes this effect (Bok et al. 2021). Zhu et al. (2011) found that estrogen enhanced CSE expression and reduced inflammation and oxidative stress in the myocardium of ovariectomized rats (Zhu et al. 2013). Homocysteine accumulation in blood, termed as hyperhomocysteinemia, is an independent risk factor for vascular dysfunction (Veeranna et al. 2011). Zhang et al. showed that homocysteine concentration was higher but estrogen level was lower in female hypertensive patients than that of healthy controls (Zhang et al. 2021). The authors further observed that estrogen inhibited homocysteine-mediated damage by promoting H_2S production via upregulating CBS and CSE expression in human umbilical vein ECs. These studies clearly point to the contribution of H_2S in estrogen functions. Mechanistically, estrogen induced H_2S synthesis via ER-mediated transcriptions of both CSE and CBS, since multiple estrogen-responsive elements were found in the proximal promoters of both CSE and CBS genes (Fu et al. 2013; Zhang et al. 2021). In addition, activated ER could induce PKG-initiated CSE phosphorylation, thus increasing CSE activity and rapidly inducing H_2S release from ECs (Xu et al. 2019).

The functional link between estrogen and H_2S has been tested in various systems. It was known that both estrogen and H_2S were able to inhibit SMC proliferation and protect from atherosclerosis development (Teoh et al. 2020; Yang and Wang 2015). Intriguingly, Shuang et al. found that estrogen stimulated the proliferation of vascular SMCs when endogenous H_2S level was lower (Shuang et al. 2021). One of the targets that estrogen inhibits SMC proliferation is attributed to insulin-like growth factor-1 (IGF-1)/IGF-1R system. A lower level of H_2S reduced IGF-1R *S*-sulfhydration and further disrupted IGF-1R/ERα hybrid, thus abolishing the inhibitory role of estrogen on SMC proliferation (Shuang et al. 2021). These studies suggest that H_2S mediates the protective roles of estrogen on cardiovascular functions, and lack of H_2S favors the interaction of estrogen with IGF-1R/ERα hybrid to stimulate SMC proliferation. In bone health, estrogen deficiency leads to decreased H_2S levels and bone loss. H_2S contributes to the beneficial effects of estrogen on bone formation and resorption. Rhein-derived thioamides, known for their H_2S-releasing properties, mitigated bone loss in estrogen deficiency-induced osteoporosis in ovariectomized mice (Jiang et al. 2019). CSE was expressed in human osteoblasts, and estrogen enhanced CSE transcription by recruiting Sp1 to the CSE promoter region (Lambertini et al. 2017). Thus, supplementation with H_2S may offer therapeutic benefits for the treatment of bone disorders associated with estrogen dysfunction.

In addition to estrogen, progesterone is also a crucial female hormone that regulates pregnancy and menstruation. There are only few studies on the interaction between H_2S and progesterone. CSE-KO mice exhibited a lower number of successful pregnancies compared to their WT littermates (Panagiotopoulos et al. 2023), but it is unclear whether CSE deficiency affects progesterone level in mice. It has been reported that CSE mediated the inhibitory role of progesterone on uterine artery blood flow in ovariectomized mice (Bok et al. 2021), suggesting progesterone may stimulate CSE activity and H_2S generation. In contrast, other studies reported that progesterone had no effect on H_2S generation in vascular tissues from male mice and endometrial stromal cells (Brancaleone et al. 2015; Qi et al. 2020). Thus, further studies are required to elucidate the interaction between H_2S and progesterone during pregnancy or the menstrual cycle.

4.4 H_2S and Testosterone

Testosterone, the principal male sex hormone produced primarily in the testes, plays a crucial role in growth, development, reproduction, and the maintenance of various organ functions (Goldman et al. 2017). The interaction between H_2S and testosterone has been studied across various systems, revealing a complex relationship that influences normal physiological functions and disease states.

Varicocele, a condition associated with testosterone deficiency, can be challenging to treat effectively. High doses of testosterone were often used, but they come with risks such as prostate cancer and cardiovascular issues (Hayden and Tanrikut 2016). By using a rat model of varicocele, Shafie et al. observed a significant lower level of serum testosterone and testicular H_2S, and a subeffective dose of testosterone combined with a subeffective dose of H_2S donor NaHS could be effective for curing varicocele by reducing testicular oxidative stress and apoptosis (Shafie et al. 2022). In addition, the combined administration of testosterone and NaHS significantly increased the percentage of progressively motile spermatozoa and decreased the percentage of non-motile spermatozoa compared to the groups of testosterones and NaHS alone. This study provides a novel approach for improving the beneficial effects of testosterone therapy and decreasing its side effects at the same time. Cyclophosphamide is a cancer drug but often causes serious side effects such as hemorrhagic cystitis and male infertility. Özatik et al. reported that H_2S had a protective and therapeutic effect on hemorrhagic cystitis and testicular dysfunction induced by cyclophosphamide (Özatik et al. 2021). The authors concluded that H_2S was a promising molecule in facilitating cancer treatment. Similarly, Azarbarz et al. also observed that H_2S protected from cisplatin-induced testicular toxicity and plasma testosterone deficiency in rats possibly by its antioxidant and anti-inflammatory properties (Azarbarz et al. 2020). Diabetes is associated with various complications including testosterone deficiency and erectile dysfunction. Adebayo et al. observed a significant decrease in H_2S and testosterone levels in diabetic rats (Adebayo et al. 2019). Almond-supplemented diet increased H_2S level and improved penile erection in diabetic rats by regulating testosterone synthesis. Srilatha et al. also noticed that H_2S supplementation markedly increased testosterone concentration and improved the compromised sexual functioning in the aging male (Srilatha et al. 2012). Bianca et al. found that H_2S contributed to the vasodilatory effect of testosterone in rat isolated aortic rings by influencing cGMP pathway and phosphodiesterase 5 activity (Bianca et al. 2015). By using rat aortic rings, Bucci et al. demonstrated that H_2S mediated the vasodilation of testosterone by opening K_{ATP} channels (Bucci et al. 2009). Wang et al. reported that CBS-derived H_2S directly induced testosterone synthesis in MLTC-1 cells via post-translational modification of phosphodiesterase followed by increased cAMP level and activation of PKA pathway (Wang et al. 2021). These studies suggest that H_2S supplement would be of great significance for maintaining normal testosterone level and male sexual health. In contrast, another study had opposite observation that H_2S caused male reproductive toxicity in mice and disrupted spermatogenesis by impairing blood-testis barrier (Wang et al. 2020). The apparent discrepancy may be due to the dose of H_2S donor used and different ways of H_2S administration.

Testosterone also stimulates H_2S generation. Brancaleone et al. observed higher blood H_2S levels in healthy males than females, pointing to the possibility of male hormone stimulation of H_2S generation (Brancaleone et al. 2015). Indeed, it was further found that testosterone stimulated H_2S generation in an androgen receptor-dependent manner, since pretreatment with androgen receptor inhibitor nilutamide prevented testosterone-induced H_2S generation and reduced its vasodilatory effect. The authors further demonstrated that androgen receptor physically associated with CSE for inducing its activation. Another study by Bucci et al. showed testosterone rapidly stimulated H_2S production in aortic rings of rats by increasing CSE activity, pointing to the potential roles of testosterone and its receptor in post-translational modification of CSE enzyme (Bucci et al. 2009).

4.5 H_2S and Glucocorticoids

Glucocorticoids, also known as corticosterones, are steroid hormones produced by the adrenal cortex that play crucial roles in regulating metabolism, stress responses, and inflammation (Adcock and Mumby 2017). By directly influencing one another, H_2S and glucocorticoids together contribute to many homeostatic functions.

Glucocorticoids, such as dexamethasone, are known to adversely affect bone health. Liao et al. observed decreased serum H_2S levels and increased glucocorticoid levels in rats with femur fractures, indicating a potential link between reduced H_2S and glucocorticoid-induced bone loss (Liao et al. 2021). Dexamethasone has been reported to lower H_2S levels by inhibiting H_2S-producing enzymes in bone marrow (Ma et al. 2019). H_2S treatment was found to counteract the negative effects of dexamethasone on bone mass. Li et al. (2021a) observed that dexamethasone reduced circulating H_2S content and CSE expression in neutrophils from lipopolysaccharide-treated rats (Li et al. 2009). However, the mechanisms by which dexamethasone alters H_2S signaling remain unclear.

H_2S inhibition of glucocorticoid levels and improvement of stress response and organ functions are often reported. Hazzaa et al. demonstrated that H_2S donor administration to acutely stressed rats lowered serum cortisol levels and ameliorated hypertension and cardiac damage (Hazzaa et al. 2021). Similarly, Russo et al. found that the H_2S precursor S-adenosyl-L-methionine reduced plasma corticosterone levels under physical stress (Russo et al. 2000). Garlic, a dietary source of H_2S, was also observed to decrease plasma corticosterone levels in rats fed a high-protein diet (Oi et al. 2001). In a model of water immersion and restraint stress, Lou et al. reported that H_2S exposure significantly reversed stress-induced increases in plasma corticosterone and protected against gastric ulcers by regulating inflammatory and oxidative stress responses (Lou et al. 2008). Furthermore, H_2S donor GYY4137 was found to inhibit glucocorticoid levels and counteracted bone loss (Liao et al. 2021). Conversely, inhibition of endogenous H_2S production with propargylglycine (PPG) increased glucocorticoid secretion and exacerbated bone loss. Further mechanism studies found that H_2S reduced glucocorticoid secretion and inhibited glucocorticoid receptor α activity by *S*-sulfhydration in the kidneys. Additionally, H_2S may enhance adrenocortical responsiveness during endotoxemia and sepsis. It has been proposed that H_2S could attenuate stress-related glucocorticoid release by targeting the hypothalamic-pituitary-adrenal axis (Russo et al. 2000). Obesity, associated with elevated perceived stress, has been linked to changes in H_2S signaling. Zheng et al. observed decreased CBS protein levels in the hypothalamus of high-fat diet-induced obese rats and db/db obese mice. Overexpression of CBS in the hypothalamus lead to reduced plasma corticosterone levels, suggesting that enhancing hypothalamic

CBS/H_2S signaling might improve stress management and glucose metabolism in diabetes by inhibiting corticosterone release (Zheng et al. 2018).

4.6 H_2S and GI Hormones

H_2S interaction with GI hormones has also been reported, including GLP-1, gastric acid, ghrelin, etc. GLP-1 is a hormone secreted primarily by enteroendocrine L-cells in the ileum and colon. GLP-1 plays a crucial role in insulin secretion and glucose metabolism, making it a target for managing type 2 diabetes and obesity (Maselli and Camilleri 2021). The relationship between H_2S and GLP-1 is complex and has been explored in several studies. Bala et al. reported that both endogenous and exogenous H_2S inhibited GLP-1 release from enteroendocrine L-cells through the bile acid receptor TGR5 via the Gs/cAMP/PLC/Ca^{2+} pathway (Bala et al. 2014). Qi et al. also observed that Desulfovibrio-derived H_2S suppressed GLP-1 production in intestinal L cells and impaired host metabolism in male mice (Qi et al. 2024). These findings indicate that H_2S exerts a negative regulatory effect on GLP-1 secretion. In contrast, Pichette et al. found that chondroitin sulfate treatment, which promotes H_2S production in the gut, was associated with increased GLP-1 and insulin secretion, leading to improved oral glucose tolerance in mice (Pichette et al. 2017). Further studies by the same group observed that a single injection of GYY4137, an H_2S donor, enhanced glucose-stimulated GLP-1 release and improved glucose clearance in Western diet-fed mice (Mezouari et al. 2020). These findings indicate that the effect of H_2S on GLP-1 secretion can vary depending on the context, including the type of H_2S donor and the physiological conditions.

Gastrin is a hormone produced by gastric parietal cells that stimulates gastric acid release and aids digestion (Ramsay and Carr 2011). The interaction between H_2S and gastrin involves both regulatory and protective mechanisms. Mard et al. found that H_2S inhibited distention-induced gastrin secretion in rats, which contributes to gastric mucosal defense and modulates gastric motility (Mard et al. 2015). Conversely, Sun et al. observed that H_2S supplementation enhanced gastrin secretion in rats through TRPV1 channel activation and NF-κB-dependent mechanisms (Sun et al. 2018). This suggests that H_2S can also stimulate gastrin release under certain conditions. *Helicobacter pylori* infection, which is linked to gastric mucosal atrophy and precancerous conditions, affected H_2S production in gastric cancer cells, suggesting that H_2S might be one of the bioactive molecules involved in gastrin secretion and mucosal functions (Kawahara et al. 2020).

Ghrelin is a hunger hormone secreted by the stomach that regulates appetite, energy metabolism, and gastric motility (Kojima and Kangawa 2008). Slade et al. reported that H_2S donor GYY4137 directly suppressed ghrelin secretion in rat primary gastric culture, in part through the activation of the protein kinase B pathway (Slade et al. 2018). Additionally, GYY4137 treatment prolonged the postprandial drop of circulating ghrelin and reduced food consumption up to 4 h after treatment in mice. Besides with ghrelin, other GI hormones, i.e., cholecystokinin and peptide YY, play essential roles in regulating appetite and facilitating food absorption. Whether H_2S can influence the biosynthesis and functions of cholecystokinin and peptide YY has never been reported. Further research is required to clarify these interactions, especially how H_2S affects appetite and metabolism via gastrointestinal hormones in humans. Understanding these dynamics could lead to novel therapeutic strategies for metabolic and gastrointestinal disorders.

4.7 H_2S and Other Hormones

In addition to the common hormones mentioned above, H_2S also interacts with other hormones, such as catecholamines and melatonin. Catecholamines are a group of hormones that play crucial roles in the body's stress response and physiological regulation, including epinephrine (adrenaline), norepinephrine (noradrenaline), and dopamine, etc. It was first demonstrated that H_2S regulated acute adrenergic stress response by stimulating the secretion of epinephrine and norepinephrine via membrane depolarization in rainbow trout (Perry et al. 2008). Differently, in isolated porcine iris-ciliary bodies, H_2S caused a concentration-dependent reduction in endogenous epinephrine and norepinephrine concentrations, thus inhibiting sympathetic neurotransmission (Kulkarni et al. 2009). Similarly, in a porcine large-animal model of severe traumatic shock, H_2S administration significantly lowered epinephrine level (Satterly et al. 2015). H_2S treatment also caused significant reduction in epinephrine levels in a chronic mild stress model in rats (Ghalwash et al. 2021). Functionally, H_2S inhibited epinephrine-induced human platelet aggregation and rat gastric damage (Fomenko et al. 2014; Zagli et al. 2007). These studies suggest that H_2S donors can ameliorate stress response by modulating catecholamine secretion or signals. There are few data currently available about the effect of catecholamines on H_2S biosynthesis. Only one paper reported that epinephrine injection induced rat myocardium damage and promoted H_2S level in blood (Ivanova et al. 2018). None of these studies explored the mechanisms underlying the interaction between H_2S and catecholamines. One possibility is that H_2S may chemically interact with epinephrine to form a new compound, thereby altering the signaling responses of both molecules (Mohamed et al. 2020).

Melatonin is a hormone secreted by the pineal gland in the brain, primarily regulating the body's circadian rhythm and sleep-wake cycle. So far, there have been few reports on the crosstalk between H_2S and melatonin in mammals. It is known that both melatonin and H_2S have protective effects on the cardiovascular system. Remote ischemic preconditioning conferred cardioprotection against ischemia reperfusion injury in rats by increasing the plasma levels of both melatonin and H_2S (Zhang et al. 2023). However, what remains unknown is whether this is merely a parallel relationship or a causal one between H_2S and melatonin. Another study reported that melatonin elevated H_2S production and promoted the apoptosis of thyroid cancer cells (Jia et al. 2022). Further studies are needed to investigate the interaction between H_2S and melatonin at multiple levels under both normal and stress conditions, beginning with their respective biosynthetic pathways, followed by signaling mechanisms and functional outcomes.

5 Is H_2S Itself a Hormone?

H_2S is increasingly recognized as a crucial signaling molecule involved in numerous physiological processes across various species and organ systems. The question of whether H_2S qualifies as a hormone hinges on its modes of action, characteristics, and regulatory mechanisms. H_2S is primarily known for its autocrine and paracrine signaling roles. H_2S can be produced by a specific cell and act on the same cell (autocrine signaling) or neighboring cells (paracrine signaling). The high lipid solubility of H_2S facilitates its diffusion across cell membranes, allowing it to exert effects on nearby cells or tissues (Hine et al. 2017; Kabil and Banerjee 2014; Wang 2012). Recent research has expanded our understanding of H_2S, revealing that it can also function through endocrine signaling mechanisms. Pharmacologically administered H_2S donors can produce effects at distant sites, suggesting that H_2S can circulate in the bloodstream and act on tissues far from its site of production (Li et al. 2011; Peng et al. 2017). This indicates that H_2S can indeed have endocrine-like effects by stabilizing and circulating in the blood, particularly in the form of bound sulfane sulfur, which provides a longer-lasting impact (Olson 2012; Takahashi et al. 2017). Similar to hormones, the levels of H_2S in cells, tissues, and circulation are tightly regulated. H_2S biosynthesis is controlled by its generating enzymes, which are subject to complex regulatory mechanisms. H_2S itself can act as a negative feedback modulator by inhibiting these enzymes, thereby regulating its own production (Corvino et al. 2016; Peleli et al. 2023). H_2S' ability to modulate its production through feedback mechanisms is akin to hormonal regulation. In addition, H_2S production can also be regulated via hypothalamic-pituitary axis (Hine et al. 2017). However, unlike many classical hormones that exert their effects through specific receptors, H_2S does not bind to a single exclusive receptor. Instead, H_2S interacts with various molecular targets, which makes its mechanism of action more versatile but less specific compared to traditional hormones [1]. There is some overlap between hormones and other signaling molecules such as cytokines and growth factors. These molecules can have hormone-like effects under certain conditions (Bahadoran et al. 2020). Overall, while H_2S may not fit neatly into the classical definition of a hormone, it demonstrates many endocrine-like properties and functions. Thus, it is reasonable to consider H_2S as an endocrine signaling molecule within the broader context of physiological regulation.

6 Prospective

The study of H_2S and its interactions with hormones has yielded significant insights into its role in cellular functions and pathological processes. However, several areas remain under-explored, and addressing these gaps will be crucial for advancing our understanding and therapeutic application of H_2S in both health and disease. First, understanding the basic mechanisms responsible for H_2S generation and removal inside and outside the cells, as well as H_2S storage and transportation in bloodstream, is very helpful for clarifying the temporal and spatial H_2S levels. Second, it will be also important to clarify the precise contribution of H_2S signals to endocrine disorders through interaction with different hormones. Understanding these interactions will definitely help in developing new therapies for a wide range of diseases. Third, developing new, slow-releasing, and organ-specific H_2S donors to treat H_2S and hormone are demonstrated by experimental and pre-clinical data, the clinical evidence from human and the detailed regulatory mechanisms are not still fully understood. By addressing the above research areas, we can gain a deeper understanding of how H_2S functions as a signaling molecule and its potential as a therapeutic target. Continued research will enhance our ability to harness H_2S for treating hormone-related diseases and other health conditions, leading to more effective and personalized medical interventions.

Acknowledgments This study was supported by a discovery grant from the Natural Sciences and Engineering Research Council of Canada (RGPIN-2022-05369) and a grand-in-aid from the Heart and Stroke Foundation of Canada (G-18-0022098).

Point-to-Point Response to Reviewer's Comments

H_2S and Insulin: The article's claims about hydrogen sulfide (H_2S) in insulin regulation are well supported by evidence. No significant inaccuracies were found in this section.

Response: I thank the reviewer for reviewing and appraising our work.

H_2S and Thyroid Hormones: The discussion on thyroid hormones is largely accurate and up-to-date. All claims here appear scientifically sound.

Response: I thank the reviewer for reviewing and appraising our work.

H_2S and Estrogen: The article's claims regarding estrogen are accurate and comprehensive.

Response: I thank the reviewer for reviewing and appraising our work.

H_2S and Testosterone: The article accurately describes known interactions between H_2S and testosterone, though this section is slightly more narrative in mixing diverse findings. Overall, the testosterone section is factually correct, balancing positive and negative findings.

Response: I thank the reviewer for reviewing and appraising our work.

H_2S and Glucocorticoids: The manuscript accurately portrays the interplay between H_2S and glucocorticoid hormones (e.g. *cortisol in humans, corticosterone in rodents). One potential over-generalization is the phrase "by targeting each other, H_2S and glucocorticoids contribute to homeostasis"—this is a bit vague and could be rephrased.*

Response: Following the suggestion, the phrase "by targeting each other, H_2S and glucocorticoids contribute to homeostasis" has been changed to "By directly influencing one another, H_2S and glucocorticoids together contribute to many homeostatic functions" (page 21).

H_2S and Gastrointestinal (GI) Hormones: The article covers three GI-related hormones—GLP-1, gastrin, and ghrelin—and its claims are accurate, with

appropriate attention to conflicting findings. A very recent study not cited in the article adds to this complexity: Qi et al. *(*2024*) showed that microbial-derived H_2S (from gut bacteria Desulfovibrio) suppresses GLP-1 secretion in mice, exacerbating metabolic syndrome (Nature Metabol, 6: 1601, 2024). While this 2024 finding wasn't included (likely due to timing), it could be included with a minor revision.*

Response: Following the suggestion, this revision has discussed the recent article by Qi et al. (2024). It was indicated that "Qi et al. also observed that Desulfovibrio-derived H_2S suppressed GLP-1 production in intestinal L cells and impaired host metabolism in male mice (Qi et al. 2024)" (page 23).

*For gastrin, the manuscript accurately reports that H_2S has a dual role. No important GI hormones are misrepresented; however, the section could be expanded in the future to other gut hormones (*e.g. *cholecystokinin or peptide YY) if data are available.*

Response: Following the suggestion, the following sentences were added to this revision: "Besides with ghrelin, other GI hormones, i.e., cholecystokinin and peptide YY, play essential roles in regulating appetite and facilitating food absorption. Whether H_2S can influence the biosynthesis and functions of cholecystokinin and peptide YY has never been reported. Further research is required to clarify these interactions, especially how H_2S affects appetite and metabolism via gastrointestinal hormones in humans" (page 24).

Is H_2S Itself a Hormone?: In a speculative section, the article ponders whether H_2S qualifies as a hormone. The manuscript accurately notes the key difference that H_2S does not signal through a dedicated receptor as classic hormones do. This assessment is scientifically reasonable. Overall, this section doesn't make firm claims but is logically consistent with known H_2S biology. There are no factual inaccuracies since this is more of a discussion.

Response: I thank the reviewer for reviewing and appraising this work.

*In summary, the scientific claims throughout the manuscript are generally accurate and well-supported. Each hormone-specific subsection reflects both historical foundational studies and recent research, giving a balanced view. Importantly, the author consistently notes instances of contradictory findings (*e.g. *H_2S both inhibiting and stimulating insulin or GLP-1, or being protective* vs. *toxic to testes), which enhances the accuracy by not presenting one-sided conclusions. The references cited in this article are, for the most part, appropriate, up-to-date, and supportive of the claims made.*

There is no evidence of mis-citation, and the author deserves credit for citing sources that disagree with each other to present a full picture. The references for these contradictory findings are up-to-date and relevant. This indicates that the author not only cited sources for facts that support a claim but also used citations to acknowledge opposing data—a hallmark of a well-referenced scientific review.

Response: I thank the reviewer for reviewing and appraising this work.

Hormones Not Covered: By necessity, the review had to select certain hormones, but this means a few areas are omitted. Notably, progesterone (another major sex hormone) is not discussed, even though the estrogen section is extensive. It's possible that the literature on H_2S–progesterone interactions is sparse, but for *completeness, a brief mention could be warranted (for example, whether progesterone affects H_2S production in pregnancy or* vice versa *is not addressed.*

Response: To address these valuable comments, one new paragraph has been added for discussing the interaction between H_2S and progesterone, as shown below:

"In addition to estrogen, progesterone is also a crucial female hormone that regulates pregnancy and menstruation. There are only few studies on the interaction between H2S and progesterone. CSE-KO mice exhibited a lower number of successful pregnancies compared to their WT littermates (Panagiotopoulos et al. 2023), but it is unclear whether CSE deficiency affects progesterone level in mice. It has been reported that CSE mediated the inhibitory role of progesterone on uterine artery blood flow in ovariectomized mice (Bok et al. 2021), suggesting progesterone may stimulate CSE activity and H2S generation. In contrast, other studies reported that progesterone had no effect on H2S generation in vascular tissues from male mice and endometrial stromal cells (Brancaleone et al. 2015; Qi et al. 2020). Thus, further studies are required to elucidate the interaction between H_2S and progesterone during pregnancy or the menstrual cycle" (page 19).

Catecholamines (adrenaline/noradrenaline) and melatonin are also hormones with potential gasotransmitter crosstalk, but they fall outside the chosen focus. Their exclusion is understandable and doesn't detract from the core content; however, the title "hormones: types and functions" (from the table of contents) suggests a broad overview, so a reader might expect mention of these.

Response: We really appreciate this comment. One new section of "4.7 H_2S and other hormones" was added to discuss the crosstalk between H_2S and catecholamines or melatonin, as shown below:

"4.7 H_2S and Other Hormones

In addition to the common hormones mentioned above, H_2S also interacts with other hormones, such as catecholamines and melatonin. Catecholamines are a group of hormones that play crucial roles in the body's stress response and physiological regulation, including epinephrine (adrenaline), norepinephrine (noradrenaline), and dopamine, etc. It was first demonstrated that H_2S regulated acute adrenergic stress response by stimulating the secretion of epinephrine and norepinephrine via membrane depolarization in rainbow trout [Perry et al., 2009]. Differently, in isolated porcine iris-ciliary bodies, H_2S caused a concentration-dependent reduction in endogenous epinephrine and norepinephrine concentrations, thus inhibiting sympathetic neurotransmission (Kulkarni et al. 2009). Similarly, in a porcine large-animal model of severe traumatic shock, H_2S administration significantly lowered epinephrine level (Satterly et al. 2015). H_2S treatment also caused significant reduction in epinephrine levels in a chronic mild stress model in rats (Ghalwash et al. 2021). Functionally, H_2S inhibited epinephrine-induced human platelet aggregation and rat gastric damage (Fomenko et al. 2014; Zagli et al. 2007). These studies suggest that H_2S donors can ameliorate stress response by modulating catecholamine secretion or signals. There are few data currently available about the effect of catecholamines on H_2S biosynthesis. Only one paper reported that epinephrine injection induced rat myocardium damage and promoted H_2S level in blood (Ivanova et al. 2018). None of these studies explored the mechanisms underlying the interaction between H_2S

and catecholamines. One possibility is that H_2S may chemically interact with epinephrine to form a new compound, thereby altering the signaling responses of both molecules (Mohamed et al. 2020).

Melatonin is a hormone secreted by the pineal gland in the brain, primarily regulating the body's circadian rhythm and sleep-wake cycle. So far, there have been few reports on the crosstalk between H_2S and melatonin on mammals. It is known that both melatonin and H_2S have protective effects on the cardiovascular system. Remote ischemic preconditioning conferred cardioprotection against ischemia reperfusion injury in rats by increasing the plasma levels of both melatonin and H_2S (Zhang et al. 2023). However, what remains unknown is whether this is merely a parallel relationship or a causal one between H_2S and melatonin. Another study reported that melatonin elevated H_2S production and promoted the apoptosis of thyroid cancer cells (Jia et al. 2022). Further studies are needed to investigate the interaction between H_2S and melatonin at multiple levels under both normal and stress conditions, beginning with their respective biosynthetic pathways, followed by signaling mechanisms and functional outcomes" (page 25–26).

*Depth of Molecular Mechanisms: Another aspect of completeness is the depth of explanation for mechanisms. The article generally provides mechanisms (S-sulfhydration of specific targets, enzyme expression changes, signaling pathways like PI3K/Akt or TGF-β pathways in fibrosis). In most cases, the mechanistic detail is adequate and backed by references. One minor gap: the insulin section cites a 2023 study on PTP1B S-sulfhydration in insulin signaling (*Lin et al. 2023*), which is a key mechanism by which H_2S can enhance insulin sensitivity. However, the text itself does not explicitly discuss protein tyrosine phosphatase 1B. Including a sentence about H_2S improving insulin signaling by sulfhydrating PTP1B (thereby enhancing insulin receptor signaling) would have been a useful detail from that recent paper. This omission is subtle and the reference is listed, suggesting the author was aware of it; thus it may have been a space or focus decision.*

Response: We are sorry for the overlook. We had followed this suggestion by adding the below sentence "More importantly, H_2S may also improve insulin sensitivity by *S*-sulfhydrating PTP1B, a major negative regulator for insulin receptor signaling (Lin et al. 2023)" (page 12).

The manuscript is written well although American and British English is sometimes mixed. For example "vasodilation" are used in place of British "vasodilatation"—although one instance of "vasodilatation" was used in a reference to estrogen's effects.

Response: In this resubmission, we have consistently used "vasodilation" instead of "vasodilatation".

Some minor spelling errors are present: "preventes" instead of "prevents" in the testosterone section. Similarly, "varicoele" is likely a typo for "varicocele". Aside from these, spelling is largely correct.

Response: We are sorry for the overlook. These spelling errors have been corrected.

One or two sentences could be rephrased for elegance. For example, "By targeting at each other, H_2S and glucocorticoids contribute to a variety of important *homeostatic functions" is slightly awkward. It could be rewritten as "By directly influencing one another, H_2S and glucocorticoids together contribute to many homeostatic functions."*

Response: We really appreciate the suggestion. The sentence is now rewritten to "By directly influencing one another, H_2S and glucocorticoids together contribute to many homeostatic functions" (page 21).

Author Disclosure Statement No competing financial interest exists.

References

Adcock IM, Mumby S (2017) Glucocorticoids. Handb Exp Pharmacol 237:171–196

Adebayo AA, Oboh G, Ademosun AO (2019) Almond-supplemented diet improves sexual functions beyond Phosphodiesterase-5 inhibition in diabetic male rats. Heliyon 5:e03035

Ali MY, Whiteman M, Low C, Moore PK (2007) Hydrogen sulphide reduces insulin secretion from HIT-T15 cells by a K_{ATP} channel-dependent pathway. J Endocrinol 195:105–112

Arif HM, Qian Z, Wang R (2022) Signaling integration of hydrogen sulfide and iron on cellular functions. Antioxid Redox Signal 36:275–293

Azarbarz N, Seifabadi ZS, Moaiedi MZ, Mansouri E (2020) Assessment of the effect of sodium hydrogen sulfide (hydrogen sulfide donor) on cisplatin-induced testicular toxicity in rats. Environ Sci Pollut Res Int 27:8119–8128

Bahadoran Z, Carlström M, Mirmiran P, Ghasemi A (2020) Nitric oxide: to be or not to be an endocrine hormone? Acta Physiol (Oxf) 229:e13443

Bai J, Qi QR, Li Y, Day R, Makhoul J, Magness RR, Chen DB (2020) Estrogen receptors and estrogen-induced uterine vasodilation in pregnancy. Int J Mol Sci 21:4349

Bai J, Lechuga TJ, Makhou J, Yan H, Major C, Hameed A, Chen D (2023) ERα/ERβ-directed CBS transcription mediates E2β-stimulated hUAEC H_2S production. J Mol Endocrinol 70:e220175

Bala V, Rajagopal S, Kumar DP, Nalli AD, Mahavadi S, Sanyal AJ, Grider JR, Murthy KS (2014) Release of GLP-1 and PYY in response to the activation of G protein-coupled bile acid receptor TGR5 is mediated by Epac/PLC-ε pathway and modulated by endogenous H_2S. Front Physiol 5:420

Bassett JH, Harvey CB, Williams GR (2003) Mechanisms of thyroid hormone receptor-specific nuclear and extra nuclear actions. Mol Cell Endocrinol 213:1–11

Bełtowski J (2019) Synthesis, metabolism, and signaling mechanisms of hydrogen sulfide: an overview. Methods Mol Biol 2007:1–8

Bettio LEB, Thacker JS, Rodgers SP, Brocardo PS, Christie BR, Gil-Mohapel J (2020) Interplay between hormones and exercise on hippocampal plasticity across the lifespan. Biochim Biophys Acta Mol basis Dis 1866:165821

Bianca REVB, Cirino G, Sorrentino R (2015) Hydrogen sulfide and urogenital tract. Handb Exp Pharmacol 230:111–136

Bitar MS, Nader J, Al-Ali W, Al Madhoun A, Arefanian H, Al-Mulla F (2018) Hydrogen sulfide donor NaHS improves metabolism and reduces muscle atrophy in Type 2 diabetes: implication for understanding sarcopenic pathophysiology. Oxidative Med Cell Longev 2018:6825452

Bok R, Guerra DD, Lorca RA, Wennersten SA, Harris PS, Rauniyar AK, Stabler SP, MacLean KN, Roede JR, Brown LD, Hurt KJ (2021) Cystathionine gamma-lyase promotes estrogen-stimulated uterine artery blood flow via glutathione homeostasis. Redox Biol 40:101827

Brancaleone V, Vellecco V, Matassa DS, Bianca REVD, Sorrentino R, Ianaro A, Bucci M, Esposito F, Cirino G (2015) Crucial role of androgen receptor in vascular H_2S biosynthesis induced by testosterone. Br J Pharmacol 172:1505–1515

Braun D, Schweizer U (2018) Thyroid hormone transport and transporters. Vitam Horm 106:19–44

Bucci M, Mirone V, Di Lorenzo A, Vellecco V, Roviezzo F, Brancaleone V, Ciro I, Cirino G (2009) Hydrogen sulphide is involved in testosterone vascular effect. Eur Urol 56:378–383

Cai J, Shi X, Wang H, Fan J, Feng Y, Lin X, Yang J, Cui Q, Tang C, Xu G, Geng B (2016) Cystathionine gamma lyase-hydrogen sulfide increases peroxisome proliferator-activated receptor gamma activity by sulfhydration at C139 site thereby promoting glucose uptake and lipid storage in adipocytes. Biochim Biophys Acta 1861:419–429

Casili G, Randi E, Panagaki T, Zuhra K, Petrosino M, Szabo C (2022) Inhibition of the 3-mercaptopyruvate sulfurtransferase-hydrogen sulfide system promotes cellular lipid accumulation. Geroscience 44:2271–2289

Chaker L, Bianco AC, Jonklaas J, Peeters RP (2017) Hypothyroidism. Lancet 390:1550–1562

Chen HJ, Ngowi EE, Qian L, Li T, Qin YZ, Zhou JJ, Li K, Ji XY, Wu DD (2021) Role of hydrogen sulfide in the endocrine system. Front Endocrinol (Lausanne) 12:704620

Chen L, Gao Y, Zhao Y, Yang G, Wang C, Zhao Z, Li S (2022) Chondroitin sulfate stimulates the secretion of H_2S by Desulfovibrio to improve insulin sensitivity in NAFLD mice. Int J Biol Macromol 213:631–638

Corvino A, Severino B, Fiorino F, Frecentese F, Magli E, Perissutti E, Santagada V, Bucci M, Cirino G, Kelly G, Servillo L, Popowicz G, Pastore A, Caliendo G (2016) Fragment-based de novo design of a cystathionine γ-lyase selective inhibitor blocking hydrogen sulfide production. Sci Rep 6:34398

Costa-e-Sousa RH, Hollenberg AN (2012) The neural regulation of the hypothalamic-pituitary-thyroid axis. Endocrinology 153:4128–4135

Donoso-Barraza C, Borquez JC, Sepúlveda C, Díaz-Castro F, Sepúlveda-Quiñenao C, Rodríguez JM, Porras O, Troncoso R (2022) Hydrogen sulfide disrupts insulin-induced glucose uptake in L6 skeletal muscle cells. Food Chem Toxicol 165:113083

Du C, Lin X, Xu W, Zheng F, Cai J, Yang J, Cui Q, Tang C, Cai J, Xu G, Geng B (2019) Sulfhydrated Sirtuin-1 increasing its deacetylation activity is an essential epigenetics mechanism of anti-atherogenesis by hydrogen sulfide. Antioxid Redox Signal 30:184–197

Dušková M (2023) The effects of different types of diets on steroid hormone concentrations. Physiol Res 72:S323–S337

Emanuele N, Emanuele MA (1997) The endocrine system: alcohol alters critical hormonal balance. Alcohol Health Res World 21:53–64

Feng X, Chen Y, Zhao J, Tang C, Jiang Z, Geng B (2009) Hydrogen sulfide from adipose tissue is a novel insulin resistance regulator. Biochem Biophys Res Commun 380:153–159

Filipovic MR (2015) Persulfidation (*S*-sulfhydration) and H_2S. Handb Exp Pharmacol 230:29–59

Fomenko I, Sklyarov A, Bondarchuk T, Biletska L, Panasyuk N, Wallace JL (2014) Effects of conventional and hydrogen sulfide-releasing non-steroidal anti-inflammatory drugs in rats with stress-induced and epinephrine-induced gastric damage. Stress 17:528–537

Font-Díaz J, Jiménez-Panizo A, Caelles C, Vivanco MD, Pérez P, Aranda A, Estébanez-Perpiñá E, Castrillo A, Ricote M, Valledor AF (2021) Nuclear receptors: lipid and hormone sensors with essential roles in the control of cancer development. Semin Cancer Biol 73:58–75

Fu X, Zhou K, Gao Q, Zheng S, Chen H, Li P, Zhang Y, Suo K, Simoncini T, Wang T (2013) 17beta-estradiol attenuates atherosclerosis development: the possible role of hydrogen sulfide. Int J Cardiol 167:1061–1063

Gadalla MM, Snyder SH (2010) Hydrogen sulfide as a gasotransmitter. J Neurochem 113:14–26

Geng B, Cai B, Liao F, Zheng Y, Zeng Q, Fan X, Gong Y, Yang J, Cui Q, Tang C, Xu G (2013) Increase or decrease hydrogen sulfide exert opposite lipolysis, but reduce global insulin resistance in high fatty diet induced obese mice. PLoS One 8:e73892

Ghalwash M, Elmasry A, Omar NMA (2021) Possible cardioprotective role of NaHS on ECG and oxidative stress markers in an unpredictable chronic mild stress model in rats. Can J Physiol Pharmacol 99:321–327

Gheibi S, Jeddi S, Kashfi K, Ghasemi A (2019) Effects of hydrogen sulfide on carbohydrate metabolism in obese type 2 diabetic rats. Molecules 24:190

Giuffrè A, Vicente JB (2018) Hydrogen sulfide biochemistry and interplay with other gaseous mediators in mammalian physiology. Oxidative Med Cell Longev 2018:e6290931
Goldman AL, Bhasin S, Wu FCW, Krishna M, Matsumoto AM, Jasuja R (2017) A reappraisal of testosterone's binding in circulation: physiological and clinical implications. Endocr Rev 38: 302–324
Hancock JT, Whiteman M (2014) Hydrogen sulfide and cell signaling: team player or referee? Plant Physiol Biochem 78:37–42
Hayden RP, Tanrikut C (2016) Testosterone and varicocele. Urol Clin North Am 43:223–232
Hazzaa SM, Elsayed Arafat ESE, Abdo Ismail AE, Eltorgoman AEA, Abdelaziz SA, Kombr YFA, Zidan RA, Assar MF (2021) H_2S releasing sodium sulfide protects from acute stress-induced hypertension by increasing the activity of endothelial nitric oxide synthase enzyme. Tissue Cell 72:101550
Hiller-Sturmhöfel S, Bartke A (1998) The endocrine system: an overview. Alcohol Health Res World 22:153–164
Hine C, Kim HJ, Zhu Y, Harputlugil E, Longchamp A, Matos MS, Ramadoss P, Bauerle K, Brace L, Asara JM, Ozaki CK, Cheng SY, Singha S, Ahn KH, Kimmelman A, Fisher FM, Pissios P, Withers DJ, Selman C, Wang R, Yen K, Longo VD, Cohen P, Bartke A, Kopchick JJ, Miller R, Hollenberg AN, Mitchell JR (2017) Hypothalamic-pituitary axis regulates hydrogen sulfide production. Cell Metab 25:1320–1333.e5
Hoffmann JP, Liu JA, Seddu K, Klein SL (2023) Sex hormone signaling and regulation of immune function. Immunity 56:2472–2491
Hu N, Dong M, Ren J (2014) Hydrogen sulfide alleviates cardiac contractile dysfunction in an Akt2-knockout murine model of insulin resistance: role of mitochondrial injury and apoptosis. Am J Physiol Regul Integr Comp Physiol 306:R761–R771
Huang C, Yao W, Wu W, Lu Y, Wan H, Wang W (2013) Endogenous CSE/H_2S system mediates TNF-α-induced insulin resistance in 3T3-L1 adipocytes. Cell Biochem Funct 31:468–475
Hughes K, Eastman C (2021) Thyroid disease: long-term management of hyperthyroidism and hypothyroidism. Aust J Gen Pract 50:36–42
Iciek M, Kowalczyk-Pachel D, Bilska-Wilkosz A, Kwiecień I, Górny M, Włodek L (2015) S-sulfhydration as a cellular redox regulation. Biosci Rep 36:e00304
Ivanova AS, Sitnikova OG, Popova IG, Nazarov SB (2018) Concentrations of gaseous transmitters during catecholamine damage to the myocardium in rats. Bull Exp Biol Med 165:725–727
Jain SK, Kahlon G, Morehead L, Lieblong B, Stapleton T, Hoeldtke R, Bass PF, Levine SN (2012) The effect of sleep apnea and insomnia on blood levels of leptin, insulin resistance, IP-10, and hydrogen sulfide in type 2 diabetic patients. Metab Syndr Relat Disord 10:331–336
Jeddi S, Gholami H, Gheibi S, Kashfi K, Ghasemi A (2019) Altered gene expression of hydrogen sulfide-producing enzymes in the liver and muscles tissues of hyperthyroid rats. J Cell Physiol 234:17937–17945
Jia H, Sun W, Li X, Xu W (2022) Melatonin promotes apoptosis of thyroid cancer cells via regulating the signaling of microRNA-21 (miR-21) and microRNA-30e (miR-30e). Bioengineered 13:9588–9601
Jiang M, Wang T, Yan X, Liu Z, Yan Y, Yang K, Qi J, Zhou H, Qian N, Zhou Q, Chen B, Xu X, Xi X, Yang C, Deng L (2019) A novel rhein derivative modulates bone formation and resorption and ameliorates estrogen-dependent bone loss. J Bone Miner Res 34:361–374
Ju Y, Zhang W, Pei Y, Yang G (2013) H(2)S signaling in redox regulation of cellular functions. Can J Physiol Pharmacol 91:8–14
Ju Y, Fu M, Stokes E, Wu L, Yang G (2017) H_2S-mediated protein S-sulfhydration: a prediction for its formation and regulation. Molecules 22:1334
Kabil O, Banerjee R (2014) Enzymology of H_2S biogenesis, decay and signaling. Antioxid Redox Signal 20:770–782
Kabil O, Motl N, Banerjee R (2014) H_2S and its role in redox signaling. Biochim Biophys Acta 1844:1355–1366

Kaneko Y, Kimura Y, Kimura H, Niki I (2006) L-cysteine inhibits insulin release from the pancreatic β-cell: possible involvement of metabolic production of hydrogen sulfide, a novel gasotransmitter. Diabetes 55:1391–1397

Kaneko Y, Kimura T, Taniguchi S, Souma M, Kojima Y, Kimura Y, Kimura H, Niki I (2008) Glucose-induced production of hydrogen sulfide may protect the pancreatic beta-cells from apoptotic cell death by high glucose. FEBS Lett 583:377–382

Kasamatsu S, Nishimura A, Morita M, Matsunaga T, Hamid HA, Akaike T (2016) Redox signaling regulated by cysteine persulfide and protein polysulfidation. Molecules 21:1721

Katsouda A, Valakos D, Dionellis VS, Bibli SI, Akoumianakis I, Karaliota S, Zuhra K, Fleming I, Nagahara N, Havaki S, Gorgoulis VG, Thanos D, Antoniades C, Szabo C, Papapetropoulos A (2022) MPST sulfurtransferase maintains mitochondrial protein import and cellular bioenergetics to attenuate obesity. J Exp Med 219:e20211894

Kawahara Y, Hirashita Y, Tamura C, Kudo Y, Sakai K, Togo K, Fukuda K, Matsunari O, Okamoto K, Ogawa R, Mizukami K, Okimoto T, Kodama M, Murakami K (2020) Helicobacter pylori infection modulates endogenous hydrogen sulfide production in gastric cancer AGS cells. Helicobacter 25:e12732

Kimura H (2015) Signaling molecules: hydrogen sulfide and polysulfide. Antioxid Redox Signal 22:362–376

Kimura H (2019) Signaling by hydrogen sulfide (H_2S) and polysulfides (H_2Sn) in the central nervous system. Neurochem Int 126:118–125

Kiyama R, Wada-Kiyama Y (2015) Estrogenic endocrine disruptors: molecular mechanisms of action. Environ Int 83:11–40

Kojima M, Kangawa K (2008) Structure and function of ghrelin. Results Probl Cell Differ 46:89–115

Kraemer WJ, Ratamess NA, Hymer WC, Nindl BC, Fragala MS (2020) Growth hormone(s), testosterone, insulin-like growth factors, and cortisol: roles and integration for cellular development and growth with exercise. Front Endocrinol (Lausanne) 11:33

Kulkarni KH, Monjok EM, Zeyssig R, Kouamou G, Bongmba ON, Opere CA, Njie YF, Ohia SE (2009) Effect of hydrogen sulfide on sympathetic neurotransmission and catecholamine levels in isolated porcine iris-ciliary body. Neurochem Res 34:400–406

Lambertini E, Penolazzi L, Angelozzi M, Grassi F, Gambari L, Lisignoli G, Bonis PD, Cavallo M, Piva R (2017) The expression of cystathionine gamma-lyase is regulated by estrogen receptor alpha in human osteoblasts. Oncotarget 8:101686–101696

Lechuga TJ, Zhang HH, Sheibani L, Karim M, Jia J, Magness RR, Rosenfeld CR, Chen DB (2015) Estrogen replacement therapy in ovariectomized nonpregnant ewes stimulates uterine artery hydrogen sulfide biosynthesis by selectively up-regulating cystathionine beta-synthase expression. Endocrinology 156:2288–2298

Lee CG, Rhee DK, Kim BO, Um SH, Pyo S (2019) Allicin induces beige-like adipocytes via KLF15 signal cascade. J Nutr Biochem 64:13–24

Li L, Whiteman M, Moore PK (2009) Dexamethasone inhibits lipopolysaccharide-induced hydrogen sulphide biosynthesis in intact cells and in an animal model of endotoxic shock. J Cell Mol Med 13:2684–2692

Li L, Rose P, Moore PK (2011) Hydrogen sulfide and cell signaling. Annu Rev Pharmacol Toxicol 51:169–187

Li H, Mani S, Wu L, Fu M, Shuang T, Xu C, Wang R (2017) The interaction of estrogen and CSE/H_2S pathway in the development of atherosclerosis. Am J Physiol Heart Circ Physiol 312: H406–H414

Li Y, Han B, Salmeron AG, Bai J, Chen DB (2021a) Estrogen-induced uterine vasodilation in pregnancy and preeclampsia. Matern Fetal Med 4:52–60

Li Y, Xie K, Chang Y, Wang C, Chen Y, Wang M, Zhu Y (2021b) S-Propargyl-Cysteine attenuates diabetic cardiomyopathy in db/db mice through activation of cardiac insulin receptor signaling. Front Cardiovasc Med 17(8):737191

Liao F, Zhu Z, Xiao C, Yuan J, Geng B, Hu J (2021) Hydrogen sulfide inhibits calcium and phosphorus loss after fracture by negatively regulating glucocorticoid/glucocorticoid receptor α. Life Sci 274:119363

Lin Y, Zeng W, Lee D (2023) H_2S- and redox-state-mediated PTP1B S-sulfhydration in insulin signaling. Int J Mol Sci 24:2898

Liu M, Li Z, Liang B, Li L, Liu S, Tan W, Long J, Tang F, Chu C, Yang J (2018) Hydrogen sulfide ameliorates rat myocardial fibrosis induced by thyroxine through PI3K/AKT signaling pathway. Endocr J 65:769–781

Lou LX, Geng B, Du JB, Tang CS (2008) Hydrogen sulphide-induced hypothermia attenuates stress-related ulceration in rats. Clin Exp Pharmacol Physiol 35:223–228

Lu A, Chu C, Mulvihill E, Wang R, Liang W (2019) ATP-sensitive K^+ channels and mitochondrial permeability transition pore mediate effects of hydrogen sulfide on cytosolic Ca^{2+} homeostasis and insulin secretion in β-cells. Pflugers Arch 471:1551–1564

Ma J, Shi C, Liu Z, Han B, Guo L, Zhu L, Ye T (2019) Hydrogen sulfide is a novel regulator implicated in glucocorticoids-inhibited bone formation. Aging (Albany NY) 11:7537–7552

Ma X, Jiang Z, Wang Z, Zhang Z (2020) Administration of metformin alleviates atherosclerosis by promoting H_2S production via regulating CSE expression. J Cell Physiol 235:2102–2112

Maldonato BJ, Russell DA, Totah RA (2021) Human METTL7B is an alkyl thiol methyltransferase that metabolizes hydrogen sulfide and captopril. Sci Rep 11:4857

Mani S, Li H, Austin RC, Dickhout JG, Lhoták S, Meng QH, Yang G, Wu L, Wang R (2013) Decreased endogenous production of hydrogen sulfide accelerates atherosclerosis. Circulation 127:2523–2534

Mard SA, Veisi A, Ahangarpour A, Gharib-Naseri MK (2015) Gastric acid induces mucosal H_2S release in rats by upregulating mRNA and protein expression of cystathionine gamma lyase. J Physiol Sci 65:545–554

Maselli DB, Camilleri M (2021) Effects of GLP-1 and its analogs on gastric physiology in diabetes mellitus and obesity. Adv Exp Med Biol 1307:171–192

Matei IV, Ii H, Yaegaki K (2017) Hydrogen sulfide enhances pancreatic β-cell differentiation from human tooth under normal and glucotoxic conditions. Regen Med 12:125–141

Mezouari A, Nangia R, Gagnon J (2020) The protective role of hydrogen sulfide against obesity-associated cellular stress in blood glucose regulation. Antioxidants (Basel) 9:1038

Mohamed A, Fahim AM, Ibrahim MA (2020) Theoretical investigation on hydrogen bond interaction between adrenaline and hydrogen sulfide. J Mol Model 26:354

Mullur R, Liu YY, Brent GA (2014) Thyroid hormone regulation of metabolism. Physiol Rev 94: 355–382

Murphy B, Bhattacharya R, Mukherjee P (2019) Hydrogen sulfide signaling in mitochondria and disease. FASEB J 33:13098–13125

Mustafa AK, Gadalla MM, Sen N, Kim S, Mu W, Gazi SK, Barrow RK, Yang G, Wang R, Snyder SH (2009) H_2S signals through protein S-sulfhydration. Sci Signal 2:ra72

Nechiporuk V, Zaichko N, Korda M, Melnyk A, Koloshko O (2017) Sulphur-containing amino acids metabolism in experimental hyper and hypothyroidism in rats. Georgian Med News 271: 96–102

Nourabadi D, Baluchnejadmojarad T, Zarch SMM, Ramazi S, Serenjeh MN, Roghani M (2021) Fetal hypothyroidism impairs aortic vasorelaxation responses in adulthood: involvement of hydrogen sulfide and nitric oxide cross talk. J Cardiovasc Pharmacol 77:238–244

Oi Y, Imafuku M, Shishido C, Kominato Y, Nishimura S, Iwai K (2001) Garlic supplementation increases testicular testosterone and decreases plasma corticosterone in rats fed a high protein diet. J Nutr 131:2150–2156

Okamoto M, Yamaoka M, Takei M, Ando T, Taniguchi S, Ishii I, Tohya K, Ishizaki T, Niki I, Kimura T (2013) Endogenous hydrogen sulfide protects pancreatic beta-cells from a high-fat diet-induced glucotoxicity and prevents the development of type 2 diabetes. Biochem Biophys Res Commun 442:227–233

Okamoto M, Ishizaki T, Kimura T (2015) Protective effect of hydrogen sulfide on pancreatic beta-cells. Nitric Oxide 46:32–36

Olson KR (2012) Mitochondrial adaptations to utilize hydrogen sulfide for energy and signaling. J Comp Physiol B 182:881–897

Özatik FY, Özatik O, Tekşen Y, Yiğitaslan S, Arı NS (2021) Protective and therapeutic effect of hydrogen sulfide on hemorrhagic cystitis and testis dysfunction induced with cyclophosphamide. Turk J Med Sci 51:1531–1543

Padiya R, Khatua TN, Bagul PK, Kuncha M, Banerjee SK (2011) Garlic improves insulin sensitivity and associated metabolic syndromes in fructose fed rats. Nutr Metab (Lond) 8:53

Panagaki T, Randi EB, Augsburger F, Szabo C (2019) Overproduction of H(2)S, generated by CBS, inhibits mitochondrial Complex IV and suppresses oxidative phosphorylation in Down syndrome. Proc Natl Acad Sci USA 116:18769–18771

Panagiotopoulos D, Andriopoulou T, Spanou VM, Droggiti DI, Gkavogianni T, Giamarellos-Bourboulis EJ, Panagopoulos P (2023) Deficiency of hydrogen sulfide production and pregnancy rate in an experimental model: association with preterm delivery. Am J Reprod Immunol 90:e13764

Patel M, Shah G (2010) Possible role of hydrogen sulfide in insulin secretion and in development of insulin resistance. J Young Pharm 2:148–151

Paul BD, Snyder SH (2015) H_2S: a novel gasotransmitter that signals by Sulfhydration. Trends Biochem Sci 40:687–700

Pektas SD, Cinar N, Pektas G, Akbaba G, Kara A, Hancer HS, Duman DD, Neselioglu S, Erel O, Aksoy DY (2022) Thiol/disulfide homeostasis and its relationship with insulin resistance in patients with rosacea. J Cosmet Dermatol 21:2655–2661

Peleli M, Antoniadou I, Rodrigues-Junior DM, Savvoulidou O, Caja L, Katsouda A, Ketelhuth DFJ, Stubbe J, Madsen K, Moustakas A, Papapetropoulos A (2023) Cystathionine gamma-lyase (CTH) inhibition attenuates glioblastoma formation. Redox Biol 64:102773

Peng H, Zhang Y, Palmer LD, Kehl-Fie TE, Skaar EP, Trinidad JC, Giedroc DP (2017) Hydrogen sulfide and reactive sulfur species impact proteome S-Sulfhydration and global virulence regulation in staphylococcus aureus. ACS Infect Dis 3:744–755

Perry SF, McNeill B, Elia E, Nagpal A, Vulesevic B (2008) Hydrogen sulfide stimulates catecholamine secretion in rainbow trout (Oncorhynchus mykiss). Am J Physiol Regul Integr Comp Physiol 296:R133–R140

Pichette J, Gagnon J (2016) Implications of hydrogen sulfide in glucose regulation: how H_2S can alter glucose homeostasis through metabolic hormones. Oxidative Med Cell Longev 2016: 3285074

Pichette J, Fynn-Sackey N, Gagnon J (2017) Hydrogen sulfide and sulfate prebiotic stimulates the secretion of GLP-1 and improves glycemia in male mice. Endocrinology 158:3416–3425

Pietri R, Román-Morales E, López-Garriga J (2011) Hydrogen sulfide and hemeproteins: knowledge and mysteries. Antioxid Redox Signal 15:393–404

Qi QR, Lechuga TJ, Patel B, Nguyen NA, Yang YH, Li Y, Sarnthiyakul S, Zhang QW, Bai J, Makhoul J, Chen DB (2020) Enhanced stromal cell CBS-H_2S production promotes estrogen-stimulated human endometrial angiogenesis. Endocrinology 161:bqaa176

Qi Q, Zhang H, Jin Z, Wang C, Xia M, Chen B, Lv B, Peres Diaz L, Li X, Feng R, Qiu M, Li Y, Meseguer D, Zheng X, Wang W, Song W, Huang H, Wu H, Chen L, Schneeberger M, Yu X (2024) Hydrogen sulfide produced by the gut microbiota impairs host metabolism via reducing GLP-1 levels in male mice. Nat Metab 6:1601–1615

Rahman MS, Hossain KS, Das S, Kundu S, Adegoke EO, Rahman MA, Hannan MA, Uddin MJ, Pang MG (2021) (2021) role of insulin in health and disease: an update. Int J Mol Sci 22:6403

Ramsay PT, Carr A (2011) Gastric acid and digestive physiology. Surg Clin North Am 91:977–982

Russo CD, Tringali G, Ragazzoni E, Maggiano N, Menini E, Vairano M, Preziosi P, Navarra P (2000) Evidence that hydrogen sulphide can modulate hypothalamo-pituitary-adrenal axis function: in vitro and in vivo studies in the rat. J Neuroendocrinol 12:225–233

Sabatino L, Iervasi G, Pingitore A (2014) Thyroid hormone and heart failure: from myocardial protection to systemic regulation. Expert Rev Cardiovasc Ther 12:1227–1236
Satterly SA, Salgar S, Hoffer Z, Hempel J, DeHart MJ, Wingerd M, Raywin H, Stallings JD, Martin M (2015) Hydrogen sulfide improves resuscitation via non-hibernatory mechanisms in a porcine shock model. J Surg Res 199:197–210
Schwartz N, Verma A, Bivens CB, Schwartz Z, Boyan BD (2016) Rapid steroid hormone actions via membrane receptors. Biochim Biophys Acta 1863:2289–2298
Shafie A, Kianian F, Ashabi G, Kadkhodaee M, Ranjbaran M, Hajiaqaei M, Lorian K, Abdi A, Seifi B (2022) Beneficial effects of combination therapy with testosterone and hydrogen sulfide by reducing oxidative stress and apoptosis: rat experimental varicocele model. Int J Reprod Biomed 20:941–954
Sheibani L, Lechuga TJ, Zhang H, Hameed A, Wing DA, Kumar S, Rosenfeld CR, Chen DB (2017) Augmented H_2S production via cystathionine-beta-synthase upregulation plays a role in pregnancy-associated uterine vasodilation. Biol Reprod 96:664–672
Shibuya N, Koike S, Tanaka M, Ishigami-Yuasa M, Kimura Y, Ogasawara Y, Fukui K, Nagahara N, Kimura H (2013) A novel pathway for the production of H_2S from D-cysteine in mammalian cells. Nat Commun 4:1366
Shoji T, Hayashi M, Sumi C, Kusunoki M, Uba T, Matsuo Y, Kimura H, Hirota K (2019) Pharmacological polysulfide suppresses glucose-stimulated insulin secretion in an ATP-sensitive potassium channel-dependent manner. Sci Rep 9:19377
Shuang T, Fu M, Yang G, Huang Y, Qian Z, Wu L, Wang R (2021) Interaction among estrogen, IGF-1, and H_2S on smooth muscle cell proliferation. J Endocrinol 248:17–30
Slade E, Williams L, Gagnon J (2018) Hydrogen sulfide suppresses ghrelin secretion in vitro and delays postprandial ghrelin secretion while reducing appetite in mice. Physiol Rep 6:e13870
Song X, Nie L, Long J, Zhao J, Liu X, Wang L, Liu D, Wang S, Liu S, Yang J (2023) Hydrogen sulfide alleviates hypothyroidism-induced myocardial fibrosis in rats through stimulating autophagy and inhibiting TGF-β1/Smad2 pathway. Korean J Physiol Pharmacol 27:1–8
Song Y, Mao C, Zhong Q, Zhang R, Jiang D, Sun X (2024) Role of hydrogen sulfide in the male reproductive system. Front Endocrinol (Lausanne) 15:1377090
Srilatha B, Muthulakshmi P, Adaikan PG, Moore PK (2012) Endogenous hydrogen sulfide insufficiency as a predictor of sexual dysfunction in aging rats. Aging Male 15:153–158
Sun HZ, Gong XY, Wu L, Wang XX, Nie YN, Shang R, Wang H, Li YC, Sun QF, Gao PF, Bi JX (2018) Hydrogen sulfide modulates gastric acid secretion in rats via involvement of substance P and nuclear factor-κB signaling. J Physiol Pharmacol 69:3
Szabo C, Coletta C, Chao C, Modis K, Szczesny B, Papapetropoulos A, Hellmich MR (2013) Tumor-derived hydrogen sulfide, produced by cystathionine-beta-synthase, stimulates bioenergetics, cell proliferation, and angiogenesis in colon cancer. Proc Natl Acad Sci USA 110:12474–12479
Tahiliani P, Kar A (2003) The combined effects of Trigonella and Allium extracts in the regulation of hyperthyroidism in rats. Phytomedicine 10:665–668
Takahashi N, Wei FY, Watanabe S, Hirayama M, Ohuchi Y, Fujimura A, Kaitsuka T, Ishii I, Sawa T, Nakayama H, Akaike T, Tomizawa K (2017) Reactive sulfur species regulate tRNA methylthiolation and contribute to insulin secretion. Nucleic Acids Res 45:435–445
Teoh J, Li X, Simoncini T, Zhu D, Fu X (2020) Estrogen-mediated gaseous signaling molecules in gardiovascular disease. Trends Endocrinol Metab 31:773–784
Tomita M, Nagahara N, Ito T (2016) Expression of 3-mercaptopyruvate sulfurtransferase in the mouse. Molecules 21:1707
Veeranna V, Zalawadiya SK, Niraj A, Pradhan J, Ference B, Burack RC, Jacob S, Afonso L (2011) Homocysteine and reclassification of cardiovascular disease risk. J Am Coll Cardiol 58:1025–1033
Wang R (2002) Two's company, three's a crowd: can H_2S be the third endogenous gaseous transmitter? FASEB J 16:1792–1798

Wang R (2012) Physiological implications of hydrogen sulfide: a whiff exploration that blossomed. Physiol Rev 92:791–896

Wang R, Li K, Wang H, Jiao H, Wang X, Zhao J, Lin H (2019) Endogenous CSE/hydrogen sulfide system regulates the effects of glucocorticoids and insulin on muscle protein synthesis. Oxidative Med Cell Longev 2019:9752698

Wang S, Chen Q, Zhang Y, Zheng F, Xue T, Ge X, Ma R, Li X, Wu R, Liang K, Qian Z, Ge Y, Ma J, Yao B (2020) Omega-3 polyunsaturated fatty acids alleviate hydrogen sulfide-induced blood-testis barrier disruption in the testes of adult mice. Reprod Toxicol 98:233–241

Wang J, Wang J, Shen T, Hong R, Tang S, Zhao X (2021) H_2S catalysed by CBS regulates testosterone synthesis through affecting the sulfhydrylation of PDE. J Cell Mol Med 25:3460–3468

Wang M, Wang Z, Miao Y, Wei H, Peng J, Zhou Y (2022) Diallyl trisulfide promotes placental angiogenesis by regulating lipid metabolism and alleviating inflammatory responses in obese pregnant mice. Nutrients 14:2230

Wiliński B, Wiliński J, Somogyi E, Piotrowska J, Opoka W (2013) Metformin raises hydrogen sulfide tissue concentrations in various mouse organs. Pharmacol Rep 65:737–742

Wu L, Yang W, Jia X, Yang G, Duridanova D, Cao K, Wang R (2009) Pancreatic islet overproduction of H_2S and suppressed insulin release in zucker diabetic rats. Lab Investig 89:59–67

Xu X, Yan Q, Liu X, Li P, Li X, Chen Y, Simoncini T, Liu J, Zhu D, Fu X (2019) 17β-Estradiol nongenomically induces vascular endothelial H_2S release by promoting phosphorylation of cystathionine γ-lyase. J Biol Chem 294:15577–15592

Xu C, He Z, Song Y, Shao S, Yang G, Zhao J (2023) Atypical pituitary hormone-target tissue axis. Front Med 17:1–17

Xue R, Hao DD, Sun JP, Li WW, Zhao MM, Li XH, Chen Y, Zhu JH, Ding YJ, Liu J, Zhu YC (2013) Hydrogen sulfide treatment promotes glucose uptake by increasing insulin receptor sensitivity and ameliorates kidney lesions in type 2 diabetes. Antioxid Redox Signal 19:5–23

Yang G, Wang R (2015) H_2S and blood vessels: an overview. Handb Exp Pharmacol 230:85–110

Yang W, Yang G, Jia X, Wu L, Wang R (2005) Activation of KATP channels by H_2S in rat insulin-secreting cells and the underlying mechanisms. J Physiol 569:519–531

Yang G, Yang W, Wu L, Wang R (2007) H_2S, endoplasmic reticulum stress, and apoptosis of insulin-secreting beta cells. J Biol Chem 282:16567–16576

Yang G, Wu L, Jiang B, Yang W, Qi J, Cao K, Meng Q, Mustafa AK, Mu W, Zhang S, Snyder SH, Wang R (2008) H_2S as a physiologic vasorelaxant: hypertension in mice with deletion of cystathionine gamma-lyase. Science 322:587–590

Yang G, Tang G, Zhang L, Wu L, Wang R (2011) The pathogenic role of cystathionine γ-lyase/hydrogen sulfide in streptozotocin-induced diabetes in mice. Am J Pathol 179:869–879

Yang G, Zhao K, Ju Y, Mani S, Cao Q, Puukila S, Khaper N, Wu L, Wang R (2013) Hydrogen sulfide protects against cellular senescence via S-sulfhydration of Keap1 and activation of Nrf2. Antioxid Redox Signal 18:1906–1919

Yang J, Minkler P, Grove D, Wang R, Willard B, Dweik R, Hine C (2019) Non-enzymatic hydrogen sulfide production from cysteine in blood is catalyzed by iron and vitamin B6. Commun Biol 2:194

Yang J, Link C, Henderson YO, Bithi N, Hine C (2021) Peripubertal Bisphenol A exposure imparts detrimental age-related changes in body composition, cognition, and hydrogen sulfide production capacities. Antioxid Redox Signal 36:1246–1267

Yusuf M, Huat BTK, Hsu A, Whiteman M, Bhatia M, Moore PK (2005) Streptozotocin-induced diabetes in the rat is associated with enhanced tissue hydrogen sulfide biosynthesis. Biochem Biophys Res Commun 333:1146–1152

Zagli G, Patacchini R, Trevisani M, Abbate R, Cinotti S, Gensini GF, Masotti G, Geppetti P (2007) Hydrogen sulfide inhibits human platelet aggregation. Eur J Pharmacol 559(1):65–68

Zhang L, Yang G, Untereiner A, Ju Y, Wu L, Wang R (2013) Hydrogen sulfide impairs glucose utilization and increases gluconeogenesis in hepatocytes. Endocrinology 154:114–126

Zhang J, Ye J, Yuan C, Fu Q, Zhang F, Zhu X, Wang L, Gao P, Shu G, Wang S, Liu Q, Jiang Q (2020) Hydrogen sulfide is a regulator of mammary gland development in prepubescent female mice. Mol Med Rep 22:4061–4069

Zhang D, Hong X, Wang J, Jiang Y, Zhang Y, Chen J, Niu X (2021) Estradiol-17β inhibits homocysteine mediated damage by promoting H_2S production via upregulating CBS and CSE expression in human umbilical vein endothelial cells. J Cell Biochem 122:915–925

Zhang H, Li S, Jin Y (2023) Remote ischemic preconditioning-induced late cardioprotection: possible role of melatonin-mitoKATP-H_2S signaling pathway. Acta Cir Bras 38:e380423

Zhao X, Cao Y, Jin H, Wang X, Zhang L, Zhang Y, Yu Y, Huang Y, Gao Y, Zhang J (2022) Hydrogen sulfide promotes thyroid hormone synthesis and secretion by upregulating Sirtuin-1. Front Pharmacol 13:838248

Zheng F, Han J, Lu H, Cui C, Yang J, Cui Q, Cai J, Zhou Y, Tang C, Xu G, Geng B (2018) Cystathionine beta synthase-hydrogen sulfide system in paraventricular nucleus reduced high fatty diet induced obesity and insulin resistance by brain-adipose axis. Biochim Biophys Acta Mol basis Dis 1864:3281–3291

Zhou K, Gao Q, Zheng S, Pan S, Li P, Suo K, Simoncini T, Wang T, Fu X (2013) 17beta-estradiol induces vasorelaxation by stimulating endothelial hydrogen sulfide release. Mol Hum Reprod 19:169–176

Zhu XY, Gu H, Ni X (2011) Hydrogen sulfide in the endocrine and reproductive systems. Expert Rev Clin Pharmacol 4:75–82

Zhu X, Tang Z, Cong B, Du J, Wang C, Wang L, Ni S, Lu J (2013) Estrogens increase cystathionine-γ-lyase expression and decrease inflammation and oxidative stress in the myocardium of ovariectomized rats. Menopause 20:1084–1091

Zhu J, Ligi S, Yang G (2021) An evolutionary perspective on the interplays between hydrogen sulfide and oxygen in cellular functions. Arch Biochem Biophys 707:108920

Zivanovic J, Kouroussis E, Kohl JB, Adhikari B, Bursac B, Schott-Roux S, Petrovic D, Miljkovic JL, Thomas-Lopez D, Jung Y, Miler M, Mitchell S, Milosevic V, Gomes JE, Benhar M, Gonzalez-Zorn B, Ivanovic-Burmazovic I, Torregrossa R, Mitchell JR, Whiteman M, Schwarz G, Snyder SH, Paul BD, Carroll KS, Filipovic MR (2019) Selective persulfide detection reveals evolutionarily conserved antiaging effects of S-sulfhydration. Cell Metab 30:1152–1170

Hydrogen Sulfide in Adipose Tissue Biology

Antonia Katsouda and Andreas Papapetropoulos

Contents

Abstract

Over the past several decades, sulfide has emerged as an important signaling molecule, playing diverse physiological and pathological roles across a broad

A. Katsouda
Clinical, Experimental Surgery and Translational Research Center, Biomedical Research Foundation of the Academy of Athens, Athens, Greece

A. Papapetropoulos (✉)
Clinical, Experimental Surgery and Translational Research Center, Biomedical Research Foundation of the Academy of Athens, Athens, Greece

Laboratory of Pharmacology, Faculty of Pharmacy, National and Kapodistrian University of Athens, Athens, Greece
e-mail: apapapet@pharm.uoa.gr

A. Papapetropoulos, C. Szabo (eds.), *Hydrogen Sulfide: A Toxicant with Healing Potential*, Handbook of Experimental Pharmacology 292,
https://doi.org/10.1007/164_2025_759

spectrum of human health conditions. Among its recently recognized systemic effects, sulfide has been identified as a key regulator of adipose tissue (AT) homeostasis; its critical involvement in numerous adipose tissue functions, including adipogenesis, lipogenesis, and lipid and glucose metabolism, has been reported. Growing evidence indicates that dysregulation of H_2S signaling within the adipose tissue contributes to the development and progression of obesity and related metabolic disorders. The pharmacological modulation of endogenous sulfide levels is being actively explored as a promising therapeutic strategy for improving metabolic health. Herein, we provide a comprehensive and critical review of the current literature on the role of sulfide in adipose tissue biology and physiology. Particular emphasis is placed on the broader implications of sulfide signaling in the prevention and treatment of obesity.

Keywords

Adipose tissue · Energy homeostasis · Lipid metabolism · Metabolic disorders · Obesity · Persulfidation · Sulfide

1 Introduction

Adipose tissue (AT), a remarkably complex organ, until the mid-twentieth century was regarded merely as a type of connective tissue containing lipid droplets, with no substantial association to metabolic regulation (David Paton Cuthbertson and Lionel Tompsett 1933; Clark and Clark 1940; Rosen and Spiegelman 2014). This perception gradually changed when it was discovered that adipose tissue plays a crucial role in energy homeostasis, serving either as an energy reservoir or as a source of free fatty acids, depending on the body's metabolic needs (Rosen and Spiegelman 2014). The identification of adipocyte-derived proteins (adipokines) in the bloodstream marked a turning point, leading to the recognition of adipose tissue as a secretory gland that was accompanied by a rising interest in understanding its physiological roles (Cook et al. 1974; Friedman and Halaas 1998; Mohamed-Ali et al. 1998). Nowadays, AT is considered the largest endocrine organ in the human body that responds to hormonal and energetic cues, affects the function of multiple other organs, and plays a central role in whole-body energy homeostasis, including food intake, glucose handling, insulin sensitivity, and thermogenesis (Kershaw and Flier 2004; Rosen and Spiegelman 2014).

Adipocytes occur in three morphologically and functionally distinct types: white, brown, and beige adipocytes (Berry et al. 2013). Unilocular white adipocytes' main function is the energy storage for the body's metabolic needs, while the multilocular brown adipocytes are highly specialized cells that dissipate stored chemical energy in the form of heat, a process known as thermogenesis (Cannon and Nedergaard 2004). Thermogenesis is driven by the uncoupling protein-1 (UCP1) that catalyzes a proton leak across the inner mitochondrial membrane, thus "uncoupling" fuel oxidation from ATP synthesis (Chouchani et al. 2019). Response to exogenous stimuli including cold exposure, catecholamines, exercise, and thiazolidinediones can trigger the appearance of clusters of $UCP1^{+}$ cells with a brown fat-like morphology within white fat depots (Van Der Lans et al. 2013). These inducible cells have been called beige adipocytes and nowadays are recognized as the third type of adipocytes. Beige cells present an overlapping, but distinct gene expression pattern compared to classic brown adipocyte, expressing a core program of thermogenic and mitochondrial genes (Wu et al. 2012; Sharp et al. 2012).

Energy imbalance involves either increasing energy uptake or decreasing energy expenditure causing adipocyte hypertrophy/hyperplasia, excessive fat expansion, and obesity (Trayhurn 2013). Overweight condition is characterized not only by an increase in AT mass but also by significant alterations in its function, including chronic low-grade inflammation, mitochondrial dysfunction, hypoxia, and impaired adipokine signaling (Rosen and Spiegelman 2014). These changes contribute to systemic metabolic dysregulation and increased disease risk. Understanding the cellular, molecular, and endocrine functions of adipose tissue is therefore critical for identifying novel therapeutic targets to combat the global rise in metabolic disease (Williams et al. 2015).

Recent research suggests that hydrogen sulfide (H_2S), a vital endogenous signaling molecule classified as a gasotransmitter, influences several aspects of adipose tissue physiology. Notably, H_2S has been implicated in adipogenesis, lipid metabolism, glucose uptake, and inflammation (Cirino et al. 2023). Protein persulfidation, a post-translational modification of cysteine residues that leads to changes in the conformation and activity of the modified proteins, is believed to mediate most of sulfide's cellular responses (Mustafa et al. 2009; Nagy 2015; Iciek et al. 2016). Understanding how H_2S regulates adipose tissue biology can unveil new therapeutic targets for metabolic diseases. This review will focus on the production and function of sulfide in adipose tissue, detailing its roles in adipocyte differentiation, metabolic homeostasis, and mitochondrial function (Fig. 1). A strong focus is directed toward the implications of sulfide signaling in obesity and related metabolic disorders, as well as the therapeutic potential of pharmacologically modulating sulfide levels to improve adipose function and systemic metabolic health (Fig. 1).

2 Endogenous Sources and Regulation of Sulfide in Adipose Tissue

2.1 H_2S Biosynthesis in Adipose Tissue

Background In the human body, hydrogen sulfide can be synthesized through enzymatically catalyzed reactions as well as non-enzymatic pathways (Szabo and Papapetropoulos 2017). Its biosynthesis occurs via the action of the enzymes: cystathionine β-synthase (CBS), cystathionine γ-lyase (CSE), and 3-mercaptopyruvate sulfurtransferase (MPST) (Szabo 2018). CBS, CSE, and MPST exhibit significant differences in terms of mechanisms through which their activity is regulated, the cofactors they require and the substrates they use, their

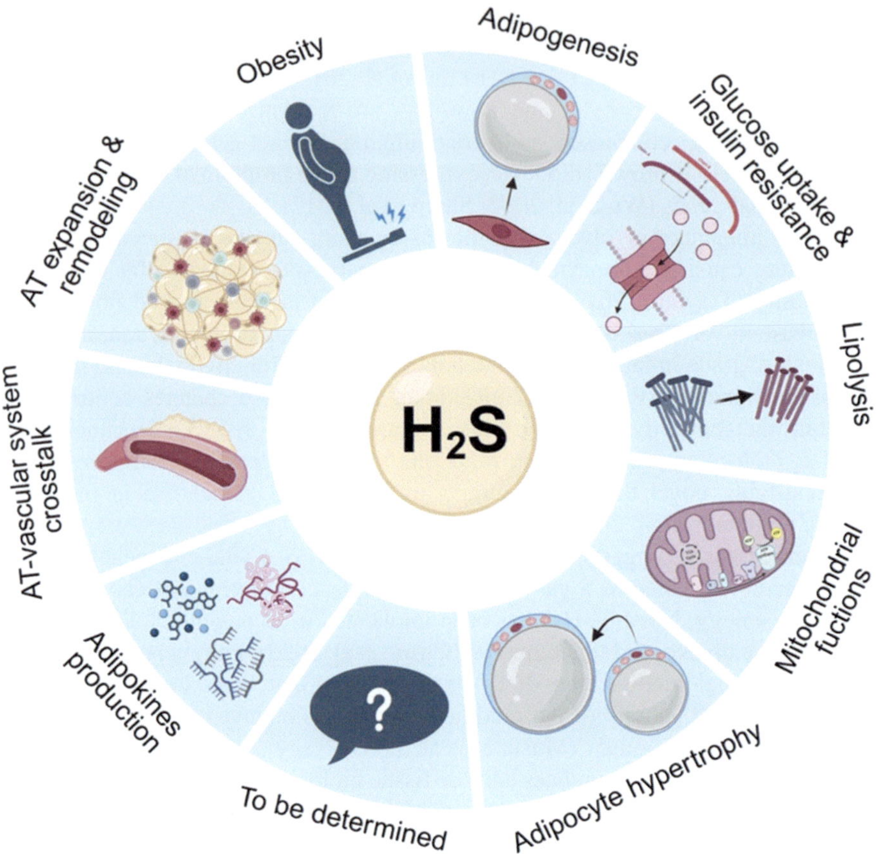

Fig. 1 The role of hydrogen sulfide in adipose tissue physiology and metabolic health. Sulfide has been identified as a key modulator of adipose tissue (AT) homeostasis, regulating adipogenesis, adipocyte lipid accumulation, lipid and glucose metabolism, and adipokine production. Dysregulation of sulfide signaling has been linked with adipose tissue expansion and weight gain. A regulatory role for H_2S in the AT-vascular system interplay has also been reported. Ongoing research is anticipated to further elucidate the modulatory effects of H_2S on adipose tissue physiology and its broader implications for metabolic health

cellular localization, and their expression profiles across tissues (Cirino et al. 2023). CBS catalyzes reactions involving cysteine, serine, and homocysteine. The most important of these include the conversion of L-cysteine and L-homocysteine into L-cystathionine and H_2S, as well as the hydrolysis of L-cysteine, which produces H_2S, L-serine, and L-lanthionine (Singh et al. 2009; Kabil and Banerjee 2014). CBS belongs to the class of pyridoxal-5′-phosphate (PLP)-dependent enzymes and is the only enzyme in this class that requires heme as a prosthetic group (Singh and Banerjee 2011). CSE uses cystathionine and L-cysteine as its primary substrates, to catalyze the formation of H_2S, ammonia, and pyruvate (Sun et al. 2009; Singh et al. 2009). Homocysteine has also been reported as a substrate for the enzyme in hydrogen sulfide production (Kabil and Banerjee 2014). Like CBS, CSE belongs to the class of PLP-dependent enzymes (Singh and Banerjee 2011). MPST requires zinc for its activity and produces H_2S from cysteine, through its coupled action with the enzyme CAT (a PLP-dependent enzyme). In the presence of CAT, cysteine is converted into 3-mercaptopyruvate (3-MP), which is subsequently metabolized by MPST to form H_2S and pyruvate (Mikami et al. 2011). The resulting hydrogen sulfide exists in a bound form and is released under reducing conditions. The catalytic mechanism of the enzyme involves the transfer of sulfur from 3-MP to nucleophilic cysteine residue 247 of MPST (Nagahara et al. 2007; Nagahara 2013). The formed disulfide releases hydrogen sulfide in the presence of reducing agents or specific enzymes (Mikami et al. 2011; Nagahara 2013).

Regulation of H_2S-Producing Enzymes in Adipose Tissue CBS and CSE were initially reported to be present in adipose tissue (Feng et al. 2009). Subsequent studies confirmed the presence of CBS and CSE in both white and brown AT and revealed that MPST is also present in AT depots (Katsouda et al. 2018; Yang et al. 2018; Liu et al. 2018). Additionally, the expression of H_2S-producing enzymes has been detected in cultured adipocytes in studies using the 3 T3-L1 cell line (Tsai et al. 2015).

Changes in the levels of H_2S-synthesizing enzymes in adipose tissue have been observed in obesity. Reduced CBS protein levels have been reported in the adipose tissue of *db/db* mice, a genetic model of obesity (Katsouda et al. 2018). In contrast, diet-induced obesity or age-related weight gain does affect CBS expression levels in adipose tissue of mice (Katsouda et al. 2018; Zampas et al. 2025). The levels of CSE in the adipose tissue of obese experimental animals have been investigated, with studies reporting increased, decreased, or unchanged expression (Feng et al. 2009; Velmurugan et al. 2015; Katsouda et al. 2018; Liu et al. 2018; Zampas et al. 2025). These discrepancies may be attributed to differences in animal models (genetic or diet-induced obesity), diet composition, and the duration of dietary interventions. In humans, calorie restriction has been shown to upregulate CSE expression in adipose tissue (Lee et al. 2025). In rodents, CSE levels remain unchanged in 6-month-old mice but increase in the visceral fat depots of 12-month-old rats (Feng et al. 2009; Katsouda et al. 2018). High-fat diet feeding leads to downregulation of MPST in inguinal and gonadal white adipose tissue, as well as in brown adipose tissue depots in mice (Katsouda et al. 2018; Zampas et al. 2025). This observation has been confirmed in visceral and subcutaneous fat from obese human subjects (Comas et al. 2021a; Katsouda et al. 2022). Moreover, mRNA levels of CSE and CBS have been shown to be reduced in the white adipose tissue of obese individuals (Comas et al. 2021a). Metabolic syndrome, which is closely associated with obesity, glucose intolerance, and hypertension, is also characterized by downregulation of MPST in adipose tissue (Zampas et al. 2025). Furthermore, aging has been shown to reduce MPST expression in both gonadal white and brown adipose tissue depots in mice (Katsouda et al. 2018).

2.2 Adipose Tissue Sulfide Pools

Background Endogenously produced sulfide exists in different forms that include free sulfide, sulfane-bound sulfur, and sulfide species. Examples of such species include thiosulfate ion ($S_2O_3^{2-}$), organic polysulfides (R-S_n-R), and elemental sulfur (S^0) (Ishigami et al. 2009; Shen et al. 2012). The release of H_2S from these pools occurs through redox reactions. Bound sulfur is also found in proteins that have been modified via persulfidation and can release hydrogen sulfide either non-enzymatically or through the action of thioredoxin (Wedmann et al. 2016).

Sulfide Levels in Fat Tissue In agreement with the findings demonstrating the MPST downregulation in the adipose tissue of mice fed a high-fat diet, H_2S concentrations were also reduced in white fat depots. Additionally, both white and brown adipose tissues exhibited decreased H_2S levels in *db/db* mice (Katsouda et al. 2018). These reductions were, at least in part, attributed to the downregulation of H_2S-producing enzymes.

2.3 H_2S Catabolism in Adipose Tissue

Background Once synthesized, H_2S does not accumulate in tissues; it undergoes catabolism primarily through mitochondrial oxidation that occurs via a multi-step pathway (Olson 2018). In detail, H_2S produced within the mitochondria or entering via diffusion from the cytosol is reacted upon by sulfide quinone oxidoreductase (SQRDL) on the inner mitochondrial membrane, forming a persulfide intermediate and transferring electrons to quinone, which contributes to ATP production (Jackson et al. 2012). The sulfur moiety bound to SQRDL can follow one of two pathways: it is either transferred to sulfite ions (SO_3^{2-}) to form thiosulfate ions ($S_2O_3^{2-}$), which subsequently reacts with glutathione via thiosulfate sulfurtransferase (TST) forming glutathione persulfide (GSSH) and SO_3^{2-}, or it directly reacts with glutathione leading to its oxidation (Libiad et al. 2014). GSSH is then reduced by persulfide dioxygenase (ETHE1), producing SO_3^{2-}, which is further oxidized by sulfite oxidase to sulfate ions (SO_4^{2}), contributing electrons to cytochrome C (Tiranti et al. 2009; Melideo et al. 2014). Alternatively, TST can also catalyze the formation of biologically active thiosulfate via interaction of sulfite and additional H_2S molecules (Kabil and Banerjee 2014).

Regulation of H2S-Degrading Enzymes in Adipose Tissue Recent studies have reported the presence of SQRDL, TST, and ETHE1 in murine white adipose tissue (Morton et al. 2016; Zampas et al. 2025). Notably, TST expression was also noted in cultured 3 T3-L1 adipocytes and in the stromal vascular fraction, which includes adipocyte precursor cells, isolated from mouse AT (Morton et al. 2016; Lainšček et al. 2020). In contrast with the extensive studies on H_2S-generating enzymes, the regulation of sulfide-catabolizing proteins in AT remains poorly understood. High-fat diet feeding does not affect the expression levels of SQRDL; however, it leads to the downregulation of both TST and ETHE1 in white adipose tissue of experimental animals (Morton et al. 2016; Lainšček et al. 2020; Zampas et al. 2025). Similar findings have been reported in a mouse model of metabolic syndrome (Zampas et al. 2025). Interestingly, TST mRNA levels in subcutaneous adipose tissue have been shown to correlate positively with insulin sensitivity and negatively with fat mass (Morton et al. 2016).

3 Sulfide, Adipogenesis, and Adipocyte Lipid Deposition

3.1 Role of H_2S in Adipogenesis

Background Adipogenesis is the biological process whereby fibroblast-like progenitor cells commit to the adipocyte lineage, accumulate nutrients, and differentiate into mature, triglyceride-filled, non-dividing fat cells. Adipogenesis involves two main stages: commitment of fibroblast-like progenitors to preadipocytes, followed by differentiation into insulin-responsive mature adipocytes (Ghaben and Scherer 2019). The transformation of preadipocytes into mature adipocytes involves the stepwise activation of multiple transcription factors, including peroxisome proliferator-activated receptor gamma (PPARγ), CCAAT/enhancer-binding protein alpha (C/EBPα), sterol regulatory element-binding protein 1 (SREBP1), and carbohydrate response element-binding protein (ChREBP) (Rosen and MacDougald 2006). Developmentally, white and brown adipocytes originate from distinct precursor populations, with brown adipocytes sharing a common lineage with muscle cells (myogenic factor 5, MYF5$^+$, paired box protein 7, PAX7$^+$), whereas white and beige adipocytes derive from MYF5$^-$, PAX7$^-$ progenitors (Seale et al. 2008; Sanchez-Gurmaches and Guertin 2014). Beige adipocyte differentiation is inducible by external stimuli and may revert to a white adipocyte phenotype in their absence (Rosenwald et al. 2013).

While the number of adipocytes in specific fat depots is largely established early in life and remains stable during adulthood, recent lineage-tracing studies in rodents reveal that prolonged caloric excess can stimulate the formation of new adipocytes from preadipocyte precursors, contributing to adipose tissue expansion (Hirsch and Han 1969; Spalding et al. 2008). The balance between the enlargement of existing adipocytes (hypertrophy) and the generation of new fat cells (hyperplasia via adipogenesis) plays a crucial role in metabolic health (Ghaben and Scherer 2019). Understanding the regulatory mechanisms that govern adipogenesis remains a subject of intensive research.

The Regulatory Role of Endogenous Sulfide in Adipocyte Differentiation The role of H_2S in adipogenesis has been extensively studied in 3T3-L1 cells during their differentiation into white adipocytes. Research on the physiological functions of endogenous sulfide has primarily focused on H_2S-generating enzymes, while the role of sulfide-degrading enzymes in adipocyte differentiation remains unknown.

CSE mRNA levels are reported to be highly expressed in preadipocytes and to increase further during differentiation, whereas CBS mRNA is barely detectable and remains unchanged (Yang et al. 2018; Comas et al. 2019). At protein level, the expression of CSE, CBS, and MPST increased during the differentiation of 3T3-L1 cells into white adipocytes. CBS protein levels peaked early, before full adipocyte maturation, whereas CSE, MPST continued to rise throughout the entire 6- or 7-day differentiation period (Tsai et al. 2015; Yang et al. 2018). Notably, upregulation of CSE during this period is linked to the binding of C/EBPβ and C/EBPδ transcription factors to the promoter region of the CSE gene, which enhances its expression (Yang et al. 2018). Upregulation of H_2S-synthesizing enzymes has also been observed during the differentiation of human preadipocytes into adipocytes (Comas et al. 2021a).

Overexpression of CSE promotes adipogenesis in 3T3-L1 cell line (Cai et al. 2016). Moreover, pharmacological inhibition of CSE using propargylglycine (PAG) results in decreased lipid accumulation during 3T3-L1 differentiation into white adipocytes. This is accompanied by downregulation of the adipogenic marker fatty acid–binding protein 4 (FABP4/aP2) and reduced mRNA levels of adipogenesis-related genes, including C/EBPα, SREBP1, and ChREBP (Tsai et al. 2015). Additionally, silencing of CBS or CSE reverses the increased expression of FABP4/aP2 observed during adipogenesis (Tsai et al. 2015). Gene silencing of CSE, CBS, and MPST in human preadipocytes resulted in a significantly decreased expression of several adipogenic markers (Comas et al. 2021a). Preadipocytes isolated from the gonadal white adipose tissue of mice with a global deletion of CSE (CSE knockout, CSE KO) and differentiated into white adipocytes exhibited reduced lipid accumulation compared to those from wild-type (WT) mice (Verma et al. 2023). Similar results were observed when primary WT preadipocytes were treated with PAG (Verma et al. 2023).

In contrast, studies using human telomerase reverse transcriptase (hTERT)-immortalized human adipose-derived mesenchymal stem cells (ASC52telo) differentiated over a 15-day protocol reported opposing results; CBS knockdown led to increased adipogenesis and upregulation of adipogenic markers, including FABP4, C/EBPα, and PPARγ (Comas et al. 2021b). Additionally, MPST deficiency has been shown to result in increased lipid accumulation (Casili et al. 2022).

In line with the upregulation of CBS, CSE, and MPST, multiple studies consistently demonstrate that sulfide levels increase during adipocyte differentiation in both 3T3-L1 cell line and human preadipocytes; this upregulation corresponds to enhanced expression and activity of key adipogenic transcription factors and lipid storage pathways (Tsai et al. 2015; Yang et al. 2018; Comas et al. 2019, 2021a).

The regulatory roles of H_2S-synthesizing enzymes in adipocyte differentiation are summarized in Fig. 2. Overall, current knowledge regarding the role of H_2S-synthesizing enzymes in adipocyte differentiation suggests that their impact may vary depending on the type of precursor cells used. It is important to note that, although 3T3-L1 cells are a well-established immortalized cell line commonly employed to study adipogenesis, primary preadipocytes, isolated directly from

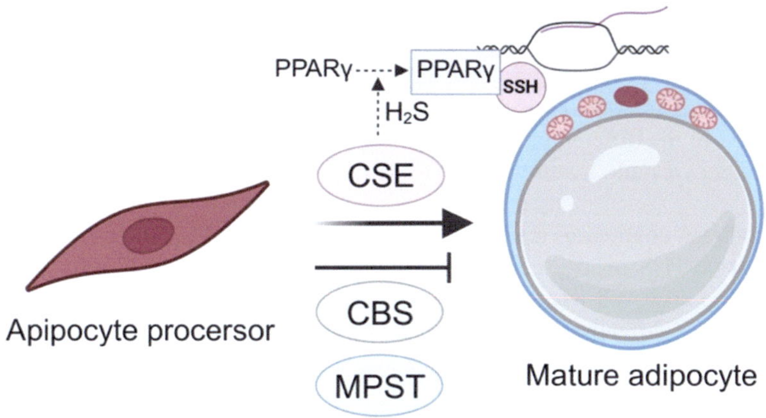

Fig. 2 Regulatory roles of H_2S-generating enzymes in adipogenesis. Cystathionine β-synthase (CBS) and 3-mercaptopyruvate sulfurtransferase (MPST) have been demonstrated to inhibit preadipocyte differentiation, whereas cystathionine γ-lyase (CSE) was shown to promote adipogenesis. CSE-derived H_2S increases peroxisome proliferator-activated receptor-γ (PPARγ) persulfidation, enhancing its activity

adipose tissue, more accurately represent the physiological characteristics of adipose cells in-vivo. As a result, 3T3-L1 and primary preadipocytes can yield differing outcomes in adipogenesis experiments, including variations in their responses to stimuli, gene expression patterns, and metabolic activity. These differences stem from their distinct origins, differentiation capacities, and molecular profiles.

Exogenous H_2S as a Regulator of Adipocyte Differentiation In 3T3-L1 cells, exogenous administration H_2S, via treatment with either inorganic sodium hydrosulfide (NaHS) or the slow-releasing donor GYY4137 (both at 50μM), promoted lipid accumulation during differentiation, as evidenced by Oil Red O staining (Tsai et al. 2015). This treatment also upregulated the expression of key early adipogenic transcription factors, PPARγ and C/EBPα, and increased protein levels of the adipocyte marker FABP4/aP2. These effects were confirmed to be H_2S-specific, as they were not replicated by the structurally related, non-H_2S-releasing compound ZYJ1122 (Tsai et al. 2015). Consistent with these findings, Cai et al. showed that treatment with H_2S-saturated buffer (200 μM) or GYY4137 during differentiation further enhanced lipid accumulation, supporting the pro-adipogenic role of H_2S. Mechanistically, CSE/H_2S activates PPARγ through persulfidation of a critical cysteine residue (C139) in its DNA-binding domain, thereby enhancing the transcription of genes involved in lipid storage (Cai et al. 2016). More recent findings indicate that H_2S can modify every cysteine residue within the DNA-binding domain of PPARγ through persulfidation, which enhances its activation and further stimulates adipogenesis (Yang et al. 2018). Moreover, data from differentiated human adipocytes also support that GYY4137 (5 μM) enhances PPARγ transcriptional activity (Comas et al. 2021a). Interestingly, GYY4137 was also found to inhibit phosphodiesterase (PDE) activity in 3T3-L1 preadipocytes in a dose-dependent manner (1–100 μM), suggesting a secondary mechanism by which H_2S may promote adipogenesis. In support of this, replacing the commonly used synthetic PDE inhibitor, 3-isobutyl-1-methylxanthine (IBMX), with either H_2S or GYY4137 in the differentiation cocktail was sufficient to induce the differentiation of preadipocytes into mature adipocytes (Cai et al. 2016).

In contrast, continuous addition of sodium sulfide (Na_2S, 50μM), which resulted in excessive H_2S exposure, inhibited adipogenesis (Comas et al. 2019). Similar results were observed when 3T3-L1 exposed to a high concentration of GYY4137 (6mM) during differentiation into white adipocytes (Casili et al. 2022). Moreover, two garlic-derived organic sulfides, ajoene and diallyl trisulfide (DATS), which also release H_2S, suppress adipogenesis in cultured 3T3-L1 cells (Ambati et al. 2009; Li et al. 2012).

Taken together, these observations suggest that the effects of exogenous H_2S on adipocyte differentiation depend on the type of sulfide donor, its concentration, and the specific stage of differentiation at which it is added to cells. Moreover, most existing studies have focused primarily on the role of sulfide in white adipocyte differentiation. Further research is needed to determine whether H_2S donors exert similar effects during brown adipocyte differentiation.

3.2 Role of H_2S in Adipocyte Lipid Deposition

Background In addition to hyperplasia, the generation of new fat cells through the differentiation of local precursor cells, adipose tissue can also expand through hypertrophy (Rosen and Spiegelman 2014). Mature adipocytes possess a remarkable capacity for hypertrophic growth, capable of enlarging to several hundred micrometers in diameter (Rutkowski et al. 2015).

During food intake, excess non-esterified fatty acids (NEFAs) enter adipocytes through specific binding and transport proteins such as platelet glycoprotein IV (CD36) and members of the fatty acid transport protein (FATP) family (Large et al. 2004). Once transported into adipocytes, NEFAs are esterified into chemically inert triglycerides (triacylglycerols, TGs) and stored in pre-existing lipid droplets, leading to their enlargement (Konige et al. 2014). Additionally, adipocytes have the ability to synthesize lipids from carbohydrates through the process of de novo lipogenesis, which is catalyzed by the enzymes fatty acid synthase (FASN) and acetyl-CoA carboxylase 1 (Herman et al. 2012). These newly synthesized lipids are also esterified into triglycerides and contribute to the growth of lipid droplets.

Under conditions of increased energy demand, TGs are hydrolyzed through the process of lipolysis into free fatty acids and glycerol (Rutkowski et al. 2015). On the surface of the lipid droplet, there is an abundance of structural proteins and metabolic enzymes that catalyze reactions involved in adipocyte homeostasis. TGs lipolysis occurs through the sequential action of the enzymes: adipose triglyceride lipase

(ATGL), hormone-sensitive lipase (HSL), and monoglyceride lipase (MGL) (Lass et al. 2011). One of the most significant activators of lipolysis is the increased secretion of catecholamines, which bind to β-adrenergic receptors (β-ARs) on adipocytes; activation of β-Ars stimulates lipolysis in adipocytes by increasing cyclic AMP (cAMP) levels, which activate protein kinase A (PKA) triggering a phosphorylation cascade that activates HSL and inactivates perilipin 1 (PLIN1) (Cannon and Nedergaard 2004). Under normal (energy-replete) conditions, PLIN1 protects triglycerides from degradation (Lass et al. 2011). The FFAs released from lipolysis are activated to form acyl-CoA and converted to acyl-carnitine by carnitine palmitoyltransferase I (CPT1) on the outer mitochondrial membrane. Acyl-carnitine is transported into the mitochondria, where CPT2 converts it back into acyl-CoA. Inside mitochondria, acyl-CoA undergoes β-oxidation, producing acetyl-CoA that enters the TCA cycle. This process generates NADH and $FADH_2$, driving ATP production through oxidative phosphorylation. (Rutkowski et al. 2015). Impaired breakdown of triglycerides or fatty acids can lead to increased lipid accumulation and adipocyte hypertrophy.

The Regulatory Role of Endogenous Sulfide in Adipocyte Hypertrophy The impact of H_2S-generating enzymes has been investigated, with most studies focusing on the roles of CSE and MPST. In contrast, knowledge about the role of H_2S-catabolizing enzymes in lipid accumulation remains largely limited to TST. Both in-vitro and in-vivo models have been used in these studies. Moreover, most research has explored the role of sulfide in lipolysis and fatty acid metabolism, while its involvement in the uptake of NEFAs by adipocytes and in lipid synthesis in mature adipocytes remains poorly understood.

Studies on the role of CSE in adipocyte hypertrophy yield controversial results. In differentiated 3T3-L1, pharmacological inhibition of CSE using PAG leads to reduced lipid deposition. PAG also enhanced the lipolytic response to the β3-adrenergic agonist CL316,243 (Tsai et al. 2015). CSE inhibition was associated with increased phosphorylation of PLIN1 and HSL, resulting in elevated basal and isoproterenol-stimulated lipolysis in primary rat adipocytes (Geng et al. 2013). Similarly, PAG alone was found to boost basal lipolysis in adipose tissue (Geng et al. 2013). Despite robust in-vitro evidence supporting a lipid-storing role for the CSE enzyme, in-vivo studies have produced inconsistent results, indicating either an inhibitory or promoting role of CSE in adipocyte hypertrophy; Yang et al. reported that CSE KO mice fed a high-fat diet (HFD) exhibited increased adipocyte size in white adipose tissue, whereas Guo et al. observed reduced adipocyte size under similar conditions. Notably, these studies used independently generated CSE KO mouse models (Yang et al. 2018; Guo et al. 2019). Additionally, PAG administration in WT mice on a HFD resulted in reduced white adipocyte size (Cai et al. 2016). The role of CBS in adipocyte hypertrophy has only been explored via its pharmacological inhibition with aminooxyacetic acid (AOAA) in differentiated 3T3-L1 adipocytes, yielding results comparable to those obtained with PAG (Tsai et al. 2015).

MPST has been shown to exert a beneficial effect against adipocyte hypertrophy. Overexpression of the enzyme in primary differentiated white adipocytes reduces lipid accumulation. Consistently, pharmacological inhibition of MPST using IMST-3 promotes lipid storage in primary cells in the presence of fatty acid supplementation (Katsouda et al. 2022). Mice with global MPST ablation (MPST knockout, MPST KO) exhibit increased adipocyte size in inguinal adipose tissue. Mechanistically, MPST downregulation disrupts mitochondrial protein import, leading to decreased fatty acid metabolism and enhanced lipid accumulation (Katsouda et al. 2022).

Mice with global TST ablation (TST knockout, TST KO) fed a high-fat diet exhibit enlarged white adipocytes (Morton et al. 2016). Preadipocytes isolated from the white adipose tissue of TST KO mice and differentiated in-vitro showed significantly greater late-stage lipid accumulation compared to those from WT mice. The mechanism by which TST regulates adipocyte lipid accumulation involves selectively enhanced mitochondrial function, along with the degradation of reactive oxygen species (ROS) and sulfide (Morton et al. 2016).

Taken together, the findings regarding the role of H_2S-synthesizing enzymes in adipocyte lipid accumulation highlight the distinct physiological effects of these enzymes. The controversial results concerning the role of CSE may arise from the use of different knockout models, variations in diet composition, and differences in the duration of dietary interventions.

Exogenous H_2S as a Regulator of Adipocyte Hypertrophy The overall impact of H_2S donors on adipocyte hypertrophy remains inconsistent across studies. Administration of SG1002 or sodium trisulfide (Na_2S_3) to WT mice fed a HFD has been shown to reduce adipocyte size in the inguinal white adipose tissue. Additionally, Na_2S_3 has been reported to decrease lipid accumulation in primary white adipocytes, particularly in the presence of fatty acid supplementation (Katsouda et al. 2022). GYY4137 reduced the glycerol content in the epididymal white adipose tissue of WT HFD-fed mice (Geng et al. 2013). In-vitro studies suggest that GYY4137 reduces both basal and isoproterenol-stimulated glycerol release in rat white adipocytes, which is associated with decreased phosphorylation of HSL at Ser639 (Geng et al. 2013). GYY4137-derived H_2S has been shown to suppress triglyceride mobilization via persulfidation of PLIN1, a modification that impairs protein's phosphorylation and thereby inhibits HSL recruitment to lipid droplets, ultimately reducing lipolysis (Ding et al. 2020). NaHS has also been reported to reduce lipolysis in the white adipose tissue of WT mice fed a HFD (Ding et al. 2020). Guo et al. demonstrated that NaHS treatment leads to either an increase or a decrease in adipocyte size in the white fat of obese mice, depending on the dosage of the donor (Guo et al. 2019). Additionally, both NaHS and GYY4137 have been shown to increase lipid accumulation in 3T3-L1 adipocytes (Tsai et al. 2015).

These conflicting findings may be attributed to variations in the type and dosage of H_2S donors, the composition of the HFD, the duration of dietary interventions, and the adipose depot. A critical factor is also the experimental design; specifically, whether the H_2S donor was administered in a preventive model, starting from the first day of HFD feeding, or in a therapeutic model, where treatment was initiated after obesity had already been established.

4 Sulfide and Adipose Tissue Insulin Sensitivity

Background Insulin plays a particularly important role in regulating adipocyte metabolism, as it triggers a cascade of events that promote the synthesis and storage of nutrients while inhibiting their breakdown. During food intake, elevated circulating insulin levels bind to the insulin receptor on adipocytes, initiating a signaling cascade that leads to the translocation of glucose transporter 4 (GLUT4) from the cytosol to the plasma membrane. This translocation allows for increased glucose uptake into the cell (Hauner et al. 1988). Once inside the adipocyte, glucose undergoes glycolysis, generating intermediates such as glyceraldehyde 3-phosphate, which serves as a backbone for triglyceride synthesis (Guan et al. 2002). Insulin also acts to suppress lipolysis through various mechanisms. One pathway involves the phosphorylation and subsequent activation of phosphodiesterase 3B (PDE3B) via protein kinase B (PKB or AKT). Through this mechanism, insulin reduces intracellular cAMP levels, thereby preventing the phosphorylation and activation of key enzymes in the lipolytic process (Choi et al. 2010). Additionally, insulin indirectly inhibits lipolysis as a result of the increased glucose influx mediated by GLUT4 activation (Ahmed et al. 2010).

The Regulatory Role of Endogenously Produced Sulfide in Adipose Tissue Insulin Sensitivity Endogenously generated hydrogen sulfide has been shown to exert both beneficial and detrimental effects on insulin sensitivity and glucose uptake. The CSE inhibitors, PAG and β-Cyanoalanine (BCA), have been found to enhance both basal and insulin-stimulated glucose uptake in primary differentiated adipocytes isolated from rat epididymal depot, suggesting that endogenous H_2S may impair glucose uptake (Feng et al. 2009). Similar findings were observed using the 3T3-L1 cell line; CSE inhibition alleviated tumor necrosis factor-α (TNFα)-induced insulin resistance; TNFα, which is overexpressed in the adipose tissue of obese individuals, contributes to insulin resistance by interfering with multiple steps of the insulin signaling cascade (Huang et al. 2013). In vivo, blocking CSE activity upregulated insulin receptor substrate (IRS) expression in adipose tissue isolated from WT mice fed a HFD (Geng et al. 2013). Furthermore, in rats fed a high-fructose diet, elevated CSE expression and increased H_2S production in adipose tissue were associated with reduced insulin-stimulated glucose uptake (Feng et al. 2009). In contrast, in 3T3-L1 adipocytes cultured under high-glucose conditions, vitamin D-induced H_2S production via CSE was shown to enhance both GLUT4 expression and its translocation to the plasma membrane, thereby increasing glucose utilization (Manna and Jain 2012). Adipose tissue isolated from mice with specific overexpression of TST in adipocytes (Ad-TST) showed elevated GLUT4 protein levels, indicating enhanced insulin sensitivity in adipose tissue. Consistently, HFD feeding reduced phospho-AKT levels in control mice but not in Ad-TST mice, further supporting the insulin-sensitizing effect of TST overexpression (Morton et al. 2016).

The roles of CBS, MPST, ETHE1, and SQRDL in regulating insulin sensitivity and glucose uptake in adipose tissue remain poorly understood. MPST-deficient mice fed a HFD have been reported to exhibit reduced whole-body glucose and insulin sensitivity; however, this observation has not been further explored specifically in adipose tissue or isolated adipocytes (Katsouda et al. 2022).

Exogenous H_2S as a Regulator of Adipose Tissue Insulin Sensitivity H_2S (10–1000 μM in solution) or NaHS (at concentrations 100–1,000 μM) reduced insulin-stimulated glucose uptake in primary culture of rat adipocytes (Feng et al. 2009). In contrast, H_2S solution or GYY4137 stimulated glucose uptake by 3T3-L1 adipocytes; H_2S facilitates the conversion of glucose into triglycerides via persulfidation of PPARγ in 3T3-L1 adipocytes (Cai et al. 2016). A protective role for H_2S has also been reported in 3T3-L1 adipocyte cultures exposed to high glucose concentrations (Manna and Jain 2011). In the presence of insulin, NaHS at concentrations of 25μM and 50μM enhanced glucose uptake in 3T3-L1 adipocytes (Xue et al. 2013).

In vivo, administration of NaHS enhanced systemic insulin sensitivity in Goto-Kakizaki rats, a model of type 2 diabetes; this improvement was accompanied by increased phosphorylation of phosphoinositide 3-kinase (PI3K) and AKT in adipose tissue (Xue et al. 2013). More studies investigating the role of GYY4137 and SG1002 in the systemic regulation of insulin resistance in obese animals have been conducted, indicating a beneficial effect. However, no insights into their impact on adipose tissue insulin resistance have been reported (Geng et al. 2013; Katsouda et al. 2022).

5 Sulfide and Adipokine Production

Background Adipocytes have the ability to produce and secrete numerous molecules that play regulatory roles in various physiological functions and enable communication between adipose tissue and other organs (Rosen and Spiegelman 2014). The rich vascularization of adipose tissue allows these secreted products to act in autocrine, paracrine, and endocrine manners. The greatest scientific interest focuses on protein-based molecules, collectively referred to as adipokines (Kershaw and Flier 2004). These substances include cytokines (such as TNFα and interleukin 6 (IL6)), complement-like factors (such as adiponectin and adipsin), chemokines (such as monocyte chemoattractant protein-1 (MCP1)), factors involved in the regulation of food intake (such as leptin and resistin), and angiogenic factors (such as vascular endothelial growth factor (VEGF)).

The Regulatory Role of Endogenously Produced Sulfide in Adipokine Production The role of endogenous sulfide in adipose tissue function has been primarily investigated through its generating enzymes, and current understanding remains limited. In 3T3-L1 cells cultured under high-glucose conditions, overexpression of CSE restored adiponectin secretion while reducing the release of proinflammatory cytokines (Pan et al. 2014). Conversely, inhibition of CSE, CBS, and MPST in human preadipocytes led to an upregulation of inflammatory gene expression (Comas et al. 2021a). In vivo, MPST KO fed a HFD exhibits upregulation of multiple genes related to inflammation and immune response in white adipose tissue. However, it remains unclear whether these changes are due to infiltrating immune cells or if adipocytes themselves (Katsouda et al. 2022).

Exogenous H_2S as a Regulator of Adipokines Production The role of exogenous sulfide in adipokine production remains unknown, with one study demonstrating that NaHS treatment in 3T3-L1 cells reduced MCP1 production while increasing adiponectin synthesis (Pan et al. 2014).

6 Sulfide, Adipose Tissue, and Obesity

Background White adipose tissue (WAT) serves as an energy reservoir that responds to hormonal and energetic cues to support the body's metabolic needs (Rosen and Spiegelman 2014). Energy imbalance involves either increasing energy uptake or decreasing energy expenditure causing adipocyte hypertrophy/hyperplasia, excessive fat expansion, and obesity (Rutkowski et al. 2015). During this expansion, adipose tissue undergoes significant remodeling, resulting in structural and functional alterations within adipose depots, including hypoxia, extracellular matrix remodeling, inflammation, and immune cell infiltration. Hypoxia arises as a critical factor in adipocyte dysfunction, due to inadequate vascularization coupled with increased oxygen demand (Trayhurn 2013; Lee et al. 2014). The activation of hypoxia-inducible factor 1α (HIF1α) contributes to the suppression of β-oxidation, increased fibrosis, and promotion of proinflammatory pathways (Halberg et al. 2009; Krishnan et al. 2012). Simultaneously, excessive deposition of extracellular matrix proteins, particularly type VI collagen, limits adipocyte expansion and exacerbates metabolic stress (Khan et al. 2009; Sun et al. 2013). Obesity also triggers a shift in macrophage phenotype within adipose tissue, from anti-inflammatory M2 to pro-inflammatory M1, thereby amplifying tissue inflammation (Chawla et al. 2011; Oh et al. 2012).

With an estimated >2 billion overweight and obese individuals worldwide, obesity is now recognized as a disease and a major risk factor for the accelerated development of cardiovascular disease, type 2 diabetes, and multiple types of cancers (Williams et al. 2015; Di Cesare et al. 2016). Given the clinical, economic, and societal impact of obesity, unravelling the mechanisms of adipose tissue expansion remains of fundamental significance.

Impact of Endogenous Sulfide on Adipose Tissue Regulation in Obesity The role of H_2S-generating enzymes in adipose tissue expansion during obesity has been investigated in rodent models primarily subjected to high-fat diet conditions. Overexpression of CBS in the paraventricular nucleus in rats fed a HFD led to reduced food uptake, body weight, and fat mass (Zheng et al. 2018). MPST also plays a protective role against white adipose tissue expansion in obesity. MPST KO mice fed a HFD exhibited an early obesity phenotype characterized by increased inguinal white adipose tissue, with no significant differences observed in gonadal adipose tissue or interscapular brown adipose tissue. The underlying mechanism involves activation of HIF-1α and a reduction in fatty acid metabolism (Katsouda et al. 2022). Interestingly, aged MPST KO mice fed a normal chow diet also showed increased body weight and white fat mass compared to age-matched WT controls (Katsouda et al. 2022). MPST deficiency has also been associated with upregulation of inflammatory markers in the adipose tissue of obese mice. Additionally, adipose tissue macrophages (ATMs) contribute to local H_2S production within adipose depots. In obese mice, ATMs exhibit elevated expression of pro-inflammatory M1 markers and a corresponding decrease in H_2S bioavailability (Velmurugan et al. 2015). Studies using genetic deletion or pharmacological inhibition of CSE alongside high-fat diet administration report either a decrease or no change in body weight gain, suggesting that H_2S-generating enzymes may play distinct roles in the pathogenesis of obesity. PAG administration in WT rats fed a HFD resulted in reduced subcutaneous and epididymal white fat mass and inhibited body weight gain (Geng et al. 2013). CSE KO mice on HFD, despite exhibiting white adipocyte hypertrophy, show no significant difference in body weight gain compared to WT controls (Guo et al. 2019). In another study, using a different CSE KO mouse model, a HFD caused a reduction in white fat mass and body weight in CSE-deficient mice (Mani et al. 2015; Yang et al. 2018).

Current knowledge about the role of H_2S-catabolizing enzymes in adipose tissue expansion during obesity is largely limited to TST. In a human cohort, TST mRNA expression was negatively correlated with fat mass (Morton et al. 2016). Transgenic overexpression of TST in adipocytes protected mice from diet-induced obesity, leading to reduced weight gain and white AT mass, without affecting the mass of brown fat. However, global ablation of TST did not affect body weight after 6 weeks of HFD feeding (Morton et al. 2016).

Taken together, these findings suggest a critical role for H_2S-generating enzymes and TST in the pathogenesis and progression of obesity (Fig. 3). The physiological effects of endogenous sulfide have been reported not only in adipose tissue expansion but also in the remodeling processes that occur during obesity, highlighting the potential of targeting H_2S signaling as a therapeutic strategy to improve metabolic health.

Exogenous Sulfide as a Modulator of Adipose Tissue in Obesity Administration of NaHS in obese mice results in either a reduction in HFD-induced body weight gain or no significant effect, depending on the dosage (Wu et al. 2016; Guo et al.

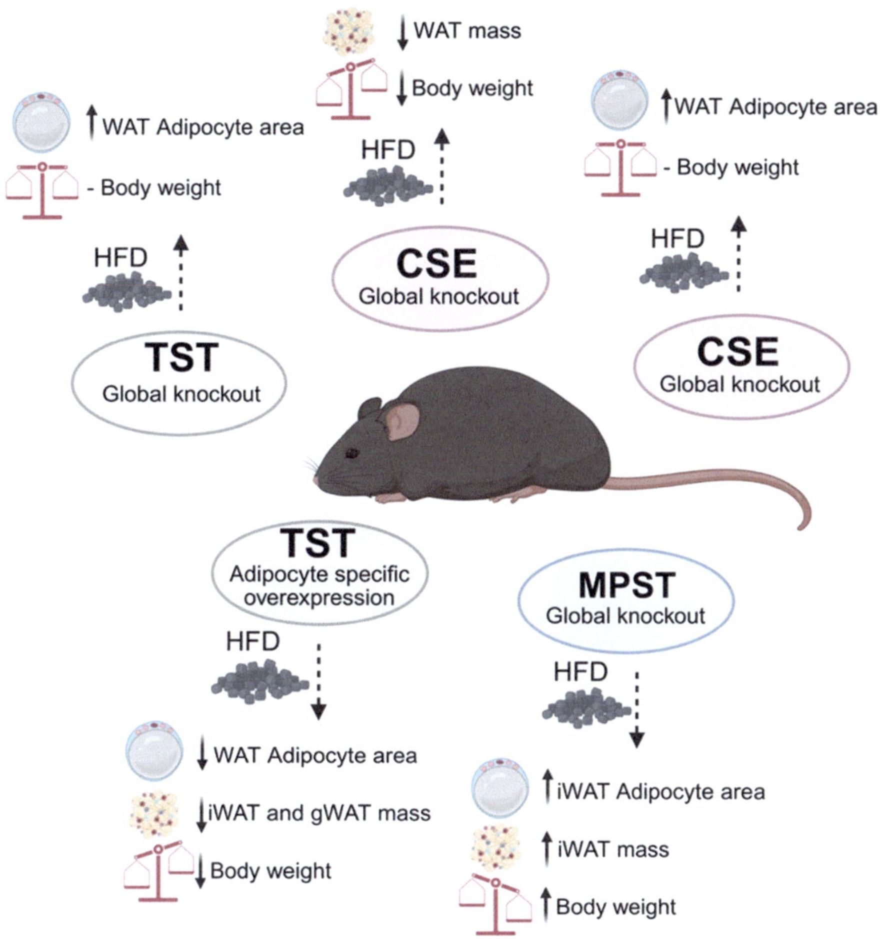

Fig. 3 Changes in the adipose tissue of genetically modified mice lacking or overexpressing CBS, CSE, MPST, and TST. Adipocyte area, adipose tissue mass, and body weight in mice fed a HFD compared to wild-type mice on HFD are depicted. CBS Cystathionine β-synthase, CSE cystathionine γ-lyase, MPST 3-mercaptopyruvate sulfurtransferase, TST thiosulfate sulfurtransferase, gWAT gonadal white adipose tissue, iWAT inguinal white adipose tissue, WAT white adipose tissue, HFD high-fat diet ↑ increased, ↓ decreased, – unchanged

2019). Another study in rats found a reduction in visceral adipose tissue mass but no change in body weight gain (Gomez et al. 2019). Administration of GYY4137 during high-fat diet (HFD) feeding does not lead to changes in body weight in preventive protocols (Geng et al. 2013; Cai et al. 2016). In agreement with this, no differences in white adipose tissue mass were observed following donor administration (Geng et al. 2013; Cai et al. 2016). Administration of SG1002 in a typical model of diet-induced obesity using a 45% HFD has been shown to inhibit white adipose tissue expansion and weight gain (Katsouda et al. 2022). However, SG1002 did not affect weight gain in a separate study using a 60% HFD administered over 24 weeks. Notably, that study primarily focused on cardiac function and did not assess the donor's impact on adipose tissue physiology (Barr et al. 2015). Additionally, the beneficial effects of Na_2S_3 and DATS against adipose tissue expansion and weight gain have also been reported in rodent models of diet-induced obesity (Katsouda et al. 2022; Hu et al. 2022).

Overall, sulfide donors, depending on their specific characteristics, appear to exert either beneficial or neutral effects on adipose tissue expansion and weight gain (Table 1). The discrepancies observed across studies may also be attributed to the selected dose/dosing regimen and/or differences in the animal models of obesity used. Further research is needed to determine whether the positive outcomes observed in preclinical models can be translated into clinical settings and pave the way for novel anti-obesity therapeutic strategies.

7 Sulfide Signaling, Adipose Tissue, and Cardiovascular Health

Background As knowledge of adipose tissue biology continues to expand, the AT is increasingly being recognized as a crucial regulator of cardiovascular health and disease. The AT impacts the cardiovascular system by the secretion of several bioactive products, which include adipocytokines, metabolites, microRNAs (miRNAs), and gaseous messengers, with a wide range of endocrine and paracrine effects on both vessels and the heart (Oikonomou and Antoniades 2019). Notably, metabolic dysregulation leads to a "shift" in AT phenotype and secretome and has been consistently associated with the development of cardiometabolic complications (Oikonomou and Antoniades 2019). Investigating the role of the adipose tissue in cardiovascular disease (CVD) pathogenesis, it has become clear that a complex landscape governs this causal relationship. The heterogeneity in both body fat distribution and the intrinsic properties of different adipose tissue depots translates into differential cardiometabolic risks (Oikonomou and Antoniades 2019). Given their anatomical proximity to the vascular wall, heart perivascular (PVAT) and epicardial (EAT) adipose tissue depots have emerged as a central regulator of vascular and cardiac function (Oikonomou and Antoniades 2019; Antoniades et al. 2023).

As the leading cause of morbidity and mortality worldwide, CVD continues to pose a major challenge to the scientific community. Better understanding the molecular mechanisms that drive this switch, governing the complex interactions of the AT-cardiovascular crosstalk, is critical in pursuing breakthrough therapeutic strategies that can target cardiovascular disease.

Table 1 Impact of sulfide administration on adipose tissue expansion and weight gain in obesity

Animal model	Therapeutic /preventive	Sulfide donor dosage	Duration in diet	Duration of treatment	Phenotype	References
HFD-fed mice (45% kcal fat)	Therapeutic	**NaHS** (50 μmol/kg/day i.p.)	16 weeks	4 weeks	↓ body weight gain	Wu et al. 2016
HFD-fed mice (60% kcal fat)	Preventive	**NaHS** (25 μmol/kg/day, i.p.)	12 weeks	12 weeks	↓ body weight gain ↓ gWAT adipocyte size	Guo et al. 2019
HFD-fed mice (60% kcal fat)	Preventive	**NaHS** (50 μmol/kg/day i.p.)	12 weeks	12 weeks	-body weight gain ↑ gWAT adipocyte size	Guo et al. 2019
HFD-fed rats (38.5% kcal fat)	Therapeutic	**NaHS** (5.6 mmol/kg/day i.p.)	16 weeks	4 weeks	↓ body weight gain ↓VAT mass	Gomez et al. 2019
HFD-fed mice (45% kcal fat)	Preventive	**GYY4137** (200 μmol/kg/day s.c.)	13 weeks	13 weeks	-body weight gain - gWAT and iWAT mass	Geng et al. 2013
HFD-fed mice (45% kcal fat)	Preventive	**GYY4137** (100 μmol/kg/ day, i.p.)	4 weeks	4 weeks	-body weight gain - gWAT and iWAT mass -gWAT adipocyte size	Cai et al. 2016
HFD-fed mice 45% kcal fat)	Therapeutic	**SG1002** (40 mg/kg/day, orally)	20 weeks	10 weeks	↓ body weight gain ↓ iWAT mass -gWAT and iBAT mass ↓ iWAT adipocyte size	Katsouda et al. 2022
HFD-fed mice (60% kcal fat)	Preventive	**SG1002** (20 mg/kg/day, orally)	24 weeks	24 weeks	-body weight gain	Barr et al. 2015
HFD-fed mice (60% kcal fat)	Therapeutic	**SG1002** (20 mg/kg/day, orally)	24 weeks	12 weeks	-body weight gain	Barr et al. 2015
HFD-fed mice (45% kcal fat)	Therapeutic	**Na_2S_3** (5 mg/kg/day, i.p.)	19 week	7 weeks	↓ body weight gain and iWAT mass ↓ adipocyte area in iWAT	Katsouda et al. 2022
HFD-fed mice (60% kcal fat)	Preventive	**DATS** 40 or 80mg/kg/day for 6days/week (g.)	12 weeks	12 weeks,	↓ body weight gain ↓ iWAT and gWAT mass, - iBAT mass ↓ iWAT and gWAT adipocyte size	Hu et al. 2022

HFD High-fat diet, *DATS* diallyl disulfide, *i.p.* intraperitoneal injection, *s.c.* subcutaneous injection, *g.* gavage, *gWAT* gonadal white adipose tissue, *iWAT* inguinal white adipose tissue, *iBAT* intrascapular brown adipose tissue, *VAT* visceral adipose tissue, ↑ increased, ↓ decreased, – unchanged

Hydrogen Sulfide Signaling in Perivascular Adipose Tissue: Implications for Vascular Health Studies investigating the role of sulfide in adipose tissue–cardiovascular system crosstalk have primarily focused on PVAT and vascular health, while the role of H_2S in EAT and its interplay with the heart remains largely unexplored. This is likely due to the fact that most studies are conducted in rodent models; mice do not develop EAT, and rats exhibit only limited development of this fat depot. Additionally, most of the current knowledge regarding the potential regulatory role of sulfide in PVAT comes from studies focusing on endogenous H_2S and the CSE enzyme.

Under normal dietary conditions, CSE expression in rat periaortic adipose tissue has been shown to regulate aortic contractility. Pharmacological inhibition of the enzyme reduces the anticontractile effects of vasoconstrictor agents in an ex vivo aortic ring model with intact adipose tissue (Fang et al. 2009). Subsequent studies confirmed that PVAT also reduces contractility in smaller arteries, such as mesenteric vessels, an effect that is abolished by CSE inhibition and reproduced by NaHS (Schleifenbaum et al. 2010). Both ATP-sensitive potassium (KATP) channels and voltage-gated potassium (KCNQ) channels have been implicated in the endogenous H_2S-mediated anticontractile effect in rodent arteries (Schleifenbaum et al. 2010; Köhn et al. 2012). Interestingly, species differences exist: CSE expression is much lower in mouse PAT compared to rats, and H_2S does not mediate PVAT's effect in mouse aortic rings (Köhn et al. 2012). Notably, CBS inhibition using AOAA did not affect the anticontractile effects of PVAT in rodent models (Köhn et al. 2012).

Obesity alters PVAT function in a time-dependent manner: short-term high-calorie diet enhances H_2S production and PVAT's anticontractile effect, while prolonged exposure impairs this function via reduced CSE expression and H_2S synthesis (Bełtowski and Jamroz-Wísniewska 2017). PVAT reduces serotonin-induced contraction in the gracilis artery of normoglycemic rats, an effect mediated by CSE-derived H_2S and reversed by CSE inhibition under high glucose conditions. In diabetic rats, this anticontractile effect is lost, with PVAT potentially promoting vasoconstriction at higher serotonin concentrations. These findings suggest a protective role of PVAT-derived H_2S in normal glucose metabolism and a detrimental shift in diabetes (Emilova et al. 2015).

Although the regulatory role of H_2S in PVAT–vascular crosstalk has been demonstrated, large parts of this interaction remain unexplored. This includes the roles of MPST, H_2S-catabolizing enzymes, and H_2S donors. Expanding studies on EAT will contribute to a better understanding of sulfide signaling in adipose tissue–cardiovascular system crosstalk. Notably, mapping the changes that H_2S induces in adipose tissue phenotype and secretome can provide clearer insights into whether H_2S in PVAT and EAT can exert beneficial effects on cardiovascular health.

8 Concluding Remarks

In recent years, hydrogen sulfide has gained increasing attention as a physiological signaling molecule involved in the regulation of both health and disease processes. In adipose tissue, H_2S has been shown to influence key processes such as adipogenesis, adipocyte hypertrophy, and insulin sensitivity. Notably, these effects have been linked to broader physiological outcomes, including adipose tissue expansion and the pathogenesis and progression of obesity.

Although significant progress has been made in elucidating the impact of endogenous H_2S system in adipose tissue, many aspects remain unresolved. Most studies have focused on the canonical H_2S-producing enzymes and the catabolizing enzyme TST; however, the physiological roles of ETHE1 and SQRDL are still poorly understood. Current evidence suggests that endogenous H_2S exerts regulatory effects in adipose tissue during obesity, highlighting the distinct and potentially divergent roles of these enzymes. Studies on exogenous H_2S administration have yielded conflicting results. These discrepancies underscore the importance of donor type, concentration, and experimental conditions in determining biological outcomes. The role of H_2S in white adipogenesis has been somewhat explored, yet its influence on brown adipocyte differentiation remains poorly understood. Moreover, data on how H_2S affects the adipokine secretory profile of adipose tissue are still limited. Despite its potential relevance to cardiovascular disease, the role of H_2S in AT-cardiovascular system crosstalk remains largely underexplored.

Given the global rise in obesity and its profound clinical, economic, and societal impact, a deeper understanding of the molecular mechanisms governing adipose tissue function and expansion is of critical importance. Investigating the role of sulfide in adipose tissue biology and physiology may uncover novel therapeutic targets for maintaining metabolic health.

References

Ahmed K, Tunaru S, Tang C et al (2010) An autocrine lactate loop mediates insulin-dependent inhibition of lipolysis through GPR81. Cell Metab 11:311–319. https://doi.org/10.1016/j.cmet.2010.02.012

Ambati S, Yang J-Y, Rayalam S et al (2009) Effects of garlic derivatives on adipocytes 513 Ajoene exerts potent effects in 3T3-L1 adipocytes by inhibiting Adipogenesis and inducing apoptosis i. Phytother Res 23:513–518. https://doi.org/10.1002/ptr

Antoniades C, Tousoulis D, Vavlukis M et al (2023) Perivascular adipose tissue as a source of therapeutic targets and clinical biomarkers. Eur Heart J 44:3827–3844

Barr LA, Shimizu Y, Lambert JP et al (2015) Hydrogen sulfide attenuates high fat diet-induced cardiac dysfunction via the suppression of endoplasmic reticulum stress. Nitric Oxide 46:145–156. https://doi.org/10.1016/j.niox.2014.12.013

Bełtowski J, Jamroz-Wísniewska A (2017) Hydrogen sulfide in the adipose tissue-physiology, pathology and a target for pharmacotherapy. Molecules 22:63. https://doi.org/10.3390/molecules22010063

Berry DC, Stenesen D, Zeve D, Graff JM (2013) The developmental origins of adipose tissue. Development (Cambridge) 140:3939–3949

Cai J, Shi X, Wang H et al (2016) Cystathionine γ lyase-hydrogen sulfide increases peroxisome proliferator-activated receptor γ activity by sulfhydration at C139 site thereby promoting glucose uptake and lipid storage in adipocytes. Biochim Biophys Acta Mol Cell Biol Lipids 1861:419–429. https://doi.org/10.1016/j.bbalip.2016.03.001

Cannon B, Nedergaard J (2004) Brown adipose tissue: function and physiological significance. Physiol Rev 84:277–359

Casili G, Randi E, Panagaki T et al (2022) Inhibition of the 3-mercaptopyruvate sulfurtransferase—hydrogen sulfide system promotes cellular lipid accumulation. Geroscience 44:2271–2289. https://doi.org/10.1007/s11357-022-00600-9

Chawla A, Nguyen KD, Goh YPS (2011) Macrophage-mediated inflammation in metabolic disease. Nat Rev Immunol 11:738–749

Choi SM, Tucker DF, Gross DN et al (2010) Insulin regulates adipocyte lipolysis via an Akt-independent signaling pathway. Mol Cell Biol 30:5009–5020. https://doi.org/10.1128/mcb.00797-10

Chouchani ET, Kazak L, Spiegelman BM (2019) New advances in adaptive thermogenesis: UCP1 and beyond. Cell Metab 29:27–37

Cirino G, Szabo C, Papapetropoulos A (2023) Physiological roles of hydrogen sulfide in mammalian cells, tissues, and organs. Physiol Rev 103:31–276. https://doi.org/10.1152/physrev.00028.2021

Clark ER, Clark EL (1940) Microscopic studies of the new formation of fat in living adult rabbits. Am J Anat 67:255–285. https://doi.org/10.1002/aja.1000670204

Comas F, Latorre J, Cussó O et al (2019) Hydrogen sulfide impacts on inflammation-induced adipocyte dysfunction. Food Chem Toxicol 131. https://doi.org/10.1016/j.fct.2019.05.051

Comas F, Latorre J, Ortega F et al (2021a) Activation of endogenous H2S biosynthesis or supplementation with exogenous H2S enhances adipose tissue Adipogenesis and preserves adipocyte physiology in humans. Antioxid Redox Signal 35:319–340. https://doi.org/10.1089/ars.2020.8206

Comas F, Latorre J, Ortega F et al (2021b) Permanent cystathionine-β-Synthase gene knockdown promotes inflammation and oxidative stress in immortalized human adipose-derived mesenchymal stem cells, enhancing their adipogenic capacity. Redox Biol 42. https://doi.org/10.1016/j.redox.2020.101668

Cook KS, Yeong Min H, Johnson D et al (1974) Adipsin: a circulating serine protease homolog secreted by adipose tissue and sciatic nerve. Science (1979) 237:402–405

David Paton Cuthbertson B, Lionel Tompsett S (1933) The degree of unsaturation of the fats of human adipose tissue in relation to depth from skin surface. Biochem J 27:1103–1106

Di Cesare M, Bentham J, Stevens GA et al (2016) Trends in adult body-mass index in 200 countries from 1975 to 2014: a pooled analysis of 1698 population-based measurement studies with 19.2 million participants. Lancet 387:1377–1396. https://doi.org/10.1016/S0140-6736(16)30054-X

Ding Y, Wang H, Geng B, Xu G (2020) Sulfhydration of perilipin 1 is involved in the inhibitory effects of cystathionine gamma lyase/hydrogen sulfide on adipocyte lipolysis. Biochem Biophys Res Commun 521:786–790. https://doi.org/10.1016/j.bbrc.2019.10.192

Emilova R, Dimitrova D, Mladenov M et al (2015) Cystathionine gamma-lyase of perivascular adipose tissue with reversed regulatory effect in diabetic rat artery. Biotechnol Biotechnol Equip 29:147–151. https://doi.org/10.1080/13102818.2014.991565

Fang L, Zhao J, Chen Y et al (2009) Hydrogen sulfide derived from periadventitial adipose tissue is a vasodilator. J Hypertens 27:2174–2185. https://doi.org/10.1097/HJH.0b013e328330a900

Feng X, Chen Y, Zhao J et al (2009) Hydrogen sulfide from adipose tissue is a novel insulin resistance regulator. Biochem Biophys Res Commun 380:153–159. https://doi.org/10.1016/j.bbrc.2009.01.059

Friedman JM, Halaas JL (1998) Leptin and the regulation of body weight in mammals. Nature 395:763–770

Geng B, Cai B, Liao F et al (2013) Increase or decrease hydrogen sulfide exert opposite lipolysis, but reduce global insulin resistance in high fatty diet induced obese mice. PLoS One 8. https://doi.org/10.1371/journal.pone.0073892

Ghaben AL, Scherer PE (2019) Adipogenesis and metabolic health. Nat Rev Mol Cell Biol 20:242–258

Gomez CB, de la Cruz SH, Medina-Terol GJ et al (2019) Chronic administration of NaHS and L-Cysteine restores cardiovascular changes induced by high-fat diet in rats. Eur J Pharmacol 863. https://doi.org/10.1016/j.ejphar.2019.172707

Guan HP, Li Y, Jensen MV et al (2002) A futile metabolic cycle activated in adipocytes by antidiabetic agents. Nat Med 8:1122–1128. https://doi.org/10.1038/nm780

Guo W, Li D, You V et al (2019) Cystathionine γ-lyase deficiency aggravates obesity-related insulin resistance via FoxO1-dependent hepatic gluconeogenesis. FASEB J 33:4212–4224. https://doi.org/10.1096/fj.201801894R

Halberg N, Khan T, Trujillo ME et al (2009) Hypoxia-inducible factor 1α induces fibrosis and insulin resistance in white adipose tissue. Mol Cell Biol 29:4467–4483. https://doi.org/10.1128/mcb.00192-09

Hauner H, Èhrig R, Spelleken M et al (1988) Development of insulin-responsive glucose uptake and GLUT4 expression in differentiating human adipocyte precursor cells. Int J Obes 22:448–453

Herman MA, Peroni OD, Villoria J et al (2012) A novel ChREBP isoform in adipose tissue regulates systemic glucose metabolism. Nature 484:333–338. https://doi.org/10.1038/nature10986

Hirsch J, Han PW (1969) Cellularity of rat adipose tissue: effects of growth, starvation, and obesity. J Lipid Res 10(1):77–82

Hu Y, Xu J, Gao R et al (2022) Diallyl Trisulfide prevents Adipogenesis and lipogenesis by regulating the transcriptional activation function of KLF15 on PPARγ to ameliorate obesity. Mol Nutr Food Res 66. https://doi.org/10.1002/mnfr.202200173

Huang CY, Yao WF, Wu W-G et al (2013) Endogenous CSE/H2S system mediates TNF-α-induced insulin resistance in 3T3-L1 adipocytes. Cell Biochem Funct 31:468–475. https://doi.org/10.1002/cbf.2920

Iciek M, Kowalczyk-Pachel D, Bilska-Wilkosz A et al (2016) S-sulfhydration as a cellular redox regulation. Biosci Rep 36. https://doi.org/10.1042/BSR20150147

Ishigami M, Hiraki K, Umemura K et al (2009) A source of hydrogen sulfide and a mechanism of its release in the brain. Antioxid Redox Signal 11:205–214. https://doi.org/10.1089/ars.2008.2132

Jackson MR, Melideo SL, Jorns MS (2012) Human sulfide:Quinone oxidoreductase catalyzes the first step in hydrogen sulfide metabolism and produces a sulfane sulfur metabolite. Biochemistry 51:6804–6815. https://doi.org/10.1021/bi300778t

Kabil O, Banerjee R (2014) Enzymology of H2S biogenesis, decay and signaling. Antioxid Redox Signal 20:770–782

Katsouda A, Szabo C, Papapetropoulos A (2018) Reduced adipose tissue H2S in obesity. Pharmacol Res 128:190–199. https://doi.org/10.1016/j.phrs.2017.09.023

Katsouda A, Valakos D, Dionellis VS et al (2022) MPST sulfurtransferase maintains mitochondrial protein import and cellular bioenergetics to attenuate obesity. J Exp Med 219. https://doi.org/10.1084/jem.20211894

Kershaw EE, Flier JS (2004) Adipose tissue as an endocrine organ. J Clin Endocrinol Metab 89:2548–2556. https://doi.org/10.1210/jc.2004-0395

Khan T, Muise ES, Iyengar P et al (2009) Metabolic dysregulation and adipose tissue fibrosis: role of collagen VI. Mol Cell Biol 29:1575–1591. https://doi.org/10.1128/mcb.01300-08

Köhn C, Schleifenbaum J, Szijártó IA et al (2012) Differential effects of cystathionine-γ-lyase-dependent vasodilatory H2S in periadventitial vasoregulation of rat and mouse aortas. PLoS One 7. https://doi.org/10.1371/journal.pone.0041951

Konige M, Wang H, Sztalryd C (2014) Role of adipose specific lipid droplet proteins in maintaining whole body energy homeostasis. Biochim Biophys Acta Mol basis Dis 1842:393–401

Krishnan J, Danzer C, Simka T et al (2012) Dietary obesity-associated hif1α activation in adipocytes restricts fatty acid oxidation and energy expenditure via suppression of the Sirt2-NAD+ system. Genes Dev 26:259–270. https://doi.org/10.1101/gad.180406.111

Lainšček D, Šuštar U, Carter RN et al (2020) Tst gene mediates protection against palmitate-induced inflammation in 3T3-L1 adipocytes. Biochem Biophys Res Commun 527:1008–1013. https://doi.org/10.1016/j.bbrc.2020.05.014

Large V, Peroni O, Letexier D et al (2004) Metabolism of lipids in human white adipocyte (Métabolisme lipidique dans le tissu adipeux blanc humain). Diabetes Metab. https://doi.org/10.1016/s1262-3636(07)70121-0

Lass A, Zimmermann R, Oberer M, Zechner R (2011) Lipolysis – a highly regulated multi-enzyme complex mediates the catabolism of cellular fat stores. Prog Lipid Res 50:14–27

Lee YS, Kim JW, Osborne O et al (2014) Increased adipocyte O2 consumption triggers HIF-1α, causing inflammation and insulin resistance in obesity. Cell 157:1339–1352. https://doi.org/10.1016/j.cell.2014.05.012

Lee AH, Orliaguet L, Youm YH et al (2025) Cysteine depletion triggers adipose tissue thermogenesis and weight loss. Nat Metab. https://doi.org/10.1038/s42255-025-01297-8

Li CK, Huang CY, Chen HW et al (2012) Diallyl trisulfide suppresses the adipogenesis of 3T3-L1 preadipocytes through ERK activation. Food Chem Toxicol 50:478–484. https://doi.org/10.1016/j.fct.2011.11.020

Libiad M, Yadav PK, Vitvitsky V et al (2014) Organization of the human mitochondrial hydrogen sulfide oxidation pathway. J Biol Chem 289:30901–30910. https://doi.org/10.1074/jbc.M114.602664

Liu M, Deng M, Su J et al (2018) Specific downregulation of cystathionine β-synthase expression in the kidney during obesity. Physiol Rep 6:10.14814/phy2.13630

Mani S, Li H, Yang G et al (2015) Deficiency of cystathionine gamma-lyase and hepatic cholesterol accumulation during mouse fatty liver development. Sci Bull (Beijing) 60:336–347. https://doi.org/10.1007/s11434-014-0722-7

Manna P, Jain SK (2011) Hydrogen sulfide and L-cysteine increase phosphatidylinositol 3,4,5-trisphosphate (PIP3) and glucose utilization by inhibiting phosphatase and tensin homolog (PTEN) protein and activating phosphoinositide 3-kinase (PI3K)/serine/threonine protein kinase (AKT)/protein kinase Cς/λ (PKCς/λ) in 3T3l1 adipocytes. J Biol Chem 286:39848–39859. https://doi.org/10.1074/jbc.M111.270884

Manna P, Jain SK (2012) Vitamin D up-regulates glucose transporter 4 (GLUT4) translocation and glucose utilization mediated by cystathionine-γ-lyase (CSE) activation and H2S formation in 3T3L1 adipocytes. J Biol Chem 287:42324–42332. https://doi.org/10.1074/jbc.M112.407833

Melideo SL, Jackson MR, Jorns MS (2014) Biosynthesis of a central intermediate in hydrogen sulfide metabolism by a novel human sulfurtransferase and its yeast ortholog. Biochemistry 53:4739–4753. https://doi.org/10.1021/bi500650h

Mikami Y, Shibuya N, Kimura Y et al (2011) Thioredoxin and dihydrolipoic acid are required for 3-mercaptopyruvate sulfurtransferase to produce hydrogen sulfide. Biochem J 439:479–485

Mohamed-Ali V, Pinkney JH, Coppack SW (1998) Adipose tissue as an endocrine and paracrine organ. Int J Obes 22:1145–1158

Morton NM, Beltram J, Carter RN et al (2016) Genetic identification of thiosulfate sulfurtransferase as an adipocyte-expressed antidiabetic target in mice selected for leanness. Nat Med 22:771–779. https://doi.org/10.1038/nm.4115

Mustafa AK, Gadalla MM, Sen N et al (2009) HS signals through protein S-Sulfhydration. Sci Signal 2. https://doi.org/10.1126/scisignal.2000464

Nagahara N (2013) Regulation of mercaptopyruvate sulfurtransferase activity via intrasubunit and intersubunit redox-sensing switches. Antioxid Redox Signal 19:1792–1802

Nagahara N, Yoshii T, Abe Y, Matsumura T (2007) Thioredoxin-dependent enzymatic activation of mercaptopyruvate sulfurtransferase: an intersubunit disulfide bond serves as a redox switch for activation. J Biol Chem 282:1561–1569. https://doi.org/10.1074/jbc.M605931200

Nagy P (2015) Mechanistic chemical perspective of hydrogen sulfide signaling. In: Methods in enzymology. Academic, pp 3–29
Oh DY, Morinaga H, Talukdar S et al (2012) Increased macrophage migration into adipose tissue in obese mice. Diabetes 61:346–354. https://doi.org/10.2337/db11-0860
Oikonomou EK, Antoniades C (2019) The role of adipose tissue in cardiovascular health and disease. Nat Rev Cardiol 16:83–99
Olson KR (2018) H2S and polysulfide metabolism: conventional and unconventional pathways. Biochem Pharmacol 149:77–90
Pan Z, Wang H, Liu Y et al (2014) Involvement of CSE/H2S in high glucose induced aberrant secretion of adipokines in 3T3-L1 adipocytes. Lipids Health Dis 13. https://doi.org/10.1186/1476-511X-13-155
Rosen ED, MacDougald OA (2006) Adipocyte differentiation from the inside out. Nat Rev Mol Cell Biol 7:885–896
Rosen ED, Spiegelman BM (2014) What we talk about when we talk about fat. Cell 156:20–44
Rosenwald M, Perdikari A, Rülicke T, Wolfrum C (2013) Bi-directional interconversion of brite and white adipocytes. Nat Cell Biol 15:659–667. https://doi.org/10.1038/ncb2740
Rutkowski JM, Stern JH, Scherer PE (2015) The cell biology of fat expansion. J Cell Biol 208:501–512. https://doi.org/10.1083/jcb.201409063
Sanchez-Gurmaches J, Guertin DA (2014) Adipocyte lineages: tracing back the origins of fat. Biochim Biophys Acta Mol basis Dis 1842:340–351
Schleifenbaum J, Köhn C, Voblova N et al (2010) Systemic peripheral artery relaxation by KCNQ channel openers and hydrogen sulfide. J Hypertens 28:1875–1882. https://doi.org/10.1097/HJH.0b013e32833c20d5
Seale P, Bjork B, Yang W et al (2008) PRDM16 controls a brown fat/skeletal muscle switch. Nature 454:961–967. https://doi.org/10.1038/nature07182
Sharp LZ, Shinoda K, Ohno H et al (2012) Human BAT possesses molecular signatures that resemble beige/Brite cells. PLoS One 7. https://doi.org/10.1371/journal.pone.0049452
Shen X, Peter EA, Bir S et al (2012) Analytical measurement of discrete hydrogen sulfide pools in biological specimens. Free Radic Biol Med 52:2276–2283
Singh S, Banerjee R (2011) PLP-dependent H 2S biogenesis. Biochim Biophys Acta Proteins Proteom 1814:1518–1527
Singh S, Padovani D, Leslie RA et al (2009) Relative contributions of cystathionine β-synthase and γ-cystathionase to H2S biogenesis via alternative trans-sulfuration reactions. J Biol Chem 284: 22457–22466. https://doi.org/10.1074/jbc.M109.010868
Spalding KL, Arner E, Westermark PO et al (2008) Dynamics of fat cell turnover in humans. Nature 453:783–787. https://doi.org/10.1038/nature06902
Sun Q, Collins R, Huang S et al (2009) Structural basis for the inhibition mechanism of human cystathionine γ-lyase, an enzyme responsible for the production of H2S. J Biol Chem 284:3076–3085. https://doi.org/10.1074/jbc.M805459200
Sun K, Tordjman J, Clément K, Scherer PE (2013) Fibrosis and adipose tissue dysfunction. Cell Metab 18:470–477
Szabo C (2018) A timeline of hydrogen sulfide (H2S) research: from environmental toxin to biological mediator. Biochem Pharmacol 149:5–19
Tiranti V, Viscomi C, Hildebrandt T et al (2009) Loss of ETHE1, a mitochondrial dioxygenase, causes fatal sulfide toxicity in ethylmalonic encephalopathy. Nat Med 15:200–205. https://doi.org/10.1038/nm.1907
Trayhurn P (2013) Hypoxia and adipose tissue function and dysfunction in obesity. Physiol Rev 93: 1–21. https://doi.org/10.1152/physrev.00017.2012.-The
Tsai CY, Peh MT, Feng W et al (2015) Hydrogen sulfide promotes adipogenesis in 3T3L1 cells. PLoS One 10. https://doi.org/10.1371/journal.pone.0119511
Van Der Lans AAJJ, Hoeks J, Brans B et al (2013) Cold acclimation recruits human brown fat and increases nonshivering thermogenesis. J Clin Invest 123:3395–3403. https://doi.org/10.1172/JCI68993

Velmurugan GV, Huang H, Sun H et al (2015) Depletion of H2S during obesity enhances store-operated Ca2+ entry in adipose tissue macrophages to increase cytokine production. Sci Signal 8. https://doi.org/10.1126/scisignal.aac7135

Verma R, Fu M, Yang G et al (2023) Hydrogen sulfide promotes adipocyte differentiation, hyperplasia, and hypertrophy. Engineering 20:36–48. https://doi.org/10.1016/j.eng.2022.09.010

Wedmann R, Onderka C, Wei S et al (2016) Improved tag-switch method reveals that thioredoxin acts as depersulfidase and controls the intracellular levels of protein persulfidation. Chem Sci 7:3414–3426. https://doi.org/10.1039/c5sc04818d

Williams EP, Mesidor M, Winters K et al (2015) Overweight and obesity: prevalence, consequences, and causes of a growing public health problem. Curr Obes Rep 6:363–370. https://doi.org/10.1007/s13679-015-0169-4

Wu J, Boström P, Sparks LM et al (2012) Beige adipocytes are a distinct type of thermogenic fat cell in mouse and human. Cell 150:366–376. https://doi.org/10.1016/j.cell.2012.05.016

Wu D, Gao B, Li M et al (2016) Hydrogen sulfide mitigates kidney injury in high fat diet-induced obese mice. Oxidative Med Cell Longev 2016. https://doi.org/10.1155/2016/2715718

Xue R, Hao DD, Sun JP et al (2013) Hydrogen sulfide treatment promotes glucose uptake by increasing insulin receptor sensitivity and ameliorates kidney lesions in type 2 diabetes. Antioxid Redox Signal 19:5–23. https://doi.org/10.1089/ars.2012.5024

Yang G, Ju Y, Fu M et al (2018) Cystathionine gamma-lyase/hydrogen sulfide system is essential for adipogenesis and fat mass accumulation in mice. Biochim Biophys Acta Mol Cell Biol Lipids 1863:165–176. https://doi.org/10.1016/j.bbalip.2017.11.008

Zampas P, Li Z, Katsouda A et al (2025) Protective role of 3-mercaptopyruvate sulfurtransferase (MPST) in the development of metabolic syndrome and vascular inflammation. Pharmacol Res 211. https://doi.org/10.1016/j.phrs.2024.107542

Zheng F, Han J, Lu H et al (2018) Cystathionine beta synthase-hydrogen sulfide system in paraventricular nucleus reduced high fatty diet induced obesity and insulin resistance by brain-adipose axis. Biochim Biophys Acta Mol basis Dis 1864:3281–3291. https://doi.org/10.1016/j.bbadis.2018.07.014

H_2S Signaling and SKM Physiopathology

Valentina Vellecco, Martina Smimmo, Veronica Casale, and Mariarosaria Bucci

Contents

Abstract

Hydrogen sulfide (H_2S) is increasingly recognized as gaseous endogenous molecule for its significant role in various physiological processes, behind its historical association with toxicity. Recent studies have highlighted H_2S's cytoprotective properties, including antioxidant, anti-inflammatory, and antifibrotic effects, particularly in the context of skeletal muscle (SKM) health. SKM disorders, such as muscular dystrophy, human malignant hyperthermia, and sarcopenia, lead to severe structural and functional impairments that adversely affect the quality of life. Although limited literature is available on the role of H_2S in SKM physiopathology, it is gaining special interest. Emerging evidence suggests that H_2S may

V. Vellecco · M. Smimmo · V. Casale · M. Bucci (✉)
Department of Pharmacy, School of Medicine and Surgery, University of Naples Federico II, Naples, Italy
e-mail: mrbucci@unina.it

A. Papapetropoulos, C. Szabo (eds.), *Hydrogen Sulfide: A Toxicant with Healing Potential*, Handbook of Experimental Pharmacology 292,
https://doi.org/10.1007/164_2025_779

have a protective role in mitigating muscle damage and dysfunction. This chapter explores the dual functions of H_2S in SKM physiology and pathophysiology, emphasizing its potential therapeutic applications. We propose that H_2S-based strategies may offer promising avenues for alleviating the progression of muscle-related disorders and warrant further investigation to fully elucidate its mechanisms of action.

Keywords

H_2S donors · Myopathies · Persulfidation

1 Introduction

Skeletal muscle (SKM) is a dynamic tissue that accounts for about 40–50% of total body weight. It is essential in supporting body structure, locomotion, oxygen (O_2) consumption, energy metabolism, and substrate turnover and storage, and it is composed of multinucleated myofibers made up of individual sarcomeres. Sarcomeres are the fundamental contractile units of SKM, representing an elegant piece of molecular machinery with a complex structure composed of two main alternating sets of protein filaments: thin filaments (primarily actin) and thick filaments (myosin). This arrangement allows for the sliding filament theory, where actin and myosin filaments slide past each other during muscle contraction (Slavin et al. 2024). Upon the generation of an action potential at the neuromuscular junction, through the binding of acetylcholine to the nicotinic receptor, the depolarization of the muscle cell's sarcolemmal membrane occurs, causing the release of calcium from the sarcoplasmic reticulum, which then leads to the breakdown of ATP, facilitating muscle contraction by shortening the sarcomere. This entire process is known as "excitation-contraction coupling". Contraction of SKM provides stability and power for all body movements. Additionally, SKM possesses metabolic properties; it contributes to basal energy metabolism, serving as the main reservoir of proteins and carbohydrates, produces heat for the maintenance of core temperature, and consumes most of the oxygen and fuel used during physical activity and exercise (Frontera and Ochala 2015). Therefore, any impairment in SKM function leads to some degree of instability or immobility. Muscle function can be impaired due to injury, disease, disuse, or old age. These damaged stimuli have been associated with changes in mitochondrial phenotypes and impaired respiration, leading to reduced energy production, excessive ROS, and increased apoptotic signaling (Adhihetty et al. 2007; Triolo et al. 2022). Such dysfunctional markers further stimulate catabolic signaling that increases muscle degradation, contributing to muscle atrophy and weakness. This, in turn, negatively impacts the quality of life, impairing the ability to perform daily activities and maintain independence. Many advances have been made in understanding the cellular and molecular mechanisms involved in SKM physiopathology to develop novel therapeutic/additive strategies to enhance muscle function and improve the overall quality of life. An essential role for the transsulfuration pathway (TSP), specifically for H_2S, has emerged in this context. This chapter summarizes our current understanding of the role of H_2S in SKM pathophysiology and describes how this new evidence is opening exciting perspective for the use of H_2S donors as a therapeutic/additive strategy for SKM disorders.

2 The Hydrogen Sulfide Pathway and SKM

2.1 H_2S Generation in SKM

The transsulfuration pathway is a sophisticated multi-step reaction system essential for cellular sulfur metabolism and redox regulation. This pathway synthesizes L-methionine from L-cysteine in prokaryotes, fungi, and plants. Conversely, in mammals, the pathway functions in reverse, converting L-methionine into L-cysteine, thus earning the designation of reverse transsulfuration pathway (TSP) (Steegborn et al. 1999). TSP facilitates sulfur transfer from homocysteine to L-cysteine through cystathionine and represents the sole route for L-cysteine biosynthesis. This non-essential sulfur-containing amino acid contributes to several processes within SKM, including protein synthesis, formation of endogenous antioxidants, and the regulation of redox signaling. TSP is ubiquitous, and its end-products are glutathione (GSH), taurine, and hydrogen sulfide (H_2S). H_2S is an endogenous gasotransmitter produced during various steps of the TSP through the enzymatic activities of cystathionine γ-lyase (CSE) and cystathionine β-synthase (CBS) (Cirino et al. 2017; Vellecco et al. 2018). CBS is a cytosolic homotetrameric enzyme of approximately 63 kDa subunits catalyzing the first step of TSP by condensing L-serine with homocysteine to generate cystathionine. Alternatively, CBS can utilize L-cysteine to produce both H_2S and cystathionine or act on two molecules of L-cysteine to form H_2S directly.

CSE uses cystathionine, generated by CBS, to produce L-cysteine (Fig. 1). This enzyme is also a cytosolic homotetramer, with a molecular weight of about 45 kDa (Cirino et al. 2023). Furthermore, H_2S can also be generated through the catabolism of L-cysteine via 3-mercaptopyruvate sulfurtransferase (MPST; 3-MST), which operates in concert with cysteine aminotransferase (CAT).

MPST is expressed in both cytosol and mitochondria, and its primary function is to catalyze the transfer of a sulfur atom from 3-mercaptopyruvate to cyanide, forming pyruvate and thiocyanate (Shibuya et al. 2009). The other sulfur metabolites produced by TSP are GSH, synthesized from L-cysteine by the action of γ-glutamylcysteine synthetase (γ-GCS) and glutathione synthase (GS) enzymes; and L-taurine (Fig. 1), produced from L-cysteine by the subsequent action of other three enzymes named cysteine dioxygenase (CDO), cysteine sulfonic acid decarboxylase (CSAD), and hypotaurine dehydrogenase (HDD). Those sulfur compounds exert several benefits in maintaining SKM health and function such as antioxidant and anti-inflammatory effects and support energy metabolism (Sbodio et al. 2019; Thirupathi et al. 2020). All the TSP enzymes and metabolites were found in the SKM of humans and rodents (Vellecco et al. 2016; Panza et al. 2021). Focusing our

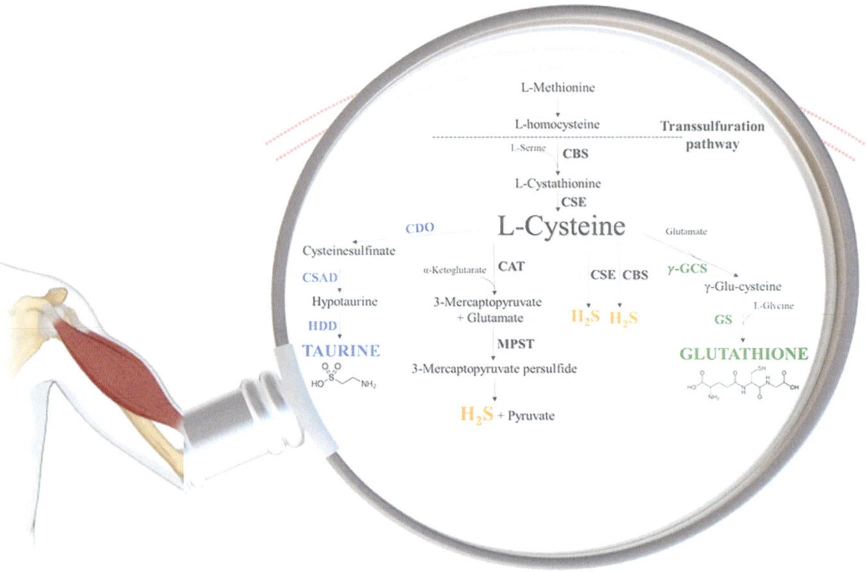

Fig. 1 Schematic overview of the transsulfuration pathway (TSP). In mammals, the amino acid L-methionine is converted into homocysteine. Homocysteine then undergoes two enzymatic reactions to be transformed into L-cysteine. L-cysteine serves as a substrate for the biosynthesis of three major final products: H_2S, taurine, and glutathione. The key enzymes involved in these processes include cysteine aminotransferase (CAT); cystathionine β-synthase (CBS); cysteine dioxygenase (CDO); cysteine sulfonic acid decarboxylase (CSAD); cystathionine-γ-lyase (CSE); glutathione synthase (GS); hypotaurine dehydrogenase (HDD); 3-mercaptopyruvate sulfurtransferase (MPST); gamma-glutamylcysteine synthetase (γ-GCS). This pathway highlights the important biochemical transformations in SKM, linking sulfur amino acid metabolism to the synthesis of key biological molecules that impact antioxidant defense, cellular signaling, and metabolism of muscle fibers

attention on H_2S, it has been demonstrated that in rat SKM homogenate, the production of this gasotransmitter was lower compared to the liver and kidney, which are well-known sources of H_2S in the body. This is due to the lower mRNA expression levels of CBS and CSE in SKM compared to those observed in the kidney and liver. In contrast, the mRNA expression of MPST in SKM is like that in the liver and is twice as high as in the kidney (Du et al. 2013). Using immunohistochemistry approaches, Du and colleagues revealed distinct sub-localizations of the three H_2S-generating enzymes within rat muscle fibers: CBS was primarily localized in the endomysium and perimysium, alongside MPST, while CSE was predominantly found in the cytosol of fibers (Du et al. 2013).

Additionally, human SKM actively produced nanomolar levels of H_2S. The quadriceps muscle expresses comparable CBS and MPST mRNA levels, while CSE mRNA is approximately three times lower than the other two H_2S-producing enzymes. Interestingly, at the protein levels, CBS and CSE are similar, whereas

MPST is about 50% lower, as determined by Western blot analysis (Vellecco et al. 2016).

2.2 H_2S Mechanisms of Action in SKM

Due to its gaseous nature, H_2S can easily move through cell membranes and activate several molecular targets in a receptor-independent manner. Many pathways and proteins are influenced by H_2S (for review see Vellecco et al. 2018; Cirino et al. 2023), among them, the potassium ATP-dependent channels (K_{ATP}), potassium voltage-dependent channels (Kv_7), and phosphodiesterases (PDEs) are the ones that have a recognized role in SKM function. Indeed, H_2S exerts its effect in SKM through (i) the activation of potassium channels leading to the outward flow of potassium ions from the cell. This efflux of positive ions makes the inside of the cell more negative, causing hyperpolarization of the membrane that in turn decreases the excitability of the muscle cell (Zhao et al. 2001; Martelli et al. 2013), (ii) non-selective inhibition of PDEs (Bucci et al. 2010, 2012) therefore increasing the half-life of cyclic nucleotides (cAMP and cGMP); (iii) persulfidation

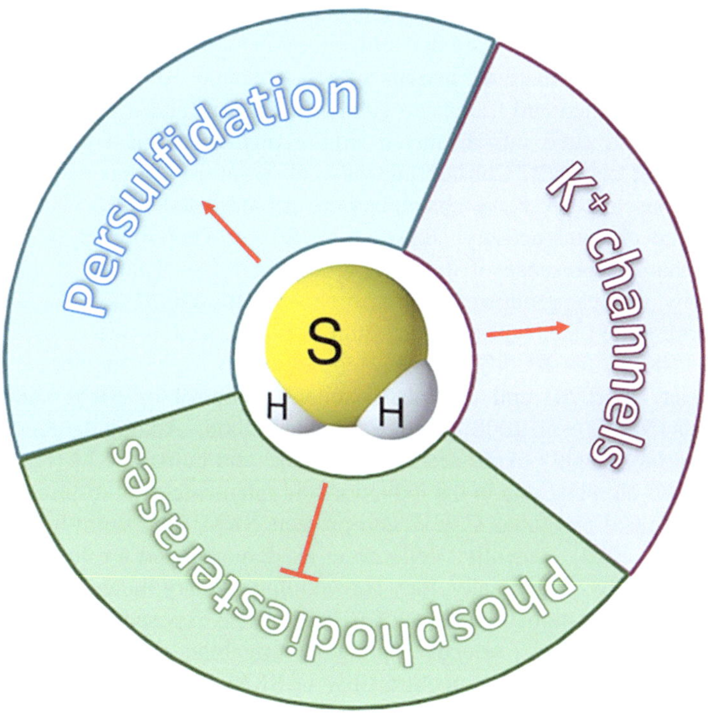

Fig. 2 Key molecular targets of hydrogen sulfide in skeletal muscle physiology. The main molecular targets of H_2S in SKM physiology include potassium channels (such as ATP-sensitive K_{ATP} and voltage-dependent K_{V7} channels), phosphodiesterases (PDEs), and persulfidation processes. Activation is denoted by a solid arrow (↑), while inhibition is indicated by a discontinued arrow (⊥)

(or S-sulfhydration) of proteins, a covalent attachment of an SH group to the thiol side chain of L-cysteine residues that modulate protein function (Vignane and Filipovic 2023). These three events activate downstream signaling cascades that shape SKM responses to H_2S (Fig. 2).

2.2.1 Potassium Channels

K_{ATP} channels are the most abundant K^+ channels expressed in the SKM sarcolemma (Flagg et al. 2010; MacIntosh et al. 2012) providing unique feedback between muscle cell metabolism and electrical activity. During muscle contraction, and especially at rest, K_{ATP} regulates the energy status in SKM. It has been demonstrated that the K_{ATP} channels affect both the basal and insulin-stimulated glucose transport in human SKM (Wasada et al. 2001) and directly influence glucose uptake in mouse SKM (Miki et al. 2002; Kristensen and Juel 2010). Therefore, they can play an important role in the control of contractility, particularly when cellular energetics are compromised, protecting the tissue against calcium overload and fiber damage. It has also been suggested that K_{ATP} channel activation is a physiologically relevant myoprotective mechanism since it protects the muscle against ATP depletion during exercise, and influences recovery of the resting muscle (Matar et al. 2000; Tricarico et al. 2016). Activation of K_{ATP}, which occurs at physiological concentrations of H_2S, is conceivably relevant in SKM although specific studies on this aspect are still missing. An interesting study demonstrated that thiol-dependent redox mechanisms regulate the activity of K_{ATP} channels of SKM fibers in young and old rats (Tricarico and Camerino 1994). Moreover, the L-cysteine exposure of K_{ATP} channels in aged rats displayed similar pharmacological properties to the young adult rat channels. Considering these observations, it has been shown that in smooth muscle cells, K_{ATP} channels were persulfidated by H_2S and this PTM increased the channel activity (Jiang et al. 2010). Therefore, it is possible to speculate that the presence of the thiol groups in K_{ATP} channels is essential for their activity, as these groups can undergo persulfidation by H_2S.

Concerning the other types of potassium channels, such as the subtype voltage-gated $Kv_{7.1}$-$Kv_{7.5}$ (or KCNQ1-5), they contribute to resting membrane potential (Vellecco et al. 2020) and electrical excitation control of SKM and promote myogenesis (Miceli et al. 2008; Roura-Ferrer et al. 2008). Among the Kv_7 channels, $Kv_{7.4}$ is the most highly expressed in both mouse and human SKM (Iannotti et al. 2013); it plays a crucial role in the induction and maintenance of differentiated state in murine skeletal myotubes C2C12 and protects SKM from statin-induced injury (Iannotti et al. 2010). Recently, Vellecco et al. demonstrated a role for $H_2S/Kv_{7.4}$ interaction in SKM. Specifically, they showed that primary human skeletal muscle cells (PHSKMC) or human biopsies of vastus muscle exposed to NAHS and then challenged with the $Kv_{7.4}$ selective activator, retigabine, displayed a paradoxical depolarizing activity, i.e., hypercontractility of SKM, suggesting that an interplay between Kv_7 channels and H_2S occurs in SKM physiopathology (Vellecco et al. 2020). In addition, it has been reported that exposure of human aortic smooth muscle cells to H_2S donors induces membrane hyperpolarization that is reversed by the Kv_7

channel blocker XE-991, reinforcing the hypothesis that the interaction between H_2S and Kv_7 channels has relevance in muscle function (Martelli et al. 2013).

2.2.2 Phosphodiesterases

PDE inhibition is another well-explored molecular mechanism underlying the biological activity of H_2S (Bucci et al. 2010, 2012; Mitidieri et al. 2021). Several PDE isoforms are expressed in SKM and modulate intracellular signaling by degrading cAMP and cGMP (Tetsi et al. 2017; Nio et al. 2017; Vellecco et al. 2022). It is widely recognized that autonomic regulation processes in SKM are predominantly mediated by cAMP-dependent signaling. In several SKM disorders such as muscular dystrophy, age-related atrophy, disuse atrophy, and cancer cachexia, cAMP-dependent pathways are frequently disrupted (Berdeaux and Stewart 2012; Stapleton et al. 2014). Even though the amount of information available on cAMP in SKM is much greater compared to cGMP, it's worth noting that cGMP-dependent signaling regulates complex intracellular communication involving calcium and sphingolipids, which are important for SKM homeostasis (De Palma and Clementi 2012). Additionally, cGMP facilitates the NO-dependent biogenesis of mitochondria by activating a pathway that involves sirtuin-1, peroxisome proliferator-activated receptor-gamma coactivator 1-alpha (PGC1α), nuclear respiratory factor-1 (NRF-1), and mitochondrial transcription factor A (TFAM). This augmented mitochondrial biogenesis increased the ability of SKM to produce ATP through mitochondrial respiration (Nisoli et al. 2004). The importance of the cAMP-cGMP/PDEs axis in SKM function was furnished by a study on mdx mice, a murine model of Duchenne muscular dystrophy (DMD), a genetic myopathy characterized by rapid weakening and functional failure of SKM. Here, it has been shown that the treatment with a combination of PDE5 and PDE4 inhibitors, sildenafil and piclamilast, respectively, resulted in an antifibrotic effect greater than the single treatments (Nio et al. 2017).

Furthermore, studies showed that sildenafil treatment displayed benefits in mdx mice by improving cardiac function, dysregulated in these models (Adamo et al. 2010), and by reducing respiratory muscle weakness (Percival et al. 2012). These positive effects were also confirmed in a clinical study involving boys with DDMD aged 8–13 years (Nelson et al. 2014).

A significant contribution to understanding the relationship between PDE and H_2S in SKM comes from research on PHSKMC obtained from human biopsies of the vastus muscle. Indeed, cells exposed to NaHS significantly increased both cAMP and cGMP levels (Vellecco et al. 2022). Furthermore, this study demonstrated that the exposure of SKM bundles to NaHS, then challenged with the selective PDE5 (sildenafil) or PDE4 (rolipram) inhibitors, significantly increased muscle contraction, leading to a pathological condition known as malignant hyperthermia (Vellecco et al. 2022). Finally, historical data dating back to 1964 by Nichol indicated a substantial reduction in SH groups in the hind leg muscles of dystrophic mice compared to healthy controls. This finding suggests that thiol deficiency may be a characteristic feature of muscular dystrophy (Nichol 1964). Jointly, these pieces of evidence highlight the complex interactions between cyclic nucleotides, H_2S, and thiol status in SKM.

2.2.3 Persulfidation

The third primary mechanism of H_2S action, persulfidation, regulates the activity of several key enzymes, channels, and transcription factors involved in SKM physiology. Indeed, Vellecco and colleagues reported that several PDEs, specifically PDE1, PDE4, PDE5, PDE7, and PDE11, were physiologically persulfidated in human vastus quadriceps (Vellecco et al. 2022). These compelling findings indicate that persulfidation is a physiological phenomenon in SKM, revealing a critical and intricate mechanism through which SKM function is regulated. Persulfidation also influences key metabolic enzymes involved in glycolysis, the citric acid cycle, and fatty acid oxidation. Therefore, by modulating the aforementioned metabolic routes, this PTM regulates energy production, particularly during exercise, and metabolic stress. Additionally, persulfidation acts as a protective mechanism that helps to maintain cellular redox status. Specifically, persulfidation enhances the antioxidant capacity of SKM, which is critical during intense physical activity, where oxygen consumption can increase up to 200-fold, leading to elevated production of reactive oxygen and nitrogen species. Supporting this hypothesis, it has been shown that higher levels of H_2O_2 induce hyperoxidation of L-cysteine, resulting in protein loss of function, and inhibit persulfidation (Zivanovic et al. 2019). Persulfidation may also influence the regenerative capacity of SKM affecting satellite cell function since their activation and differentiation are influenced by redox modifications.

3 The Importance of H_2S in SKM Function: Insights from CSE and CBS Deficiency Studies

The first example of the relevance of H_2S in SKM function has been furnished by $CSE^{-/-}$ mice generated by Ishii and colleagues, as an animal model of cystathioninemia/cystathioninuria. This study shows that mice fed with a diet poor in L-cysteine from 3 weeks of age developed an acute myopathy, associated with a reduction in glutathione levels and intracellular accumulation of two autophagic hallmarks: microtubule-associated protein 1 light chain 3 (LC3) and p62, in skeletal myofibers (Ishii et al. 2010). Indeed, $CSE^{-/-}$ mice displayed progressive paralysis of their lower extremities and severe atrophy in the abdominal regions, the trapezius and rectus femoris muscles. Thereafter, the mice become lame, paralyzed in the upper extremities, and eventually die (Ishii et al. 2010).

Further important confirmation of the pivotal role of CSE-derived H_2S in SKM development and/or function comes from the study by Zhang et al. (2021). Here, it has been demonstrated that CSE acts as the major H_2S-producing enzyme in SKM since the H_2S production rate in $CSE^{-/-}$ mice was reduced by more than 90% compared with WT. These in vivo results were further supported by in vitro experiments on C2C12 cells. Specifically, by using propargylglycine (PAG, a CSE inhibitor), H_2S content was inhibited by more than 86% in undifferentiated C2C12 cells compared to untreated cells. Moreover, this study reported a progressive downregulation of CSE/H_2S signaling in SKM of aged mice. This evidence is coupled with a reduced SKM weight and SKM fiber cross-sectional area, two typical clinical signs of sarcopenia. An age-dependent decrease in CSE expression was observed by comparing WT mice at 10, 26, and 51 weeks. Similarly, H_2S levels, as well as L-cysteine amount, were also significantly lower in SKM from 51-week-old to 10-week-old mice; this decrease was enhanced in $CSE^{-/-}$ mice in both 10 and 51 weeks old (Zhang et al. 2021). The role of CSE-derived H_2S in protecting SKM from age-associated sarcopenia was also confirmed by the finding that the intracellular thiols were much lower in old mice and further decreased in $CSE^{-/-}$ mice in both young and old mice. In addition, the levels of L-cysteine oxidation, an index of protein inactivation, were severe in aged mice and again CSE deficiency promoted the L-cysteine residue oxidation within proteins (Zhang et al. 2021). This latter finding strongly suggests that: (i) being an antioxidant molecule, H_2S exerts a protective effect on SKM maintaining the physiological redox status; (ii) the impairment of the CSE/H_2S signaling in SKM disturbed the redox homeostasis resulting in protein oxidation. This event damages protein activity leading to the generation of ER stress that further ends with cell death and muscle wasting. This hypothesis is also confirmed by the finding that: (i) dietary supplementation of cystine ameliorated muscle structure and function in old mice (Sinha-Hikim et al. 2013); (ii) glucoraphanin, a natural H_2S-releasing compound derived from Brassicaceae, improves sarcopenia in vitro in C2C12-derived myotubes exposed to dexamethasone (DEX). In detail, it has been reported that glucoraphanin reduced the DEX-induced rise in O^{2-} levels and protein carbonylation, the major hallmarks of oxidative-stress-related disorders (Micheli et al. 2022). Furthermore, the cytoprotective effects of the H_2S-releasing molecule were also highlighted in myotubes since DEX treatment leads to an impairment of morphology, decreasing the diameter and the formation of mature myotubes, whereas glucoraphanin was able to restore the alterations observed in myotube diameter and the number of multinucleated cells (Micheli et al. 2022).

Regarding the contribution of CBS-derived H_2S in preserving SKM function, most of the published data has been furnished by hyperhomocysteinemia (HHcy). HHcy is characterized by CBS deficit/lack with a consequent reduction of homocysteine conversion into L-cysteine, thereby resulting in an impairment of TSP by-end products (H_2S, GSH, and taurine) coupled with an increase of homocysteine. Indeed, inefficient H_2S biosynthesis was found in patients with CBS gene mutation (Beard and Bearden 2011). Similarly, heterozygous CBS ($CBS^{+/-}$) mice displayed lower H_2S levels in their blood and showed oxidative stress-mediated damage (Ishihara et al. 2008; John et al. 2017). HHcy represents a risk factor for several human diseases, and patients with HHcy manifest many characteristic features (early thrombotic events, cognitive decline) including a functional deficit in SKM (Veeranki and Tyagi 2013; Morris et al. 2017). Specifically, the SKM morphological abnormalities caused by HHcy are the disruption of Z-disks and disorganized myofilaments with excessive collagen deposition in the basal lamina (Kanwar et al. 1976; Kalra et al. 1985; Veeranki and Tyagi 2015). These findings are also confirmed in a preclinical setting, where it was found that chronic administration of homocysteine to rats caused a decrease in cellular metabolic activity and led to an energy imbalance in the gastrocnemius (Kolling et al. 2013).

In this context, studies performed on heterozygous CBS-deficient ($CBS^{+/-}$) mice are of particular interest. These animals represent a useful model of HHcy since they have less pronounced hyperhomocysteinemia than $CBS^{-/-}$ mice, thereby living longer (Familtseva et al. 2014; Nandi and Mishra 2017; Majumder et al. 2019). In the swim test, $CBS^{+/-}$ mice exhibited lower performance, showing less fatigue resistance and reduced contractile force. This effect was not due to any noticeable changes in muscle structure, but rather to a decrease in the number of large muscle fibers (Veeranki et al. 2015). In addition, SKM harvested from $CBS^{+/-}$ displayed lower ATP levels, dystrophin, and mitochondrial transcription factor A (mtTFA) compared to WT littermates (Veeranki et al. 2015). Furthermore, CBS deficiency caused redox imbalance and ER stress in SKM in vivo as reported by Majumder et al. (2018). In this study, it was demonstrated that HHcy had negative effects on SKM, especially the gastrocnemius and quadriceps muscles, leading to muscle weight loss and atrophy. Based on the available data, it is unclear whether the damaging effect of HHcy on SKM is caused by the direct toxic impact of high levels of Hcy or it could be attributed to lower levels of H_2S, leading to a minimal activation of H_2S molecular targets. Part of the same report, using an exogenous source of H_2S, answered the question. Indeed, the administration of an H_2S donor, like NaHS, was able to reduce SKM injuries caused by HHcy-induced oxidative and ER stress responses, as it notably enhanced grip strength and swimming performance (Majumder et al. 2018). By translating the in vivo results to an in vitro setting using the C2C12 myoblast cell line, it was found that the increased ER stress and the activation of the JNK pathway were involved in the development of muscle atrophy observed when CBS was absent (Majumder et al. 2018). These findings were also confirmed by a recent report where it has been demonstrated that the treatment of $CBS^{+/-}$ with exogenous H_2S improves SKM performance. Specifically, the creatine kinase, a marker of injury, was higher in $CBS^{+/-}$. In contrast, musclin, a myokine produced by SKM to increase the mitochondrial oxidative capacity during exercise augmenting physical endurance, was significantly abated. The H_2S treatment ameliorated the multi-organ damage in $CBS^{+/-}$ mice increasing the musclin levels. Additionally, the H_2S treatment reversed the increased level of MMP2 and collagen fibrosis detected in $CBS^{+/-}$ mice resulting in an amelioration of muscle strength (Singh et al. 2021).

4 H_2S and Therapy of SKM Disorders

H_2S possesses anti-inflammatory, anti-apoptotic, and antioxidant properties, making it a promising candidate for treating various SKM diseases. Given the wide range of protective effects associated with H_2S, it is surprising that its role in SKM pathology has not been more thoroughly investigated. Indeed, only a few studies detail the molecular mechanisms underlying the H_2S activity in SKM disorders. Recently, it has been shown that the H_2S pathway is defective in both murine and human DMD (Ellwood et al. 2021; Panza et al. 2021). Specifically, Panza and coauthors demonstrated a dysregulated TSP pathway in terms of both mRNA and metabolites in the mdx mouse model of DMD; meanwhile, the study performed by Ellwood and colleagues reported a potent decrease in total sulfide, 3-MST, and CSE levels in dystrophin/utrophin double knockout mice (a more severe model of DMD). Additionally, in an in vitro model of sarcopenia, both H_2S content and CBS and CSE expression were significantly decreased (Micheli et al. 2022). In line with this observation, the muscles of elderly mice displayed a significant reduction of L-cysteine content (Zhang et al. 2021). Moreover, a diminished concentration of endogenous H_2S was observed in the SKM of patients with critical limb ischemia, along with a decrease in NRF2, an essential cytoprotective factor, leading to increased oxidative stress (Islam et al. 2015).

In contrast, CBS-derived H_2S biosynthesis is increased in malignant hyperthermia (MH), a human syndrome marked by abnormal hypercontractility of SKM following exposure to volatile anesthetics (Vellecco et al. 2016). The augmented H_2S content, persulfidating $K_{v7.4}$ channel and PDEs within SKM of susceptible patients (MHS), contributes to the pathological hypercontractility observed in this condition (Vellecco et al. 2020, 2022). Specifically, Vellecco and colleagues, by performing functional studies (IVCT) on human SKM, reported that exposing MHS biopsies to retigabine, a selective activator of K_{v7} channels, increased the SKM contracture beyond 0.2 g of tension, which is the threshold of tension used for diagnosing MH. Additionally, electrophysiological experiments on MHS-derived PHSKM revealed anomalous depolarizing activity of the $K_{v7.4}$ channel due to H_2S-derived persulfidation that changes the Kv7.4 activity from hyperpolarizing to depolarizing one (Vellecco et al. 2020). The results concerning the role of phosphodiesterases (PDEs) in MH clearly align with those observed for Kv7. Indeed, in MHS biopsies an increased level of cyclic nucleotides was found, coupled with an augmented PDEs persulfidation. Furthermore, the IVCT bioassay conducted on MHS samples demonstrated that both sildenafil and rolipram induce a remarkable increase in muscle tension, surpassing the physiological threshold of 0.2 g (Vellecco et al. 2022). Despite the two studies evidence the crucial role of H_2S-derived persulfidation in MH, it is essential to highlight a critical difference between the two targets of persulfidation: in healthy human SKM, $K_{v7.4}$ is not persulfidated indicating that this specific PTM is directly tied to the pathological condition, i.e., MHS. In contrast, in physiological conditions, persulfidation regulates PDEs activity and this PTM is markedly enhanced in pathological states. This evidence highlights the role of persulfidation, not only as a physiological PTM essential in modulating protein function but also as a potential biomarker in SKM pathologies. Based on these findings, we can conclude that the H_2S pathway is impaired in SKM diseases characterized by atrophy and weakness, such as DMD, sarcopenia, and aging. Conversely, in SKM conditions associated with hypercontractility, such as MH, the pathway is upregulated.

Most studies have focused on the role of H_2S in metabolic-related SKM diseases such as obesity and diabetes. It is well-established that SKM plays a crucial role in glucose uptake and storage, accounting for 80% of insulin-mediated glucose disposal. Consequently, insulin resistance (IR) in SKM significantly contributes to the progression of metabolic disturbances. For example, Goto Kakizaki (GK) rats (an animal model of type-2 diabetes) displayed decreased systemic and muscle H_2S bioavailability due to a diminished expression of the mRNA encoding for the three H_2S-producing enzymes, when compared to their corresponding control (Bitar et al. 2018). The H_2S impairment was associated with a state of heightened oxidative stress in SKM of GK rats together with an upregulation in myostatin expression and the ubiquitin/proteasome system typified by the FoxO1/MuRF1/atrogin-1 signaling pathway. These biochemical aberrations were associated with muscle loss and deficit in muscle function (i.e., sarcopenia) that were identified as a complication in diabetic patients (Bitar et al. 2018). NaHS-treated diabetic animals displayed increased muscle mass and strength, along with reduced myostatin levels compared to diabetic rats. Furthermore, NaHS administration led to lower levels of superoxide and H_2O_2 in the SKM membrane and mitochondrial fractions, as well as improved antioxidant capacity indicated by the GSH/GSSG ratio (Bitar et al. 2018). To confirm the molecular mechanisms proposed in the in vivo study, C2C12 t cells were exposed to high glucose (HG), and a significant reduction in CSE expression (but not of CBS or 3-MST) was found. Following exogenous H_2S administration (i.e., NaHS), an increase in GSH levels and a decrease in ROS generation were observed (Parsanathan and Jain 2018). Conversely, the knockdown of CSE with siRNA resulted in opposing effects, showing increased H_2O_2 generation and reduced expression of enzymes involved in the GSH biosynthesis pathway. Thus, CSE-derived H_2S appears to be a modulator of oxidative balance in myoblasts and SKM. Another study conducted in a murine model of type II diabetes (db/db mice) postulated that H_2S provides SKM-protective effects via the persulfidation of muscle RING finger 1 (MuRF1) (Lu et al. 2020). MuRF1 is an E3 ubiquitin ligase, and its elevated expression contributes to muscle mass loss in diabetes. db/db mice treated with NaHS displayed diminished SKM atrophy, ROS production, and degradation of myomesin-1 (MYOM1) and myosin heavy chain 4 (MYH4). In vitro studies on C2C12 myoblasts exposed to HG and palmitate showed that H_2S, by persulfidating the Cys44, modified MuRF1, decreasing the ubiquitination of MYOM1 and MYH4. This mechanism in turn resulted in diminished SKM atrophy (Lu et al. 2020).

Recently, the role of CSE-derived H_2S has been investigated in metabolic disorders of SKM induced by a high-fat diet (HFD). In SKM harvested from mice fed with HFD, a downregulation of CSE was found (Wang et al. 2017; Parsanathan and Jain 2018, 2019; Lu et al. 2024). An interesting observation that links CSE, SKM, and HFD was made using SKM-specific $CSE^{-/-}$ mice ($Cse^{\Delta skm}$) and their littermate controls with floxed Cse ($Cse^{f/f}$). $Cse^{\Delta skm}$ mice fed with a standard diet exhibited similar patterns to those observed in $Cse^{f/f}$ mice fed with HFD, in terms of energy metabolism (glycolysis, tricarboxylic acid cycle, fatty acid β-oxidation, amino acid metabolism), mitochondrial morphology, oxidative stress, and signaling pathways linked to insulin resistance. Notably, in $Cse^{\Delta skm}$ mice fed with HFD, exacerbated oxidative stress, mitochondrial dysfunction, and lipid peroxidation were observed. Furthermore, $Cse^{\Delta skm}$ mice fed with an HFD showed increased body weight gain, elevated fasting blood glucose, higher plasma insulin levels, greater IR, and reduced GLUT-4 expression in SKM compared to their littermate controls. Therefore, the CSE deficiency in SKM contributed to more severe obesity and IR after 12 weeks of HFD, which was attributed to significant disturbances in energy metabolism, mitochondrial dysfunction, and processes related to ferroptosis (Lu et al. 2024).

Based on these findings, H_2S donors show promising potential as therapeutic options for treating various SKM diseases, as well as other SKM dysfunctions related to conditions such as diabetes and obesity, cancer-related cachexia, statin-induced myopathies, multiple sclerosis, and corticosteroid-induced myopathies, among others. Indeed, it has been discovered that H2S treatment could alleviate SKM damage due to ischemia-reperfusion injury, diabetes, and DMD (Du et al. 2013; Lu et al. 2020; Yang et al. 2020), and DMD (Ellwood et al. 2021; Panza et al. 2021) even though the underlying molecular mechanisms have yet to be elucidated. Interestingly, in a murine model of SKM contusion, it was demonstrated a beneficial effect induced by an H_2S donor (Zhao et al. 2020). Mice were treated daily with NaHS (28 mmol/kg) and after 2 weeks of treatment, the gastrocnemius muscles were collected. Molecular analysis revealed a noticeable reduction in SKM fibrosis in NaHS-treated mice, evidenced by decreased mRNA levels of Col1a1, Col1a3, and TGFβ, as well as reduced collagen content measured through trichrome staining and the regulation of MMPs. Additionally, H_2S treatment led to a decrease in the expression of TGF-β, pro-inflammatory cytokines, chemokines, and markers of oxidative stress, as indicated by decreased expression of the key NADPH oxidase subunit, and gp91phox. This effect was linked to a reduction in M1 CD68 macrophages counterbalanced by an increase in M2 CD206 (Zhao et al. 2020). It's worth noting that this study, despite highlighting the multifaceted role of H_2S in mitigating SKM injury, is mainly descriptive without any hypothesis of mechanism of action. The protective effect of H_2S against fibrosis was also observed in the diaphragms of a rat model of streptozotocin-induced diabetes (Yang et al. 2020). NaHS treatment reduced collagen deposition and slightly decreased the content of pro-fibrotic markers, such as Col1a1, Col1a3, and TGFβ, along with a reduction in pro-inflammatory cytokines (IL-1β, IL-6, IL-18, and TNF-α). Additionally, NaHS also suppressed the activation of the NLRP3 inflammasome. These effects contributed to improving the biomechanical properties of the diaphragm (Yang et al. 2020). The anti-fibrotic effect of H_2S supplementation was also described in other SKM disorders. For example, Cain et al. found that the H_2S donor, SG1002, attenuated fibrosis in the gastrocnemius and diaphragm in humanized dystrophic mice (Cain et al. 2019). Similar results were obtained by using NaHS in mdx mice (Panza et al. 2021). Indeed, NaHS administration, in both short-term (2 weeks) and long-term (12 weeks) treatments, reduced fibrosis, necrosis, and inflammatory cell infiltration. Furthermore, the H_2S supplementation reactivated the autophagy process. Specifically, in the quadriceps of mdx mice, a decrease in the levels of autophagy regulators (Atg3, Atg7, Atg12, and Ulk1) was observed, alongside an elevated expression of pro-inflammatory cytokines (IL-1β, IL-6, and Tnfα) and the pro-fibrotic factor TGF-β. Following NaHS treatment, these markers were largely restored to physiological levels. These biochemical improvements resulted in the prevention of locomotor activity loss, as demonstrated by the rotarod and weight tests (Panza et al. 2021). To further confirm the therapeutic potential of H_2S donors in SKM diseases, an interesting study performed on the C. elegans model of DMD reported that the supplementation with H_2S donors improved motor skills, contractile force, and the structure of muscle mitochondria (Ellwood et al. 2021). Although the supplement with H_2S did not improve the overall lifespan of the mutated worms, it delayed muscle cell death, indicating an enhancement in health span. Interestingly, in this study, two different H_2S donors were used: the slow-release sodium compound, GYY4137 (NaGYY), and the mitochondria-targeted donor, AP39. Even though both donors ameliorated DMD, GYY4137 exhibited its SKM benefits at a significantly higher concentration (100 μM) compared to AP 39 (100 pM), suggesting that the positive effects observed are mitochondria-based (Ellwood et al. 2021).

The treatment with either slow (GYY4137) or rapid (NaHS) H_2S-releasing donors showed protective effects also in mice against immobilization-induced muscle fibrosis and atrophy. Both compounds inhibited inflammation and oxidative stress associated with immobilization in SKM, highlighting its potential as a therapeutic agent in managing muscle degeneration due to immobilization, therefore supporting its use in patients with disuse-associated muscle atrophy.

An additional interesting aspect of these H_2S donors in SKM comes from the study conducted by Zhang et al. (2021). This intriguing study showed that H_2S can stimulate SKM regeneration. Specifically, they found that cardiotoxin-induced injury and age-related sarcopenia were exacerbated in $CSE^{-/-}$ mice, while NaHS treatment enhanced myogenesis. The mechanism underlying this effect involved the promotion of heterodimer formation between myogenic factors, specifically myocyte enhancer factor 2 (MEF2c) and myogenic regulatory factor-4 (MRF4), facilitating their binding to the myogenin promoter. Therefore, increasing H_2S levels could be considered a potential treatment for muscle injuries and a preventive measure against age-related sarcopenia (Zhang et al. 2021). Overall, these findings indicate that various H_2S donors hold promising potential for H_2S-based therapies. However, further research is needed to fully clarify the molecular mechanisms through which H_2S exerts its beneficial effects in different SKM diseases. A deeper understanding of these pathways will be essential for developing targeted therapies. Although preclinical studies have shown promising results on H_2S donors in SKM dysfunctions, their effectiveness and safety in human patients have not yet been evaluated therefore clinical trials are necessary to validate these findings in clinical practice (Table 1).

Table 1 Different H_2S donors: natural and synthetic sources with varying release kinetics used in SKM disorders

H_2S donor	Full name	Formula	Chemical features	Release kinetics
NaHS	Sodium hydrosulfide	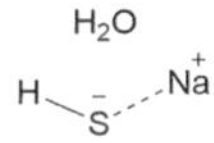	Sulfide salt	Fast
GYY4137	(4Methoxyphenyl) (morpholino) phosphinodithioic acid	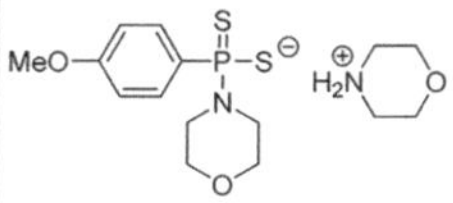	Derivative of Lawesson's reagent	Slow
AP39	10-oxo-10-(4-(3-thioxo-3*H*-1,2-dithiol-5yl)phenoxy)decyl) triphenylphosphonium bromide	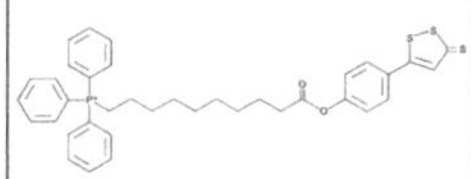	1,2-Dithiole-3-thione (DTT) derivative	Slow
SG1002	Sodium polysulthionate	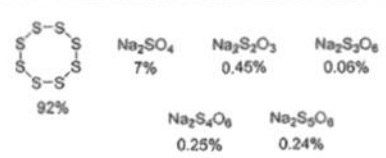	Polysulfur mixture	Slow
Glucoraphanin	4-methylsulfinylbutyl glucosinolate	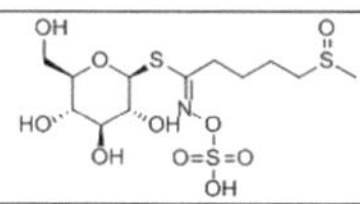	Natural	Slow

This table presents various H_2S donors, categorized into fast and slow groups based on their release kinetics, as used in studies on SKM. The donors differ in their mechanisms and sources of H_2S release, classified as either natural or synthetic. Natural donors include compounds such as glucoraphanin, while synthetic donors consist of substances like NaHS, AP39, GYY4137, and SG1002. Understanding the release kinetics of each donor is essential for assessing their suitability in different experimental settings and SKM biological contexts

5 Concluding Remarks

The discovery of H_2S as a key player in SKM pathophysiology and, thus, the possibility of using H_2S-based approaches in SKM diseases, opens new avenues for its application as a therapeutic molecule. The potential use of H_2S donors not only in muscular dystrophy but all SKM diseases characterized by wasting and atrophy raises several important considerations. Firstly, H_2S is vital for various physiological functions in SKM, influencing muscle contraction and enhancing mitochondrial biogenesis and function, both critical for sustaining energy production. Its antioxidant and anti-inflammatory properties benefit SKM function, muscle repair, and recovery. Additionally, H_2S offers protective effects against muscle atrophy by regulating pathways involved in protein synthesis and degradation, including the ubiquitin-proteasome pathway and autophagy. Theoretically, H_2S, due to its pleiotropic actions and gaseous nature, could serve as an example of systems pharmacology. This emerging alternative approach to traditional pharmacology is based on the concept that various stimuli can activate multiple signaling pathways that converge on the same downstream targets, leading to similar cellular functions (Martelli et al. 2023).

In conclusion, despite the numerous variables involved in human clinical studies, H_2S-based therapies have the potential to be valuable tools for managing complex diseases, including skeletal muscle disorders, cardiovascular diseases, and cancer. These therapies may be effective on their own or in combination with existing treatments, offering enhanced therapeutic benefits or reduced risks.

References

Adamo CM, Dai D-F, Percival JM et al (2010) Sildenafil reverses cardiac dysfunction in the mdx mouse model of Duchenne muscular dystrophy. Proc Natl Acad Sci U S A 107:19079–19083. https://doi.org/10.1073/pnas.1013077107

Adhihetty PJ, O'Leary MFN, Chabi B et al (2007) Effect of denervation on mitochondrially mediated apoptosis in skeletal muscle. J Appl Physiol (1985) 102:1143–1151. https://doi.org/10.1152/japplphysiol.00768.2006

Beard RS, Bearden SE (2011) Vascular complications of cystathionine β-synthase deficiency: future directions for homocysteine-to-hydrogen sulfide research. Am J Physiol Heart Circ Physiol 300:H13–H26. https://doi.org/10.1152/ajpheart.00598.2010

Berdeaux R, Stewart R (2012) cAMP signaling in skeletal muscle adaptation: hypertrophy, metabolism, and regeneration. Am J Physiol Endocrinol Metab 303:E1–E17. https://doi.org/10.1152/ajpendo.00555.2011

Bitar MS, Nader J, Al-Ali W et al (2018) Hydrogen sulfide donor NaHS improves metabolism and reduces muscle atrophy in type 2 diabetes: implication for understanding sarcopenic pathophysiology. Oxid Med Cell Longev 2018:6825452. https://doi.org/10.1155/2018/6825452

Bucci M, Papapetropoulos A, Vellecco V et al (2010) Hydrogen sulfide is an endogenous inhibitor of phosphodiesterase activity. Arterioscler Thromb Vasc Biol 30:1998–2004. https://doi.org/10.1161/ATVBAHA.110.209783

Bucci M, Papapetropoulos A, Vellecco V et al (2012) cGMP-dependent protein kinase contributes to hydrogen sulfide-stimulated vasorelaxation. PLoS One 7:e53319. https://doi.org/10.1371/journal.pone.0053319

Cain C, Devarakonda T, Thompson J et al (2019) Prevention and treatment of Duchenne cardiomyopathy with hydrogen sulfide-donor therapy. FASEB J 33. https://doi.org/10.1096/fasebj.2019.33.1_supplement.831.5

Cirino G, Vellecco V, Bucci M (2017) Nitric oxide and hydrogen sulfide: the gasotransmitter paradigm of the vascular system. Br J Pharmacol 174:4021–4031. https://doi.org/10.1111/bph.13815

Cirino G, Szabo C, Papapetropoulos A (2023) Physiological roles of hydrogen sulfide in mammalian cells, tissues, and organs. Physiol Rev 103:31–276. https://doi.org/10.1152/physrev.00028.2021

De Palma C, Clementi E (2012) Nitric oxide in myogenesis and therapeutic muscle repair. Mol Neurobiol 46:682–692. https://doi.org/10.1007/s12035-012-8311-8

Du J, Li W, Yang J et al (2013) Hydrogen sulfide is endogenously generated in rat skeletal muscle and exerts a protective effect against oxidative stress. Chin Med J 126:930–936

Ellwood RA, Hewitt JE, Torregrossa R et al (2021) Mitochondrial hydrogen sulfide supplementation improves health in the C. elegans Duchenne muscular dystrophy model. Proc Natl Acad Sci USA 118:e2018342118. https://doi.org/10.1073/pnas.2018342118

Familtseva A, Kalani A, Chaturvedi P et al (2014) Mitochondrial mitophagy in mesenteric artery remodeling in hyperhomocysteinemia. Physiol Rep 2:e00283. https://doi.org/10.14814/phy2.283

Flagg TP, Enkvetchakul D, Koster JC, Nichols CG (2010) Muscle KATP channels: recent insights to energy sensing and myoprotection. Physiol Rev 90:799–829. https://doi.org/10.1152/physrev.00027.2009

Frontera WR, Ochala J (2015) Skeletal muscle: a brief review of structure and function. Calcif Tissue Int 96:183–195. https://doi.org/10.1007/s00223-014-9915-y

Iannotti FA, Panza E, Barrese V et al (2010) Expression, localization, and pharmacological role of Kv7 potassium channels in skeletal muscle proliferation, differentiation, and survival after myotoxic insults. J Pharmacol Exp Ther 332:811–820. https://doi.org/10.1124/jpet.109.162800

Iannotti FA, Barrese V, Formisano L et al (2013) Specification of skeletal muscle differentiation by repressor element-1 silencing transcription factor (REST)-regulated Kv7.4 potassium channels. Mol Biol Cell 24:274–284. https://doi.org/10.1091/mbc.E11-12-1044

Ishihara H, Kubota H, Lindberg RLP et al (2008) Endothelial cell barrier impairment induced by glioblastomas and transforming growth factor β_2 involves matrix metalloproteinases and tight junction proteins. J Neuropathol Exp Neurol 67:435–448. https://doi.org/10.1097/NEN.0b013e31816fd622

Ishii I, Akahoshi N, Yamada H et al (2010) Cystathionine gamma-Lyase-deficient mice require dietary cysteine to protect against acute lethal myopathy and oxidative injury. J Biol Chem 285:26358–26368. https://doi.org/10.1074/jbc.M110.147439

Islam KN, Polhemus DJ, Donnarumma E et al (2015) Hydrogen sulfide levels and nuclear factor-erythroid 2-related factor 2 (NRF2) activity are attenuated in the setting of critical limb ischemia (CLI). J Am Heart Assoc 4:e001986. https://doi.org/10.1161/JAHA.115.001986

Jiang B, Tang G, Cao K et al (2010) Molecular mechanism for H(2)S-induced activation of K(ATP) channels. Antioxid Redox Signal 12:1167–1178. https://doi.org/10.1089/ars.2009.2894

John ASP, Kundu S, Pushpakumar S et al (2017) GYY4137, a hydrogen sulfide donor modulates miR194-dependent collagen realignment in diabetic kidney. Sci Rep 7:10924. https://doi.org/10.1038/s41598-017-11256-3

Kalra BR, Ghose S, Sood NN (1985) Homocystinuria with bilateral absolute glaucoma. Indian J Ophthalmol 33:195–197

Kanwar YS, Manaligod JR, Wong PW (1976) Morphologic studies in a patient with homocystinuria due to 5, 10-methylenetetrahydrofolate reductase deficiency. Pediatr Res 10:598–609. https://doi.org/10.1203/00006450-197606000-00008

Kolling J, Scherer EBS, Siebert C et al (2013) Homocysteine induces energy imbalance in rat skeletal muscle: is creatine a protector? Cell Biochem Funct 31:575–584. https://doi.org/10.1002/cbf.2938

Kristensen M, Juel C (2010) Potassium-transporting proteins in skeletal muscle: cellular location and fibre-type differences. Acta Physiol (Oxf) 198:105–123. https://doi.org/10.1111/j.1748-1716.2009.02043.x

Lu F, Lu B, Zhang L et al (2020) Hydrogen sulphide ameliorating skeletal muscle atrophy in db/db mice via muscle RING finger 1 S-sulfhydration. J Cell Mol Med 24:9362–9377. https://doi.org/10.1111/jcmm.15587

Lu J, Tang Z, Xu M et al (2024) Skeletal muscle cystathionine γ-lyase deficiency promotes obesity and insulin resistance and results in hyperglycemia and skeletal muscle injury upon HFD in mice. Redox Rep 29:2347139. https://doi.org/10.1080/13510002.2024.2347139

MacIntosh BR, Holash RJ, Renaud J-M (2012) Skeletal muscle fatigue--regulation of excitation-contraction coupling to avoid metabolic catastrophe. J Cell Sci 125:2105–2114. https://doi.org/10.1242/jcs.093674

Majumder A, Singh M, Behera J et al (2018) Hydrogen sulfide alleviates hyperhomocysteinemia-mediated skeletal muscle atrophy via mitigation of oxidative and endoplasmic reticulum stress injury. Am J Phys Cell Phys 315:C609–C622. https://doi.org/10.1152/ajpcell.00147.2018

Majumder A, Singh M, George AK, Tyagi SC (2019) Restoration of skeletal muscle homeostasis by hydrogen sulfide during hyperhomocysteinemia-mediated oxidative/ER stress condition 1. Can J Physiol Pharmacol 97:441–456. https://doi.org/10.1139/cjpp-2018-0501

Martelli A, Testai L, Breschi MC et al (2013) Vasorelaxation by hydrogen sulphide involves activation of Kv7 potassium channels. Pharmacol Res 70:27–34. https://doi.org/10.1016/j.phrs.2012.12.005

Martelli A, di Villa d'E, Bianca R, Cirino G et al (2023) Hydrogen sulfide and sulfaceutic or sulfanutraceutic agents: classification, differences and relevance in preclinical and clinical studies. Pharmacol Res 196:106947. https://doi.org/10.1016/j.phrs.2023.106947

Matar W, Nosek TM, Wong D, Renaud J (2000) Pinacidil suppresses contractility and preserves energy but glibenclamide has no effect during muscle fatigue. Am J Physiol Cell Physiol 278: C404–C416. https://doi.org/10.1152/ajpcell.2000.278.2.C404

Miceli F, Soldovieri MV, Martire M, Taglialatela M (2008) Molecular pharmacology and therapeutic potential of neuronal Kv7-modulating drugs. Curr Opin Pharmacol 8:65–74. https://doi.org/10.1016/j.coph.2007.10.003

Micheli L, Mitidieri E, Turnaturi C et al (2022) Beneficial effect of H2S-releasing molecules in an in vitro model of sarcopenia: relevance of glucoraphanin. Int J Mol Sci 23:5955. https://doi.org/10.3390/ijms23115955

Miki T, Minami K, Zhang L et al (2002) ATP-sensitive potassium channels participate in glucose uptake in skeletal muscle and adipose tissue. Am J Physiol Endocrinol Metab 283:E1178–E1184. https://doi.org/10.1152/ajpendo.00313.2002

Mitidieri E, Vellecco V, Brancaleone V et al (2021) Involvement of 3′,5′-cyclic inosine monophosphate in cystathionine γ-lyase-dependent regulation of the vascular tone. Br J Pharmacol 178:3765–3782. https://doi.org/10.1111/bph.15516

Morris AAM, Kožich V, Santra S et al (2017) Guidelines for the diagnosis and management of cystathionine beta-synthase deficiency. J Inherit Metab Dis 40:49–74. https://doi.org/10.1007/s10545-016-9979-0

Nandi SS, Mishra PK (2017) H2S and homocysteine control a novel feedback regulation of cystathionine beta synthase and cystathionine gamma lyase in cardiomyocytes. Sci Rep 7: 3639. https://doi.org/10.1038/s41598-017-03776-9

Nelson MD, Rader F, Tang X et al (2014) PDE5 inhibition alleviates functional muscle ischemia in boys with Duchenne muscular dystrophy. Neurology 82:2085–2091. https://doi.org/10.1212/WNL.0000000000000498

Nichol CJ (1964) Sulfhydryl and disulfide concentrations in dystrophic mouse muscle. Can J Biochem 42:1643–1645. https://doi.org/10.1139/o64-176

Nio Y, Tanaka M, Hirozane Y et al (2017) Phosphodiesterase 4 inhibitor and phosphodiesterase 5 inhibitor combination therapy has antifibrotic and anti-inflammatory effects in mdx mice with Duchenne muscular dystrophy. FASEB J 31:5307–5320. https://doi.org/10.1096/fj.201700249R

Nisoli E, Falcone S, Tonello C et al (2004) Mitochondrial biogenesis by NO yields functionally active mitochondria in mammals. Proc Natl Acad Sci USA 101:16507–16512. https://doi.org/10.1073/pnas.0405432101

Panza E, Vellecco V, Iannotti FA et al (2021) Duchenne's muscular dystrophy involves a defective transsulfuration pathway activity. Redox Biol 45:102040. https://doi.org/10.1016/j.redox.2021.102040

Parsanathan R, Jain SK (2018) Hydrogen sulfide increases glutathione biosynthesis, and glucose uptake and utilisation in C2C12 mouse myotubes. Free Radic Res 52:288–303. https://doi.org/10.1080/10715762.2018.1431626

Parsanathan R, Jain SK (2019) Hydrogen sulfide regulates circadian-clock genes in C2C12 myotubes and the muscle of high-fat-diet-fed mice. Arch Biochem Biophys 672:108054. https://doi.org/10.1016/j.abb.2019.07.019

Percival JM, Whitehead NP, Adams ME et al (2012) Sildenafil reduces respiratory muscle weakness and fibrosis in the mdx mouse model of Duchenne muscular dystrophy. J Pathol 228:77–87. https://doi.org/10.1002/path.4054

Roura-Ferrer M, Solé L, Martínez-Mármol R et al (2008) Skeletal muscle Kv7 (KCNQ) channels in myoblast differentiation and proliferation. Biochem Biophys Res Commun 369:1094–1097. https://doi.org/10.1016/j.bbrc.2008.02.152

Sbodio JI, Snyder SH, Paul BD (2019) Regulators of the transsulfuration pathway. Br J Pharmacol 176:583–593. https://doi.org/10.1111/bph.14446

Shibuya N, Tanaka M, Yoshida M et al (2009) 3-Mercaptopyruvate sulfurtransferase produces hydrogen sulfide and bound sulfane sulfur in the brain. Antioxid Redox Signal 11:703–714. https://doi.org/10.1089/ars.2008.2253

Singh M, George AK, Eyob W et al (2021) High-methionine diet in skeletal muscle remodeling: epigenetic mechanism of homocysteine-mediated growth retardation. Can J Physiol Pharmacol 99:56–63. https://doi.org/10.1139/cjpp-2020-0093

Sinha-Hikim I, Sinha-Hikim AP, Parveen M et al (2013) Long-term supplementation with a cystine-based antioxidant delays loss of muscle mass in aging. J Gerontol A Biol Sci Med Sci 68:749–759. https://doi.org/10.1093/gerona/gls334

Slavin MB, Khemraj P, Hood DA (2024) Exercise, mitochondrial dysfunction and inflammasomes in skeletal muscle. Biom J 47:100636. https://doi.org/10.1016/j.bj.2023.100636

Stapleton DI, Lau X, Flores M et al (2014) Dysfunctional muscle and liver glycogen metabolism in mdx dystrophic mice. PLoS One 9:e91514. https://doi.org/10.1371/journal.pone.0091514

Steegborn C, Clausen T, Sondermann P et al (1999) Kinetics and inhibition of recombinant human cystathionine gamma-lyase. Toward the rational control of transsulfuration. J Biol Chem 274: 12675–12684. https://doi.org/10.1074/jbc.274.18.12675

Tetsi L, Charles A-L, Paradis S et al (2017) Effects of cyclic nucleotide phosphodiesterases (PDEs) on mitochondrial skeletal muscle functions. Cell Mol Life Sci 74:1883–1893. https://doi.org/10.1007/s00018-016-2446-0

Thirupathi A, Pinho RA, Baker JS et al (2020) Taurine reverses oxidative damages and restores the muscle function in overuse of exercised muscle. Front Physiol 11:582449. https://doi.org/10.3389/fphys.2020.582449

Tricarico D, Camerino DC (1994) ATP-sensitive K+ channels of skeletal muscle fibers from young adult and aged rats: possible involvement of thiol-dependent redox mechanisms in the age-related modifications of their biophysical and pharmacological properties. Mol Pharmacol 46:754–761

Tricarico D, Selvaggi M, Passantino G et al (2016) ATP sensitive potassium channels in the skeletal muscle function: involvement of the KCNJ11(Kir6.2) gene in the determination of mechanical Warner Bratzer shear force. Front Physiol 7:167. https://doi.org/10.3389/fphys.2016.00167

Triolo M, Slavin M, Moradi N, Hood DA (2022) Time-dependent changes in autophagy, mitophagy and lysosomes in skeletal muscle during denervation-induced disuse. J Physiol 600:1683–1701. https://doi.org/10.1113/JP282173

Veeranki S, Tyagi SC (2013) Defective homocysteine metabolism: potential implications for skeletal muscle malfunction. Int J Mol Sci 14:15074–15091. https://doi.org/10.3390/ijms140715074

Veeranki S, Tyagi SC (2015) Mechanisms of hyperhomocysteinemia induced skeletal muscle myopathy after ischemia in the CBS−/+ mouse model. Int J Mol Sci 16:1252–1265. https://doi.org/10.3390/ijms16011252

Veeranki S, Winchester LJ, Tyagi SC (2015) Hyperhomocysteinemia associated skeletal muscle weakness involves mitochondrial dysfunction and epigenetic modifications. Biochim Biophys Acta 1852:732–741. https://doi.org/10.1016/j.bbadis.2015.01.008

Vellecco V, Mancini A, Ianaro A et al (2016) Cystathionine β-synthase-derived hydrogen sulfide is involved in human malignant hyperthermia. Clin Sci (Lond) 130:35–44. https://doi.org/10.1042/CS20150521

Vellecco V, Armogida C, Bucci M (2018) Hydrogen sulfide pathway and skeletal muscle: an introductory review. Br J Pharmacol 175:3090–3099. https://doi.org/10.1111/bph.14358

Vellecco V, Martelli A, Bibli IS et al (2020) Anomalous Kv 7 channel activity in human malignant hyperthermia syndrome unmasks a key role for H2 S and persulfidation in skeletal muscle. Br J Pharmacol 177:810–823. https://doi.org/10.1111/bph.14700

Vellecco V, Panza E, Bibli S-I et al (2022) Phosphodiesterases S-sulfhydration contributes to human skeletal muscle function. Pharmacol Res 177:106108. https://doi.org/10.1016/j.phrs.2022.106108

Vignane T, Filipovic MR (2023) Emerging chemical biology of protein Persulfidation. Antioxid Redox Signal 39:19–39. https://doi.org/10.1089/ars.2023.0352

Wang B, Zeng J, Gu Q (2017) Exercise restores bioavailability of hydrogen sulfide and promotes autophagy influx in livers of mice fed with high-fat diet. Can J Physiol Pharmacol 95:667–674. https://doi.org/10.1139/cjpp-2016-0611

Wasada T, Yano T, Ohta M et al (2001) ATP-sensitive potassium channels modulate glucose transport in cultured human skeletal muscle cells. Endocr J 48:369–375. https://doi.org/10.1507/endocrj.48.369

Yang R, Jia Q, Li Y, Mehmood S (2020) Protective effect of exogenous hydrogen sulfide on diaphragm muscle fibrosis in streptozotocin-induced diabetic rats. Exp Biol Med (Maywood) 245:1280–1289. https://doi.org/10.1177/1535370220931038

Zhang Y, Masters L, Wang Y et al (2021) Cystathionine gamma-lyase/H2 S signaling facilitates myogenesis under aging and injury condition. FASEB J 35:e21511. https://doi.org/10.1096/fj.202002675R

Zhao W, Zhang J, Lu Y, Wang R (2001) The vasorelaxant effect of H(2)S as a novel endogenous gaseous K(ATP) channel opener. EMBO J 20:6008–6016. https://doi.org/10.1093/emboj/20.21.6008

Zhao L, Liu X, Zhang J et al (2020) Hydrogen sulfide alleviates skeletal muscle fibrosis via attenuating inflammation and oxidative stress. Front Physiol 11:533690. https://doi.org/10.3389/fphys.2020.533690

Zivanovic J, Kouroussis E, Kohl JB et al (2019) Selective Persulfide detection reveals evolutionarily conserved antiaging effects of S-Sulfhydration. Cell Metab 30:1152–1170.e13. https://doi.org/10.1016/j.cmet.2019.10.007

Hydrogen Sulfide in Experimental Organ Transplantation: From Bench to Bedside

George J. Dugbartey and Alp Sener

Contents

Abstract

Organ transplantation is the treatment of choice for patients with organ failure. However, the long-term success of this complex life-saving procedure is severely challenged by several inherent factors such as ischemia-reperfusion injury (IRI), an unavoidable pathological condition which occurs due to temporary cessation of blood supply to the donor organ during procurement and cold preservation, and subsequent blood restoration upon engraftment. IRI decreases organ graft quality and function and increases the incidence of post-transplant complications. While the transplant community has adopted the use of sub-optimal organ grafts from

G. J. Dugbartey (✉) · A. Sener (✉)
Department of Surgery, Division of Urology, London Health Sciences Center, University of Western Ontario, London, ON, Canada

Western University, LHSC University Hospital, London, ON, Canada
e-mail: gdugbart@uwo.ca; alp.sener@lhsc.on.ca

A. Papapetropoulos, C. Szabo (eds.), *Hydrogen Sulfide: A Toxicant with Healing Potential*, Handbook of Experimental Pharmacology 292,
https://doi.org/10.1007/164_2025_778

extended criteria donors and deceased donors in addition to viable organs from healthy living donors to expand the pool of transplantable organs with the aim to overcome the global organ shortage crisis, the sub-optimal organs are more susceptible to IRI. Also worrying is the fact that organ preservation techniques have not changed for over the last 50 years, suggesting the need to optimize existing preservation techniques or develop effective alternative preservation solutions to limit ischemic injury during organ graft preservation. Among several pharmacological agents being tested in preservation solutions in the pretext of improving transplantation outcomes, hydrogen sulfide (H_2S), the third established member of a family of gaseous signaling molecules, is emerging as an excellent candidate, with therapeutic properties such as antioxidant, anti-inflammatory, anti-apoptotic, and vasodilating properties. In this chapter, we discussed the therapeutic benefits of H_2S and its donor compounds against IRI in various experimental models of organ transplantation. In fact, one of the H_2S donor compounds, sodium thiosulfate, which our research team is currently investigating in experimental kidney transplantation, is already in clinical use. Therefore, it is a question of repositioning it for clinical organ transplantation after substantial evidence of its protective effects in experimental organ transplantation.

1 Introduction

Organ transplantation is the treatment of choice for patients with organ failure. It provides the only chance for their survival and improves their quality of life at a relatively lower cost. It is the newest and the most dynamic revolutionary field and one of the most successful advances in modern medicine, which began in the mid-twentieth century with the kidney as the first human organ to be transplanted successfully between identical twins in 1954 (Morris 2004). This ground-breaking success was against the background of several failed attempts, and demonstrates the evolution of a challenging life-saving intervention from its experimental origin. However, despite a lot of advances made in the field of organ transplantation, which have so far improved patient outcomes significantly, many challenges remain. These challenges include global donor organ shortage crisis due to disproportion between organ demand and availability, optimization of immunosuppression protocols, infections, and ischemia-reperfusion injury (an inevitable pathological condition due to cessation and restoration of blood to transplanted organ), delayed graft function (a manifestation of acute kidney injury commonly associated with kidneys transplanted from deceased donors), primary graft dysfunction (a major cause of morbidity and mortality in lung transplant patients), all of which contribute to poor graft quality, and post-transplant complications, and thereby limiting long-term success of organ transplantation (Abouna 2008; De Martin et al. 2022; Dugbartey 2024a; Ye et al. 2022; Van Slambrouck et al. 2022; Snydman 2001). These challenges have provided a fertile ground for research into new techniques with potential to expand the pool of transplantable organs, mitigate transplant-induced graft injury, and ensure improved graft quality and function, as well as long-term transplant recipient survival.

Strategies to overcome the global donor organ shortage crisis and expand the pool of transplantable organs include the use of sub-optimal organ grafts from deceased donors (DCD; donation-after-cardiac-death, DBD; donation-after-brain-death, DBCD; donation-after-brain-death followed by cardiac-death), in addition to extended criteria donors (ECD) and organ grafts from healthy living donors. Interestingly, DCD currently represents ~25% of all transplants done globally (Saat et al. 2016; Sun et al. 2018a; Troppmann et al. 2018). Due to lack of perfusion and prolonged warm ischemic time, it is important to note that organs from deceased donors, regardless of the category, are associated with increased ischemic damage. Nonetheless, organ grafts from deceased donors still represent a viable option to meet the increasing demand for organ transplantation and shorten the transplant recipient waitlist. This will help improve utilization of thousands of organs from higher-risk and deceased donors that are discarded annually due to increased likelihood of post-transplant complications. This chapter begins with static cold storage as the gold standard of organ graft preservation with reference to renal grafts. Next, the chapter will introduce hydrogen sulfide and its protective effects in various experimental models of organ transplantation. Finally, the chapter will conclude with a translation of the therapeutic properties of hydrogen sulfide from experimental organ transplantation to clinical organ transplantation.

2 Static Cold Storage as Gold Standard of Organ Preservation

Since the inception of organ transplantation in the 1960s, organ grafts have been stored in cold preservation solution with the goal of maintaining long-term organ graft quality and function during storage and reperfusion. The most used solutions in preserving donor organs include histidine-tryptophan-ketoglutarate (HTK) solution, low-potassium dextran (LPD) solution, University of Wisconsin (UW) solution, Marshall's solution, Euro-Collins solution, hypertonic citrate adenine solution (HC-A), Celsior solution, and Institute Georges Lopez-1 (IGL-1) solution (Chen et al. 2019). These preservation solutions contain osmotic and oncotic buffering components and other supplements intended to attenuate IRI. Considering that hypothermia is the first line of defense against hypoxic injury and that at 4°C temperature, oxygen consumption reduces significantly to about 5% of normoxic value (Guibert et al. 2011), the gold standard of donor organ preservation clinically is static cold storage (SCS) in these preservation solutions on ice at 4°C. This is a simple and inexpensive approach to reducing donor organ metabolism, slowing cell death and tissue injury, reducing production of toxic metabolites, and preserving the quality and function of the organ graft prior to transplantation (Bon et al. 2012). Despite its simplicity and cost-effectiveness, challenges exist in the quest for optimizing organ preservation by SCS. Using the kidney as an ideal organ that is particularly prone to IRI due to its high metabolic activities and oxygen consumption, for example, we and other researchers have reported in pre-clinical models of kidney transplantation that prolonged SCS of donor kidneys decreases renal blood flow (RBF) and glomerular filtration rate (GFR), and induces significant kidney injury (Zhao et al. 2013; Dugbartey et al. 2023; DiRito et al. 2021), leading to delayed graft function (DGF), loss of graft function, and shortened recipient survival (Debout et al. 2015; Grenda 2017). In fact, it is important to note that every 6 h of SCS is associated with increased risk of DGF by 23%, with acute renal graft rejection and early graft loss (Ojo et al. 1997). To address the challenges associated with SCS, researchers are investigating the outcome of organ graft preservation by machine perfusion at different preservation temperatures (hypothermic, normothermic, and sub-normothermic). However, organ preservation by machine perfusion involves a complex and expensive set-up. Besides, there is no clinically approved protocol. Therefore, SCS remains the global primary preservation technique.

3 Hydrogen Sulfide and Its Sources of Production

Hydrogen sulfide (H_2S), a colorless, flammable, water-soluble, and membrane-permeable gas, which was originally considered as toxic gas with an obnoxious smell of rotten egg, has risen above its historic negative image, and has now been established as the third member of a family of gaseous signaling molecules referred to as *"gasotransmitters"* (Polhemus and Lefer 2014; Dugbartey 2021, 2024b). It is involved in cellular homeostasis, and regulates many physiological and pathological processes at low nanomolar to micromolar concentrations. Several studies have shown that this homeostatic gas contributes to pathogenesis of many diseases when its production is dysregulated. However, at the right concentration, H_2S been identified to exert multiple therapeutic effects in various experimental models of human diseases, including in the field of organ transplantation by targeting multiple molecular pathways (Dugbartey et al. 2023). It is enzymatically synthesized in all mammalian cells through four metabolic pathways that use sulfur-containing amino acids as substrates: (i) desulfhydration of L-cysteine by the cytosolic enzyme cystathionine β-synthase (CBS); (ii) desulfhydration of L-cysteine or L-homocysteine by another cytosolic enzyme cystathionine γ-lyase (CSE); (iii) transamination of L-cysteine by cysteine aminotransferase to 3-mecaptopyruvate, followed by its desulfhydration to pyruvate catalyzed by the mitochondrial enzyme 3-mecaptopyruvate sulfurtransferase (3-MST); (iv) H_2S production from D-cysteine by the peroxisomal enzyme D-amino acid oxidase (DAO) (Xia et al. 2009; Mikami et al. 2011; Modis et al. 2013; Shibuya et al. 2013; Kimura 2014) (Fig. 1). While the expression of these H_2S-producing enzymes is subcellular and tissue-specific, all four enzymes are predominantly found in the kidney, CBS, CSE, and 3-MST are also localized in heart, liver, brain, pancreas, and other tissues (Xia et al. 2009; Mikami et al. 2011; Modis et al. 2013; Shibuya et al. 2013; Kimura 2014). Under hypoxic conditions, for example, CBS and CSE undergo mitochondrial

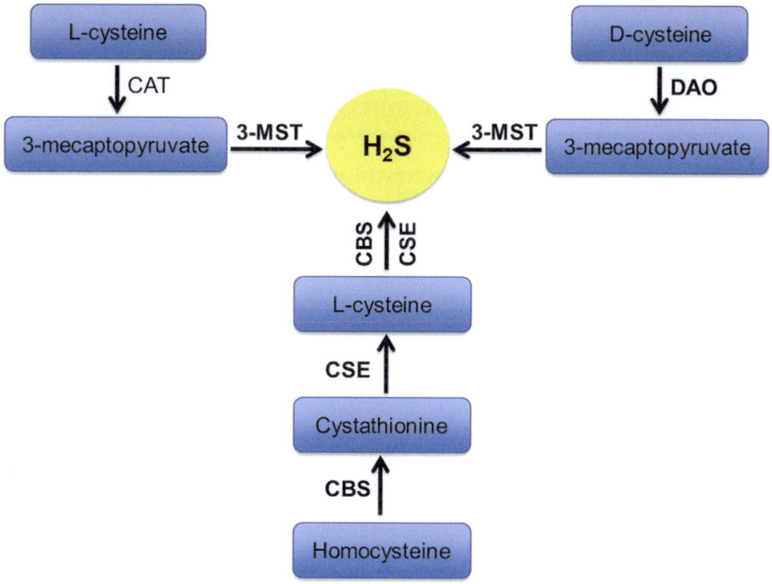

Fig. 1 Simplistic view of enzymatic production of hydrogen sulfide (H_2S). H_2S is produced enzymatically by cystathionine β-synthase (CBS) and cystathionine γ-lyase (CSE) using homocysteine and L-cysteine as substrates. L-cysteine can also be converted to 3-mecaptopyruvate as an intermediate product in a reaction catalyzed by cysteine aminotransferase (CAT). The enzyme 3-mercaptopyruvate sulfurtransferase (3-MST) then produces H_2S from 3-mecaptopyruvate, and coupled with D-amino acid oxidase (DAO), 3-MST produces H_2S from exogenously administered D-cysteine

translocation, where they metabolize cysteine to augment mitochondrial H_2S production and increase adenosine triphosphate (ATP) production by 3-MST (Fu et al. 2012).

Besides the experimental upregulation of endogenously produced H_2S, gaseous H_2S has been used as an exogenous source of H_2S in various preclinical studies. Unfortunately, administration of gaseous H_2S through inhalation is limited by its toxic effects to non-participants. This drawback has led to the development of H_2S donor compounds such as sodium hydrosulfide (NaHS) and sodium sulfide (Na_2S) as exogenous sources of H_2S by releasing H_2S into cells upon their exogenous administration. These are fast-releasing H_2S donor compounds and have been used in many experimental models of human diseases, although the release of H_2S by these compounds is short-lasting and sometimes fails to reach the mitochondria and other targeted organelles (Caliendo et al. 2010; Kashfi and Olso 2013). Considering the downside of these inorganic sulfide salts, compounds such as GYY4137 have been synthesized and offer sustained and longer-lasting H_2S release (Li et al. 2008). Remarkably, a research team led by Prof. Matthew Whiteman also synthesized mitochondria-targeted slow-releasing H_2S donor compounds (AP39 and AP123), which have so far shown therapeutic benefits by augmenting mitochondrial production of H_2S in many in vitro and in vivo models, including our work on ex vivo kidney transplantation (Dugbartey et al. 2023; Zhao et al. 2016; Juriasingani et al. 2021). In addition to these synthetic sources of exogenous H_2S, organic compounds such as garlic-derived polysulfides (e.g., diallyl trisulfide, diallyl disulfide, and diallyl sulfide) have been identified as natural sources of H_2S donors with potent cytoprotective and cardioprotective properties (Benavides et al. 2007; Ginter and Simko 2010). Recently, our research team also identified sodium thiosulfate (STS), a drug already in clinical use, as another exogenous source of H_2S with renal protective properties (Zhang et al. 2022; Nelson et al. 2024; Abou Taka et al. 2024).

4 H_2S in Organ Transplantation

In the face of the organ graft injury induced by the transplant protocol, including during SCS even with reduced preservation time, the quest for pharmacological modification of the transplant protocol with novel agents such as H_2S to reduce IRI in transplantation of various solid organs, has attracted significant experimental attention (Dugbartey et al. 2021). The success of such preclinical investigations could translate into pharmacological modification of clinical organ transplantation protocols, thereby improving organ preservation techniques which could allow programs to extend preservation times with minimal costs; all of this could ultimately reduce post-transplant complications and prolong transplant recipient survival.

4.1 H_2S in Experimental Heart Transplantation

In light of the ongoing investigations on the therapeutic properties of H_2S, the transplant community has also investigated and reported the impact of H_2S in experimental models of heart transplantation. In an in vitro model of cold hypoxia/reoxygenation, which mimics the in vivo version of cold ischemia during organ preservation and warm reperfusion after implantation, we observed significant reduction in cardiomyocyte apoptosis and pro-inflammatory gene expression, which corresponded with markedly increased cardiomyocyte viability following 16 h of primary cardiomyocyte preservation at 10°C in UW solution supplemented with 100 nmol/L of AP39 and 6 h of reoxygenation after 24 h of hypoxia in normal culture medium at 10°C (Zhu et al. 2019). Using the same concentration of AP39 in the same study, our in vitro observation was supported by our in vivo murine model of syngeneic heterotopic heart transplantation in which preservation of cardiac grafts by SCS in AP39-supplemented UW solution for 24 h at 4°C significantly reduced cardiac tissue injury and fibrosis and protected the transplanted graft against IRI and preserved cardiac graft function as revealed by quantitative ultrasound scan (Zhu et al. 2019).

A previous study from another research group also investigated cardioprotective effect of H_2S in a rat model of heart transplantation, in which 6 h of SCS of cardiac grafts in UW solution at 4°C with diallyl trisulfide (a natural source of H_2S) carried by mesoporous silica nanoparticles and followed by 1 h of reperfusion resulted in greater cardiac allograft performance, with long-term graft survival and function after 8 weeks of transplantation when compared to control group without H_2S treatment (Sun et al. 2018b). The cardioprotection was evidenced by a significant increase in left ventricular developed pressure, reduced levels of cardiac damage markers and pro-inflammatory markers, along with increased antioxidant activities as well as decreased expression of cardiomyocyte apoptotic protein and preserved mitochondrial ultrastructure (Sun et al. 2018b). Interestingly, in comparing the cardioprotective effect of diallyl trisulfide to traditional H_2S donors such as NaHS and GYY4137 at corresponding concentrations in the same study, the authors also reported that diallyl trisulfide carried by mesoporous silica nanoparticles provided a superior cardioprotection than these traditional H_2S donors (Sun et al. 2018b). In the same study, the authors also confirmed their in vivo result with an in vitro model of hypoxia/reoxygenation in which diallyl trisulfide carried by mesoporous silica nanoparticles protected cardiomyocytes against hypoxic injury by inhibiting inflammatory pathways such as Toll-like receptor 4/nucleotide-binding oligomerization domain NOD-like receptor containing pyrin domain 3 (TLR4/NLRP3) signaling pathway (Sun et al. 2018b) (Fig. 2). The greater protection by diallyl trisulfide carried by mesoporous silica nanoparticles in their transplantation model is due to the fact that it is a long-term, slow, and controlled-releasing H_2S system compared to NaHS and GYY4137 whose effects sometimes do not include the mitochondria, the major targets in IRI. Besides, it is known that glutathione, the most abundant naturally occurring antioxidant, stimulates H_2S-controlled release by diallyl trisulfide, which in turn, activates glutathione and other antioxidants (Chatzianastasiou et al. 2016; Karwi et al. 2017). Furthermore, diallyl trisulfide has been reported to increase the production of nitric oxide (a member of the gasotransmitter family) through its H_2S release, leading to cardioprotection in a murine model of myocardial IRI (Predmore et al. 2012). These explanations also suggest a possible interaction among members of the gasotransmitter family, leading to the observed superior cardioprotection over the traditional H_2S donors after transplantation.

Recently, a systematic review and meta-analysis of 16 experimental studies on cardiac graft preservation revealed that postconditioning of cardiac grafts with slow-releasing H_2S donors markedly decreased the expression of cardiac damage markers and oxidative stress, which positively correlated with significantly reduced myocardial infarct size and improved cardiac graft function (Ertugrul et al. 2021). In addition to these experimental models of heart transplantation, several preclinical models of myocardial IRI also showed cardioprotection by H_2S donors, including gaseous H_2S, and revealed some protective mechanisms (Johansen et al. 2006; Ji et al. 2008; Snijder et al. 2013). In rat models of myocardial IRI, for example, ischemia and reperfusion of isolated rat hearts decreased myocardial H_2S production and downregulated CSE expression, contributing to significant cardiomyocyte damage, reduced cardiac function, increased myocardial infarct size and mitochondrial cytochrome c release due to opening of mitochondrial permeability transition pores

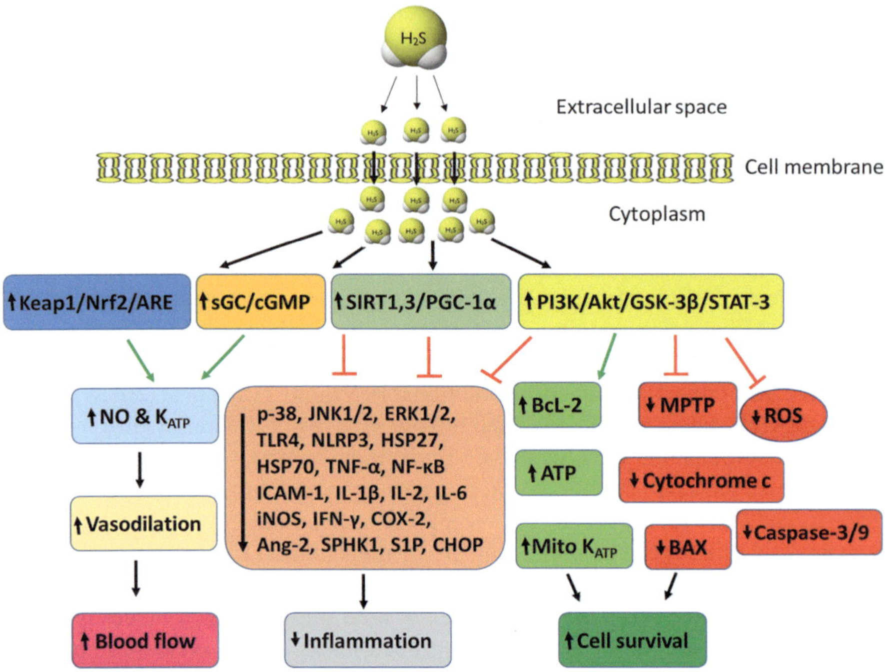

Fig. 2 Protective mechanism of hydrogen sulfide (H_2S) against ischemia-reperfusion injury (IRI) in organ transplantation. Administration of H_2S donors increases endogenous H_2S level, leading to inhibition of several pathological molecular pathways while activating protective signaling pathways against IRI to improve organ quality and function in transplantation

(MPTPs), markedly upregulated pro-apoptotic genes (e.g., caspase-3 and caspase-9 mRNA), and enhanced oxidative stress while downregulating myocardial expression of phosphorylated phosphoinositide 3-kinase (PI3K), protein kinase B (Akt), and glycogen synthase kinase-3 beta (GSK-3β) (Ji et al. 2008; Yao et al. 2010; Li et al. 2015; Hu et al. 2016; Testai et al. 2016). Interestingly, postconditioning of isolated rat hearts with 10 μmol/L of NaHS following induction of ischemia and reperfusion increased H_2S production and myocardial CSE expression and reversed these pathological pathways, and in addition, opened mitochondrial ATP-sensitive potassium (K_{ATP}) channels (cardiomyocyte-protective channels and regulators of vascular tone) and activated SIRT1/PGC-1α pathway, a signaling pathway that attenuates oxidative damage and promotes mitochondrial biogenesis and function (Ji et al. 2008; Yao et al. 2010; Li et al. 2015; Hu et al. 2016; Testai et al. 2016) (Fig. 2). Hence, inhibition of these pathological pathways and activation of protective pathways in the ischemic and reperfused rat heart resulted in decreased myocardial infarct size and myocardial enzyme release, and improved cardiac function as well as increased the levels of ATP and mitochondrial antioxidant enzymes (Ji et al. 2008; Yao et al. 2010; Li et al. 2015; Hu et al. 2016; Testai et al. 2016). In addition to H_2S postconditioning, other signaling pathways that are activated by H_2S

preconditioning to mitigate myocardial IRI and promote cardioprotection include ERK1/2, Keap1/Nrf2/ARE, and STAT-3 (Yao et al. 2010; Calvert et al. 2009; Peake et al. 2013; Lambert et al. 2014) (Fig. 2). Collectively, pharmacological modification of heart transplant protocols with H_2S in experimental setting protects against myocardial IRI and preserves cardiac graft quality and function, and prolongs transplant recipient survival by activating several cardioprotective signaling pathways while inhibiting pathological pathways. Therefore, the addition of H_2S to the transplant protocol should be considered as a novel approach for cardiac allograft preservation for clinical transplantation. However, the long-term effects of H_2S should be further investigated, especially in larger animals (e.g., in porcine models) before its translation to clinical heart transplantation.

4.2 H_2S in Experimental Kidney Transplantation

As seen in experimental models of heart transplantation, various H_2S donors have also been used to mitigate IRI and improve graft quality and function in animal models of kidney transplantation. Hosgood and Nicholson were among the first group of researchers to report on the therapeutic benefits of H_2S in experimental kidney transplantation. In their porcine model of donation-after-cardiac-death (DCD) kidney transplantation, 0.5 mmol/L of H_2S infusion at 10 min before and after reperfusion significantly improved renal hemodynamics and renal graft function, which was characterized by markedly reduced serum creatinine level and increased glomerular filtration rate, and correlated with reduced oxidative damage compared to the control group with saline infusion (Hosgood and Nicholson 2010). Following this study, and using rat models of allogeneic and syngeneic kidney transplantation, we also observed that modification of UW solution with NaHS and GYY4137 respectively during 24 h of renal allograft preservation by SCS at 4°C significantly mitigated renal IRI, reduced acute tubular and glomerular necrosis scores, graft neutrophil and macrophage infiltration, apoptotic injury, and markedly improved renal allograft function and prolonged recipient rat survival compared to control allografts preserved in UW solution without NaHS and GYY4137 supplementation (Lobb et al. 2012, 2017). In the same study, we supported our in vivo findings with in vitro model of hypoxia/reoxygenation in which AP39 prevented cold hypoxic injury in rat renal tubular epithelial cells (Lobb et al. 2012, 2017).

We have also shown that H_2S preserves renal graft function in large animals. For example, in a swine model of DCD renal autotransplantation in which the autograft was clamped for 1 h and preserved in UW solution for 24 h under hypothermic perfusion supplemented with 150 μM NaHS, we observed that renal autografts preserved in NaHS-supplemented UW solution had markedly lower acute tubular necrosis score and serum creatinine level, significantly improved renal hemodynamics, and upregulated expression of protective genes, all of which culminated in faster recovery of renal autograft function and prolonged recipient survival relative to control group without NaHS supplementation (Juriasingani et al. 2022). In non-survival ex vivo porcine models of kidney transplantation, we also observed the same salutary effect when UW solution at 4°C and 21°C was supplemented with 200 nM AP39 for 4 h and 24 h respectively (Dugbartey et al. 2023; Juriasingani et al. 2018, 2021). To make our research findings more clinically translatable, we replaced these non-clinically viable H_2S donors with sodium thiosulfate (STS), an FDA-approved H_2S donor drug for the treatment of calciphylaxis in end-stage renal disease patients and as an antidote to acute cyanide poisoning. In rat models of syngeneic orthotopic kidney transplantation, administration of donor rats with 150 μM of STS at 30 min prior to donor kidney procurement, and supplementation of UW solution with the same concentration of STS during 24 h of renal syngraft preservation by SCS at 4°C produced similar renoprotective effect and prolonged recipient rat survival (Zhang et al. 2022; Nelson et al. 2024; Abou Taka et al. 2024).

A number of molecular mechanisms have been studied in various experimental models of IRI to explain the renoprotective effects of H_2S in kidney transplantation. In a mouse model of renal IRI, for example, induction of 30 min of ischemia and 24 h of reperfusion activated NLRP3-mediated inflammation and apoptosis, which partly contributed to renal injury. However, NaHS administration protected against renal IRI by downregulating renal NLRP3 protein expression and reducing pro-inflammatory cytokine release (Su et al. 2021). In a rat model of renal IRI, intraperitoneal administration of 150 μmol/L NaHS prior to 45 min of ischemia followed by 6 h of reperfusion downregulated renal expression of molecular chaperones such as heat shock protein 70 (HSP70) and HSP27, and major inflammatory mediators such as nuclear factor-kappaB (NF-κB) and TLR4, leading to suppressed release of pro-inflammatory cytokines and improved renal antioxidant status, and ultimately promoted renal protection. Remarkably, the opposite effect was observed in the control group without NaHS pretreatment (Du et al. 2019) (Fig. 2). A similar renoprotective effect of H_2S was observed in another rat model of renal IRI (45 min of ischemia and 6 h of reperfusion) with an additional effect by inhibiting the phosphorylation of mitogen-activated protein kinases (MAPK; a family of pro-inflammatory proteins) such as p-38, c-JUN N-terminal protein kinase 1/2 (JNK1/2), and extracellular signal-regulated kinase 1/2 (ERK1/2), and further downregulated the expression of NF-κB-dependent proteins such as inducible nitric oxide synthase (iNOS), cyclo-oxygenase-2 (COX-2), and intercellular adhesion molecule-1 (ICAM-1) (Tripatara et al. 2008) (Fig. 2). Altogether, various H_2S donors have shown therapeutic benefits against IRI in experimental kidney transplantation, with well-documented protective mechanisms. Considering that STS is currently the only identified H_2S donor in clinical use, it is important for other research groups to also investigate the therapeutic benefits of STS in experimental kidney transplantation as well as its effects in large animals, and its long-term effects in order to facilitate its use in clinical kidney transplantation.

4.3 H_2S in Experimental Lung Transplantation

In addition to experimental heart and kidney transplantation, the protective effect of H_2S has also been reported in experimental models of lung transplantation. In an ex vivo rabbit model of lung transplantation, for example, administration of 150 ppm gaseous H_2S mixed with room air (21% oxygen) to donor rabbits for 2 h prior to external cooling to 34°C and procurement and preservation of inflated donor lungs in Perfadex solution at 4°C for 18 h followed by 2 h of reperfusion resulted in better oxygenation and ventilation, with significant improvement in pulmonary artery pressure during reperfusion in comparison with control lungs from donor rabbits that did not receive H_2S pre-treatment (George et al. 2012a). Mechanistically, the authors observed that H_2S-treated lungs showed markedly reduced level of reactive oxygen species (ROS; a destructive mediator of IRI) and preserved the activity of cytochrome c in the mitochondria compared to control lungs (George et al. 2012a). Using the same ex vivo model in another experiment by the same authors in which lung grafts were flushed with Perfadex solution and preserved by SCS in LPD solution at 4°C for 18 h, a bolus injection of 100 μg/kg NaHS through the reperfusion circuit at the start of a 2-h ex vivo reperfusion and followed by continuous infusion at 1 mg/kg/h produced the same beneficial effects, with high level of cyclic guanosine monophosphate (cGMP; a pulmonary vasodilator) (George et al. 2012b) (Fig. 2). These empirical findings suggest that traditional lung transplantation protocol could be improved with H_2S for protection of cellular and subcellular structures against cold IRI to produce better lung graft quality and function, including regulation of pulmonary hypertension after lung transplantation. The findings also suggest that improvement of conventional lung transplantation techniques with H_2S could be extended to those of other solid organ grafts to reduce post-transplant complications and improve post-transplant outcomes.

In a rat model of DCD orthotopic lung transplantation in which the effects of H_2S on lung inflation during warm ischemia was investigated, Meng and colleagues (Meng et al. 2017) reported that inflation of lung grafts with 80 ppm of gaseous H_2S for 2 h during warm ischemia followed by 3 h of SCS in LPD solution at 4°C and 3 h of reperfusion showed significant improvement in mitochondrial structure and downregulated expression of pro-apoptotic (TUNEL stain) and pro-inflammatory (e.g., NF-κB) proteins as well as decreased oxidative damage (by upregulating Nrf2) and lactic acid level relative to oxygen-inflated lung grafts and deflated control grafts. These results were confirmed by a recent study, and with an additional experimental group, which showed that H_2S-inflated lung grafts had improved lung oxygenation and stable pulmonary surfactant level compared to H_2S-ventilated lung grafts and deflated control grafts (Yan et al. 2023). To accentuate the therapeutic benefits of exogenous H_2S administration and the involvement of endogenous H_2S in these lung transplantation models, Wu et al. (2013) pharmacologically inhibited endogenous H_2S production while administering exogenous H_2S in a rat model of orthotopic lung transplantation. In their study, intraperitoneal administration of 37.5 mg/kg of the CSE inhibitor, propargylglycine, at 15 min after 3 h of graft preservation in LPD solution at 4°C was associated with significantly decreased H_2S level and downregulated CSE protein and anti-inflammatory cytokines (e.g., IL-10) in the lung graft, and markedly increased ROS production as well as level of pro-inflammatory cytokines (e.g., IL-1β) and myeloperoxidase activity, all of which correlated with severe lung IRI and poor pulmonary function after transplantation. Interestingly, intraperitoneal injection of NaHS (14 μmol/kg) produced the opposite effect, with attenuated lung IRI and improved pulmonary function (Wu et al. 2013) (Fig. 2). These results suggest that activation of endogenous H_2S with exogenous H_2S donors in the recipient animal activates several protective pathways that limit cold IRI to preserve graft integrity. The results further suggest that the cells that produce CSE are perhaps more sensitive than other cells in the lung graft since their pharmacological inhibition resulted in significant pathological changes within such a short time. One could also surmise that the dose of propargylglycine may be too high and therefore might have led to the immediate death of the CSE-producing cells in the lung graft.

In a miniature porcine model of DCD lung transplantation, intravenous administration of Na_2S in the donor pig at 3 h prior to graft procurement and in the recipient pig before 2 h of reperfusion decreased lung graft IRI and prolonged graft and recipient survival. Compared to control grafts, improved quality of H_2S-treated grafts was shown by significantly higher oxygen partial pressure, reduced lung neutrophil infiltration, intra-alveolar edema, and downregulated expression of pro-inflammatory cytokines (Sahara et al. 2015). From a mechanistic perspective, H_2S (via GYY4137) has been shown to reduce lung IRI by inhibiting the release of angiopoietin-2 (Ang-2; a pro-inflammatory vascular growth factor that increases vascular leak) while activating the signaling pathway of silent information regulator type-1 and 3 (SIRT1 and 3; proteins that have anti-inflammatory and anti-apoptotic properties and reduce lung endothelial injury and oxidative stress and improve mitochondrial function) (Tang et al. 2017; Jiang et al. 2019, 2020) (Fig. 2). Although Ang-2, SIRT1 and SIRT3 have not been studied experimentally in the context of lung transplantation, it is possible that the attenuation of IRI in the lung grafts and the anti-inflammatory and anti-apoptotic effects of H_2S on the lung grafts and the consequent improvement in lung function after transplantation could be partly attributed to the influence of H_2S on these proteins and perhaps other unidentified protective proteins. However, this hypothesis warrants future investigations. Taken together, these fascinating pieces of laboratory evidence on the beneficial role of H_2S in experimental lung transplantation suggest that a focus on H_2S and its modulation may be a potential therapeutic approach to improving lung transplantation in the clinical setting.

4.4 H_2S in Experimental Liver Transplantation

While several experimental investigations on the effects of H_2S have been conducted in the transplantation of other solid organ grafts, not many studies have been done on liver transplantation. Only three of such experimental studies have been reported in the literature so far. In a rat model of ex vivo DCD liver transplantation, preservation of liver grafts in diallyl disulfide (3.4 mM)-supplemented UW solution by SCS at 4° C for 48 h followed by 90 min of ex vivo reperfusion (normothermic machine perfusion) improved parameters of liver function as seen in markedly reduced intrahepatic vascular resistance and levels of lactate dehydrogenase (LDH; a tissue damage marker), improved bile production, oxygen consumption and microcirculation, and increased ability for uptake for bromsulfthalein (BSP; an indicator of improved hepatobiliary function) compared to control liver grafts. The improvement in hepatobiliary function in H_2S-treated grafts corresponded histologically with preserved liver structures (Balaban et al. 2011). A similar result was obtained in a later study by the same group in another rat model ex vivo DCD liver transplantation when liver grafts were preserved in Na_2S (10 μM)-supplemented HTK solution by SCS at 4°C for 24 h after which the grafts were reperfused ex vivo by machine perfusion at 37°C for 90 min. Additional results from this study showed improved viability parameters with reduced ROS production, downregulated the expression of heme oxygenase-1 (HO-1) and tumor necrosis factor-alpha (TNF-α) genes and upregulated the expression of the anti-apoptotic gene, B-cell lymphoma-2 (Bcl-2), and markedly elevated the levels of antioxidants (glutathione and catalase) in H_2S-treated liver grafts (Balaban et al. 2015). Most recently, another rat model of ex vivo liver transplantation was studied by a different research group in which SCS of liver grafts in AP39 (200 nM)-supplemented UW solution at 4°C for 72 h followed by 6 h of normothermic machine perfusion at 37°C also showed improved liver graft viability, reduced portal congestion, hepatic vascular resistance, transaminitis (high AST and ALT levels) and apoptosis, with improved bile production, bile glucose and energy charge, and preserved endothelial structures when compared to control grafts without AP39 treatment (McLean et al. 2024).

Several mechanistic studies explained the hepatoprotective effects of H_2S after liver transplantation. For example, pretreatment of liver cell line BRL-3A (normal hepatocytes in rats) with H_2S was recently reported to downregulate the expression of hypoxia-induced CHOP, SPHK1, and S1P proteins, and effectively protected the cells against hypoxia/reoxygenation injury, the in vitro version of IRI in organ transplantation. Interestingly, overexpression of these proteins, especially CHOP, significantly reduced H_2S level in the cells and abrogated the protective effects of H_2S by increasing inflammatory responses and endoplasmic reticulum stress-induced apoptosis (Chen et al. 2023). In in vivo mouse and rat models of hepatic IRI, H_2S preconditioning prior to induction of ischemia and reperfusion markedly reduced hypertransaminasemia, prevented hepatocyte necrosis, reduced oxidative damage by improving hepatic antioxidant status, attenuated autophagy, and downregulated pro-apoptotic genes by inhibiting the opening of MPTPs through upregulated expression of phosphorylated Akt, GSK-3β, HSP90 and Bcl-2, and in addition, reduced the release of cytochrome c from the mitochondria and prevented the activation of initiator and executioner caspases (Zhang et al. 2013; Cheng et al. 2014; Jha et al. 2008) (Fig. 2). Furthermore, H_2S also downregulated the phosphorylation of JNK1 and ERK, two principal members of the MAPK family, leading to reduced pro-inflammatory cytokine release (Zhang et al. 2013; Cheng et al. 2014; Jha et al. 2008) (Fig. 2). While these and several other mechanistic studies are needed to expand the molecular basis of the hepatoprotective effects of H_2S, the findings, so far, provide experimental support for the therapeutic use of H_2S in preclinical and clinical liver transplantation and the broader field of transplant medicine.

4.5 H_2S in Experimental Pancreas/Pancreatic Islet Transplantation

Like other solid organ grafts, IRI and its associated complications represent a major challenge after pancreas transplantation, necessitating the need for experimental studies toward improvement in pancreas transplantation protocol, including pharmacological modification of pancreas preservation and pancreatic islet isolation techniques in order to enable unstable type 1 diabetic patients restore normoglycemia and avoid diabetic complications. However, as with liver transplantation, there is a dearth of information in the literature on the role of H_2S in pancreas/pancreatic islet transplantation. One study reported that preconditioning of pancreatic islets with H_2S improved graft function in type 1 diabetic mice (Wang et al. 2016). A follow-up study by another group also demonstrated in a porcine model that SCS of porcine pancreata in AP39 (400 nM)-supplemented UW solution for 18 h at 4°C followed by islet isolation resulted in significantly higher islet recovery after culture when compared to the control group without AP39 supplementation (Nishime et al. 2021). Compared to the control group, the AP39-treated group was associated with substantial reduction in ROS production, marked increase in mitochondrial membrane potential (an indicator of mitochondrial health) and ATP production, improved islet transplantation outcome, and subsequently inhibited the occurrence of graft pancreatitis in type 1 diabetic mice (Nishime et al. 2021) (Fig. 2). Considering that this area of research is still in its infantile stage, more studies from different research groups with other H_2S donors, especially with STS (clinically viable H_2S donor drug) are required to affirm the therapeutic impact of H_2S in pancreas/pancreatic islet transplantation and its protective mechanisms, to facilitate clinical translation. Nonetheless, the observations from these experimental studies suggest that pharmacological modification of the preservation protocol with H_2S may be a simple, inexpensive, yet novel technique to increase islet yield and protect mitochondrial integrity and prevent pancreatic complications in pancreatic islet transplantation.

5 Concluding Remarks and Clinical Translation

Organ transplantation is a complex life-saving intervention that gives patients with organ failure a second chance to live. It requires optimal conditions to ensure long-term graft and patient survival. However, IRI during organ procurement, cold preservation, and reperfusion poses an inevitable challenge that limits the long-term success of this important technical procedure, leading to increased incidence of post-transplant complications and eventual loss of graft function. Modification of the transplant protocol (including the preservation techniques) with H_2S has been shown in various experimental models of organ transplantation as a novel pharmacological approach to mitigating the transplant surgeon's unavoidable "enemy" to improve graft quality and ensure long-term survival, and therefore, should be considered for human clinical trials. However, most of these experiments were performed using non-clinically viable H_2S donor compounds, and hence, hamper their translation to the clinic. To surmount this hurdle, we recently identified STS as an FDA-approved H_2S donor drug already in clinical use as an antidote to acute cyanide poisoning and for the treatment of calcific uremic arteriolopathy in patients with end-stage renal disease and cisplatin-induced organ toxicity in cancer patients. As an innovative approach, we have repositioned this drug in our rodent models of kidney transplantation with compelling success. While the clinically promising results in our rodent models merit more investigations by other research groups, including the dose of STS, timing of treatment in the organ donor and/or organ recipient as well the duration in the preservation solutions, we are currently investigating whether our observation of the renoprotective effects of STS in our rodent models is reproducible in pigs, as they are large animals and serve as an attractive model that is closely relevant to humans, with genetic, embryological and morphological similarity as well as similar responses to injury with human kidneys. Should STS produce the same therapeutic benefits in our porcine model as observed in our rodent models, and also be reproduced by other research groups, its translation from bench to bedside can lay the foundation for STS to be used in clinical organ transplantation in the future.

References

Abou Taka M, Dugbartey GJ, Richard-Mohamed M, McLeod P, Jiang J, Major S, Arp J, O'Neil C, Liu W, Gabril M, Moussa M, Luke P, Sener A (2024) Evaluating the effects of kidney preservation at 10 °C with Hemopure and sodium thiosulfate in a rat model of syngeneic orthotopic kidney transplantation. Int J Mol Sci 25(4):2210

Abouna GM (2008) Organ shortage crisis: problems and possible solutions. Transplant Proc 40(1): 34–38

Balaban CL, Rodriguez JV, Guibert EE (2011) Delivery of the bioactive gas hydrogen sulfide during cold preservation of rat liver: effects on hepatic function in an ex vivo model. Artif Organs 35(5):508–515

Balaban CL, Rodríguez JV, Tiribelli C, Guibert EE (2015) The effect of a hydrogen sulfide releasing molecule (Na2S) on the cold storage of livers from cardiac dead donor rats. A study in an ex vivo model. Cryobiology 71(1):24–32

Benavides GA, Squadrito GL, Mills RW, Patel HD, Isbell TS, Patel RB et al (2007) Hydrogen sulfide mediates the vasoactivity of garlic. Proc Natl Acad Sci USA 104:17977–17982

Bon D, Chatauret N, Giraud S, Thuillier R, Favreau, Hauet T (2012) New strategies to optimize kidney recovery and preservation in transplantation. Nat Rev Nephrol 8:339–347

Caliendo G, Cirino G, Santagada V, Wallace JL (2010) Synthesis and biological effects of hydrogen sulfide (H_2S): development of H_2S-releasing drugs as pharmaceuticals. J Med Chem 53:6275–6286

Calvert JW, Jha S, Gundewar S, Elrod JW, Ramachandran A, Pattillo CB, Kevil CG, Lefer DJ (2009) Hydrogen sulfide mediates cardioprotection through Nrf2 signaling. Circ Res 105(4): 365–374

Chatzianastasiou A, Bibli SI, Andreadou I, Efentakis P, Kaludercic N, Wood ME, Whiteman M, Di Lisa F, Daiber A, Manolopoulos VG, Szabó C, Papapetropoulos A (2016) Cardioprotection by H2S donors: nitric oxide-dependent- and independent mechanisms. J Pharmacol Exp Ther 358: 43140

Chen Y, Shi J, Xia TC, Xu R, He X, Xia Y (2019) Preservation solutions for kidney transplantation: history, advances and mechanisms. Cell Transplant 28(12):1472–1489

Chen L, Lin B, Yang J, Zhong L, Xiong X, Wang X (2023) Hydrogen sulfide alleviates ischemia induced liver injury by repressing the SPHK1/S1P pathway. Ann Transl Med 11(2):73

Cheng P, Wang F, Chen K, Shen M, Dai W, Xu L, Zhang Y, Wang C, Li J, Yang J, Zhu R, Zhang H, Zheng Y, Lu J, Zhou Y, Guo C (2014) Hydrogen sulfide ameliorates ischemia/reperfusion-induced hepatitis by inhibiting apoptosis and autophagy pathways. Mediators Inflamm 2014:935251

De Martin E, Londoño MC, Emamaullee J, Lerut J, Potts J, Aluvihare V, Spiro M, Raptis DA, McCaughan G, ERAS4OLT.org Working Group (2022) The optimal immunosuppression management to prevent early rejection after liver transplantation: A systematic review of the literature and expert panel recommendations. Clin Transpl 36(10):e14614

Debout A, Foucher Y, Trebern-Launay K, Legendre C, Kreis H, Mourad G, Garrigue V, Morelon E, Buron F, Rostaing L, Kamar N, Kessler M, Ladrie're M, Poignas A, Blidi A, Soulillou JP, Giral M, Dantan E (2015) Each additional hour of cold ischemic time significantly increases the risk of graft failure and mortality following renal transplantation. Kidney Int 87:343–349

DiRito JR, Hosgood SA, Reschke M, Albert C, Bracaglia LG, Ferdinand JR, Stewart BJ, Edwards CM, Vaish AG, Thiru S, Mulligan DC, Haakinson DJ, Clatworthy MR, Saltzman WM, Pober JS, Nicholson ML, Tietjen GT (2021) Lysis of cold-storage-induced microvascular obstructions for ex vivo revitalization of marginal human kidneys. Am J Transplant 21(1):161–173

Du Y, Liu XH, Zhu HC, Wang L, Wang ZS, Ning JZ, Xiao CC (2019) Hydrogen sulfide treatment protects against renal ischemia-reperfusion injury via induction of heat shock proteins in rats. Iran J Basic Med Sci 22(1):99–105

Dugbartey GJ (2021) Emerging role of carbon monoxide in intestinal transplantation. Biomed Pharmacother 143:112237

Dugbartey GJ (2024a) Cellular and molecular mechanisms of cell damage and cell death in ischemia-reperfusion injury in organ transplantation. Mol Biol Rep 51(1):473

Dugbartey GJ (2024b) Gasotransmitters in organ transplantation: a new era in transplant medicine. Springer Nature. https://doi.org/10.1007/978-3-031-48067-6

Dugbartey GJ, Juriasingani S, Zhang MY, Sener A (2021) H_2S donor molecules against cold ischemia-reperfusion injury in preclinical models of solid organ transplantation. Pharmacol Res 172:105842

Dugbartey GJ, Juriasingani S, Richard-Mohamed M, Rasmussen A, Levine M, Liu W, Haig A, Whiteman M, Arp J, Luke PPW, Sener A (2023) Static cold storage with mitochondria-targeted hydrogen sulfide donor improves renal graft function in an ex vivo porcine model of controlled donation-after-cardiac-death kidney transplantation. Int J Mol Sci 24(18):14017

Ertugrul IA, van Suylen V, Damman K, de Koning MLY, van Goor H, Erasmus ME (2021) Donor heart preservation with hydrogen sulfide: a systematic review and meta-analysis. Int J Mol Sci 22(11):5737

Fu M, Zhang W, Wu L, Yang G, Li H, Wang R (2012) Hydrogen sulfide (H2S) metabolism in mitochondria and its regulatory role in energy production. Proc Natl Acad Sci USA 109(8): 2943–2948

George TJ, Arnaoutakis GJ, Beaty CA, Jandu SK, Santhanam L, Berkowitz DE, Shah AS (2012a) Inhaled hydrogen sulfide improves graft function in an experimental model of lung transplantation. J Surg Res 178(2):593–600

George TJ, Arnaoutakis GJ, Beaty CA, Jandu SK, Santhanam L, Berkowitz DE, Shah AS (2012b) Hydrogen sulfide decreases reactive oxygen in a model of lung transplantation. J Surg Res 178(1):494–501

Ginter E, Simko V (2010) Garlic (Allium sativum L.) and cardiovascular diseases. Bratisl Lek Listy 111:452–456

Grenda R (2017) Delayed graft function and its management in children. Pediatr Nephrol 32(7): 1157–1167

Guibert EE, Petrenko AY, Balaban CL et al (2011) Organ preservation: current concepts and new strategies for the next decade. Transfus Med Hemother 38:125–142

Hosgood, Nicholson (2010) Hydrogen sulphide ameliorates ischaemia-reperfusion injury in an experimental model of non-heart-beating donor kidney transplantation. Br J Surg 97(2): 202–209

Hu MZ, Zhou B, Mao HY, Sheng Q, Du B, Chen JL, Pang QF, Ji Y (2016) Exogenous hydrogen sulfide postconditioning protects isolated rat hearts from ischemia/reperfusion injury through Sirt1/PGC-1alpha signaling pathway. Int Heart J 57(4):477–482

Jha S, Calvert JW, Duranski MR, Ramachandran A, Lefer DJ (2008) Hydrogen sulfide attenuates hepatic ischemia-reperfusion injury: role of antioxidant and antiapoptotic signaling. Am J Physiol Heart Circ Physiol 295(2):H801–H806

Ji Y, Pang QF, Xu G, Wang L, Wang JK, Zeng YM (2008) Exogenous hydrogen sulfide postconditioning protects isolated rat hearts against ischemia-reperfusion injury. Eur J Pharmacol 587(1–3):1–7

Jiang T, Liu Y, Meng Q, Lv X, Yue Z, Ding W, Liu T, Cui X (2019) Hydrogen sulfide attenuates lung ischemia-reperfusion injury through SIRT3-dependent regulation of mitochondrial function in type 2 diabetic rats. Surgery 165(5):1014–1026

Jiang T, Yang W, Zhang H, Song Z, Liu T, Lv X (2020) Hydrogen sulfide ameliorates lung ischemia-reperfusion injury through SIRT1 signaling pathway in type 2 diabetic rats. Front Physiol 11:596

Johansen D, Ytrehus K, Baxter GF (2006) Exogenous hydrogen sulfide (H2S) protects against regional myocardial ischemia-reperfusion injury--Evidence for a role of K ATP channels. Basic Res Cardiol 101(1):53–60

Juriasingani S, Akbari M, Chan JY, Whiteman M, Sener A (2018) H2S supplementation: A novel method for successful organ preservation at subnormothermic temperatures. Nitric Oxide 81: 57–66

Juriasingani S, Jackson A, Zhang MY, Ruthirakanthan A, Dugbartey GJ, Sogutdelen E, Levine M, Mandurah M, Whiteman M, Luke P, Sener A (2021) Evaluating the effects of subnormothermic perfusion with AP39 in a novel blood-free model of ex vivo kidney preservation and reperfusion. Int J Mol Sci 22(13):7180

Juriasingani S, Vo V, Akbari M, Grewal J, Zhang M, Jiang J, Haig A, Sener A (2022) Supplemental hydrogen sulfide in models of renal transplantation after cardiac death. Can J Surg 65(2):E193–E202

Karwi QG, Bornbaum J, Boengler K, Torregrossa R, Whiteman M, Wood ME, Schulz R, Baxter GF (2017) AP39, a mitochondria-targeting hydrogen sulfide (H_2S) donor, protects against myocardial reperfusion injury independently of salvage kinase signalling. Br J Pharmacol 174:287301

Kashfi K, Olso KR (2013) Biology and therapeutic potential of hydrogen sulfide and hydrogen sulfide-releasing chimeras. Biochem Pharmacol 85:689–703

Kimura H (2014) The physiological role of hydrogen sulfide and beyond. Nitric Oxide 41:4–10

Lambert JP, Nicholson CK, Amin H, Amin S, Calvert JW (2014) Hydrogen sulfide provides cardioprotection against myocardial/ischemia reperfusion injury in the diabetic state through the activation of the RISK pathway. Med Gas Res 4(1):20

Li L, Whiteman M, Guan YY, Neo KL, Cheng Y, Lee SW et al (2008) Characterization of a novel, water-soluble hydrogen sulfide-releasing molecule (GYY4137): new insights into the biology of hydrogen sulfide. Circulation 117:2351–2360

Li H, Wang Y, Wei C, Bai S, Zhao Y, Li H, Wu B, Wang R, Wu L, Xu C (2015) Mediation of exogenous hydrogen sulfide in recovery of ischemic post-conditioning-induced cardioprotection via down-regulating oxidative stress and up-regulating PI3K/Akt/GSK-3beta pathway in isolated aging rat hearts. Cell Biosci 5:11

Lobb I, Mok A, Lan Z, Liu W, Garcia B, Sener A (2012) Supplemental hydrogen sulphide protects transplant kidney function and prolongs recipient survival after prolonged cold ischaemia-reperfusion injury by mitigating renal graft apoptosis and inflammation. BJU Int 110(11 Pt C):E1187–E1195

Lobb I, Jiang J, Lian D, Liu W, Haig A, Saha MN, Torregrossa R, Wood ME, Whiteman M, Sener A (2017) Hydrogen sulfide protects renal grafts against prolonged cold ischemia-reperfusion injury via specific mitochondrial actions. Am J Transplant 17(2):341–352

McLean ST, Holkup S, Tchir A, Mojoudi M, Hassan M, Taveras C, Ozge SO, James FM, Yeh H, Uygun K, Longchamp A (2024) UW supplementation with AP39 improves liver viability following static cold storage. Res Sq. rs.3.rs-4487319

Meng C, Cui X, Qi S, Zhang J, Kang J, Zhou H (2017) Lung inflation with hydrogen sulfide during the warm ischemia phase ameliorates injury in rat donor lungs via metabolic inhibition after cardiac death. Surgery 161(5):1287–1298

Mikami Y, Shinuya N, Kimura Y, Nagahara N, Ogasawara Y, Kimura H (2011) Thioredoxin and dihydrolipoic acid are required for 3-mercaptopyruvate sulfurtransferase to produce hydrogen sulfide. Biochem J 439:479–485

Modis K, Coletta C, Erdelyi K, Papapetropoulos A, Szabo C (2013) Intramitochondrial hydrogen sulfide production by 3-mecaptopyruvate sulfurtransferase maintains mitochondrial electron transport flow and supports cellular biogenesis. FASEB J 27:601–611

Morris PJ (2004) Transplantation—a medical miracle of the 20th century. N Engl J Med 351:2678–2680

Nelson P, Dugbartey GJ, McFarlane L, McLeod P, Major S, Jiang J, O'Neil C, Haig A, Sener A (2024) Effect of sodium thiosulfate pre-treatment on renal ischemia-reperfusion injury in kidney transplantation. Int J Mol Sci 25(17):9529

Nishime K, Miyagi-Shiohira C, Kuwae K, Tamaki Y, Yonaha T, Sakai-Yonaha M, Saitoh I, Watanabe M, Noguchi H (2021) Preservation of pancreas in the University of Wisconsin solution supplemented with AP39 reduces reactive oxygen species production and improves islet graft function. Am J Transplant 21(8):2698–2708

Ojo AO, Wolfe RA, Held PJ, Port FK, Schmouder RL (1997) Delayed graft function: risk factors and implications for renal allograft survival. Transplantation 63:968

Peake BF, Nicholson CK, Lambert JP, Hood RL, Amin H, Amin S, Calvert JW (2013) Hydrogen sulfide preconditions the db/db diabetic mouse heart against ischemia-reperfusion injury by activating Nrf2 signaling in an Erk-dependent manner. Am J Physiol Heart Circ Physiol 304(9): H1215–H1224

Polhemus DJ, Lefer DJ (2014) Emergence of hydrogen sulfide as an endogenous gaseous signaling molecule in cardiovascular disease. Cir Res 114:730–737

Predmore BL, Kondo K, Bhushan S et al (2012) The polysulfide diallyl trisulfide protects the ischemic myocardium by preservation of endogenous hydrogen sulfide and increasing nitric oxide bioavailability. Am J Physiol - Heart Circ Physiol 302(11):H2410–H2418

Saat TC, van den Akker EK, IJzermans JNMI, Dor FJMF, de Bruin RWF (2016) Improving the outcome of kidney transplantation by ameliorating renal ischemia reperfusion injury: lost in translation? J Transl Med 14:20

Sahara H, Sekijima M, Waki S, Ichinari Y, Shimizu A, Yamada K (2015) Hydrogen sulfide (H2S) prevents ischemia-reperfusion injury (IRI) and prolongs survival of the fully MHC-disparate lung graft in CLAWN miniature swine. J Heart Lung Transplant 34(4):S149

Shibuya N, Koike S, Tanaka M, Ishigami-Yuasa M, Kimura Y, Ogasawara Y et al (2013) A novel pathway for the production of hydrogen sulfide from D-cysteine in mammalian cells. Nat Commun 4:1366

Snijder PM, de Boer RA, Bos EM, van den Born JC, Ruifrok WP, Vreeswijk-Baudoin I, van Dijk MC, Hillebrands JL, Leuvenink HG, van Goor H (2013) Gaseous hydrogen sulfide protects against myocardial ischemia-reperfusion injury in mice partially independent from hypometabolism. PLoS One 8(5):e63291

Snydman DR (2001) Epidemiology of infections after solid-organ transplantation. Clin Infect Dis 33(Suppl 1):S5–S8

Su Y, Wang Y, Liu M, Chen H (2021) Hydrogen sulfide attenuates renal I/R-induced activation of the inflammatory response and apoptosis via regulating Nrf2-mediated NLRP3 signaling pathway inhibition. Mol Med Rep 24(1):518

Sun Q, Zhou H, Cao R, Lin M, Hua X, Hong L, Huang Z, Na N, Cai R, Wang G, Meng F, Sun Q (2018a) Donation after brain death followed by circulatory death, a novel donation pattern, confers comparable renal allograft outcomes with donation after brain death. BMC Nephrol 19(1):164

Sun X, Wang W, Dai J, Huang J, Shi M, Chu X, Wang F, Guo C, Wang C, Pang L, Wang Y (2018b) Donor heart preservation with a novel long-term and slow-releasing hydrogen sulfide system. Nitric Oxide 81:1–10

Tang B, Ma L, Yao X, Tan G, Han P, Yu T, Liu B, Sun X (2017) Hydrogen sulfide ameliorates acute lung injury induced by infrarenal aortic cross-clamping by inhibiting inflammation and angiopoietin 2 release. J Vasc Surg 65(2):501–508.e1

Testai L, Marino A, Piano I, Brancaleone V, Tomita K, Di Cesare ML, Martelli A, Citi V, Breschi MC, Levi R, Gargini C, Bucci M, Cirino G, Ghelardini C, Calderone V (2016) The novel H(2)S-donor 4-carboxyphenyl isothiocyanate promotes cardioprotective effects against ischemia/reperfusion injury through activation of mitoK(ATP) channels and reduction of oxidative stress. Pharmacol Res 113(Pt A):290–299

Tripatara P, Patel NS, Collino M, Gallicchio M, Kieswich J, Castiglia S, Benetti E, Stewart KN, Brown PA, Yaqoob MM, Fantozzi R, Thiemermann C (2008) Generation of endogenous hydrogen sulfide by cystathionine gamma-lyase limits renal ischemia/reperfusion injury and dysfunction. Lab Investig 88(10):1038–1048

Troppmann C, Santhanakrishnan C, Fananapazir G, Troppmann K, Perez R (2018) Pediatric en bloc kidney transplantation from very small (≤10 kg) donation after circulatory death (versus brain death) donors: single-center matched-pair analysis of 130 transplants. Am J Transplant 18(11):2811–2817

Van Slambrouck J, Van Raemdonck D, Vos R, Vanluyten C, Vanstapel A, Prisciandaro E, Willems L, Orlitová M, Kaes J, Jin X, Jansen Y, Verleden GM, Neyrinck AP, Vanaudenaerde BM, Ceulemans LJ (2022) A focused review on primary graft dysfunction after clinical lung transplantation: A multilevel syndrome. Cells 11(4):745

Wang D, Ding M, Xue J, Zheng J, Feng S, Tian H, Li Y (2016) Transplantation of islet preconditioning with hydrogen sulfide enhances the graft function in diabetic mice. Am J Transplant (Suppl 3)

Wu J, Wei J, You X, Chen X, Zhu H, Zhu X, Liu Y, Xu M (2013) Inhibition of hydrogen sulfide generation contributes to lung injury after experimental orthotopic lung transplantation. J Surg Res 182(1):e25–e33

Xia M, Chen L, Muh RW, Li PL, Li N (2009) Production and action of hydrogen sulfide, a novel gaseous bioactive substance in the kidneys. J Pharmacol Exp Ther 329:1056–1062

Yan J, Zhang B, Yu J, Wang X, Zhao H, Li G, Yin Y, Meng C (2023) Comparison of inflation and ventilation with hydrogen sulfide during the warm ischemia phase on ischemia-reperfusion injury in a rat model of non-heart-beating donor lung transplantation. Biomed Res Int 2023: 3645304

Yao LL, Huang XW, Wang YG, Cao YX, Zhang CC, Zhu YC (2010) Hydrogen sulfide protects cardiomyocytes from hypoxia/reoxygenation-induced apoptosis by preventing GSK-3beta-dependent opening of mPTP. Am J Physiol Heart Circ Physiol 298(5):H1310–H1319

Ye BM, Hyeon JM, Kim SJ, Kim MJ, Kim SR, Kim IY, Lee SB, Lee DW (2022) Analysis of risk factors for delayed graft function after kidney transplantation. Transplant Proc 54(8):2154–2158

Zhang Q, Fu H, Zhang H, Xu F, Zou Z, Liu M, Wang Q, Miao M, Shi X (2013) Hydrogen sulfide preconditioning protects rat liver against ischemia/reperfusion injury by activating Akt-GSK-3beta signaling and inhibiting mitochondrial permeability transition. PLoS One 8(9):e74422

Zhang MY, Dugbartey GJ, Juriasingani S, Akbari M, Liu W, Haig A, McLeod P, Arp J, Sener A (2022) Sodium thiosulfate-supplemented UW solution protects renal grafts against prolonged cold ischemia-reperfusion injury in a murine model of syngeneic kidney transplantation. Biomed Pharmacother 145:112435

Zhao H, Watts HR, Chong M, Huang H, Tralau-Stewart C, Maxwell PH, Maze M, George AJT, Ma D (2013) Xenon treatment protects against cold ischemia associated delayed graft function and prolongs graft survival in rats. Am J Transplant 13(8):2006–2018

Zhao FL, Fang F, Qiao PF, Yan N, Gao D, Yan Y (2016) AP39, a mitochondria-targeted hydrogen sulfide donor, supports cellular bioenergetics and protects against Alzheimer's disease by preserving mitochondrial function in APP/PS1 mice and neurons. Oxidative Med Cell Longev 2016:8360738

Zhu C, Su Y, Juriasingani S, Zheng H, Veramkovich V, Jiang J, Sener A, Whiteman M, Lacefield J, Nagpal D, Alotaibi F, Liu K, Zheng X (2019) Supplementing preservation solution with mitochondria-targeted H_2S donor AP39 protects cardiac grafts from prolonged cold ischemia-reperfusion injury in heart transplantation. Am J Transplant 19(11):3139–3148

Correction to: The Role of H_2S in Atherosclerosis and Associated Cardiometabolic Comorbidities

Odysseia Savvoulidou, Turtushikh Damba, Daniel F. J. Ketelhuth, and Maria Peleli

Correction to:
Chapter 760 in: Handbook of Experimental Pharmacology,
https://doi.org/10.1007/164_2025_760

This chapter was originally published without open access. Following the authors' decision to opt for open access, the copyright of the chapter changed on February, 2026 to © The Author(s) 2026 and the chapter is now distributed under the terms of the Creative Commons Attribution-NonCommercial-NoDerivatives 4.0 International License (http://creativecommons.org/licenses/by-nc-nd/4.0/), which permits any noncommercial use, sharing, distribution and reproduction in any medium or format, as long as you give appropriate credit to the original author(s) and the source, provide a link to the Creative Commons license and indicate if you modified the licensed material. You do not have permission under this license to share adapted material derived from this chapter or parts of it.

The images or other third party material in this chapter are included in the chapter's Creative Commons license, unless indicated otherwise in a credit line to the material. If material is not included in the chapter's Creative Commons license and your intended use is not permitted by statutory regulation or exceeds the permitted use, you will need to obtain permission directly from the copyright holder.

The updated version of this chapter can be found at https://doi.org/10.1007/164_2025_760

A. Papapetropoulos, C. Szabo (eds.), *Hydrogen Sulfide: A Toxicant with Healing Potential*, Handbook of Experimental Pharmacology 292,
https://doi.org/10.1007/164_2026_795

MIX
Papier aus verantwortungsvollen Quellen
Paper from responsible sources
FSC® C105338

If you have any concerns about our products,
you can contact us on
ProductSafety@springernature.com

In case Publisher is established outside the EU,
the EU authorized representative is:
Springer Nature Customer Service Center GmbH
Europaplatz 3, 69115 Heidelberg, Germany

Printed by Libri Plureos GmbH
in Hamburg, Germany